Lecture Notes in Physics

Lecture Notes in Physics

Edited by H. Araki, Kyoto, J. Ehlers, München, K. Hepp, Zürich
R. Kippenhahn, München, H. A. Weidenmüller, Heidelberg
and J. Zittartz, Köln

178

Detectors in Heavy-Ion Reactions

Proceedings of the Symposium
Commemorating the 100th Anniversary
of Hans Geiger's birth
Held at the Hahn-Meitner-Institut
für Kernforschung Berlin
October 6 – 8, 1982

Edited by W. von Oertzen

Springer-Verlag
Berlin Heidelberg GmbH 1983

Editor

W. von Oertzen
Hahn-Meitner-Institut für Kernforschung Berlin GmbH
Bereich Kern- und Strahlenphysik
Postfach 39 01 28, D-1000 Berlin 39

ISBN 978-3-540-12001-8   ISBN 978-3-540-39475-4 (eBook)
DOI 10.1007/978-3-540-39475-4

2153/3140-543210

HANS   GEIGER

30.9.1882 - 24.9.1945

Symposium on Detectors in Heavy Ion Reactions
Hahn-Meitner-Institut für Kernforschung Berlin
October 6 - 8, 1982

K. Bethge, Frankfurt	V. Metag, Heidelberg
R. Bock, Darmstadt	D. Möhl, Genf
M. Brauner, Heidelberg	K.-M. Mutterer, Darmstadt
A. Breskin, Rehovot	R. Pengo, München
G. Charpak, Genf	J. Poitou, Saclay
J. Cahuvin, Grenoble	U. Quade, München
H. Crawford, Berkeley	E. Roeckl, Darmstadt
M. Dumail, Orsay	D. Schüll, Darmstadt
J. Gastebois, Saclay	S. Skorka, München
C.R. Gruhn, Genf	M. Steck, Heidelberg
H. Gutbrod, Berkeley	H. Stelzer, Darmstadt
R. Gyufko, Heidelberg	R. Stock, Berkeley
O. Haxel, Heidelberg	P. Volkov, Orsay
B. Heck, Heidelberg	W. Wagner, München
S.D. Hoath, Birmingham	C.-A. Wiedner, Heidelberg
K. Kilian, Genf	S. Zagromski, Karlsruhe
P. Martin, Grenoble	

PARTICIPANTS FROM HMI

D. Alber	U. Jahnke	R. Sielemann
A. Berger	M. Janke	G. Schiwietz
H.-G. Bohlen	H. Kluge	T. Schneider
W. Bohne	A. Kyanowski	U. Stettner
M. Bürgel	H.-E. Mahnke	N. Stolterfoht
J.G. Cramer	K.-H. Maier	R. Ulrich
C. Egelhaaf	A. Maj	W. Weller
B. Gebauer	A. Miczaika	
K. Grabisch	G. Nolte	
H. Grawe	W. von Oertzen	
O. Häusser	H. Orf	
D. Hilscher	Z. Roller	
G. Ingold	H. Rossner	
A. Itoh	N. Roy	

PREFACE

The one-hundredth anniversary of Hans Geiger's birth on September 29,
1882 in Neustadt gives us occasion to commemorate two historic events
which stood at the cradle of modern scientific research. The first was
the scattering experiment which Geiger conducted together with Marsden
in Rutherford's laboratory and which led to the discovery of the atomic
nucleus. The second was the development of gaseous detectors, better
known as Geiger counters. Geiger, who lived for a long time in Berlin,
worked first at the Physikalisch-Technische-Reichsanstalt (today
Physikalisch-Technische-Bundesanstalt, PTB), and then at the Technicall
University (TU), where his lectures became very famous.

A Festkolloquium with speeches by Professor K.-H. Lindenberger
(Hahn-Meitner-Institut), Professor Kind (PTB) and Professor Starnik
(TU) was held at the Technical University on October 8, 1982.
Professor O. Haxel honored the life and work of Hans Geiger, and
Professor R. Stock gave a review of the extensive development of
gaseous detectors which followed the first steps taken by Geiger and
his collaborators at the beginning of the century.

From October 6 to October 7 a symposium on detectors in heavy-ion
reactions was held at the HMI. It reviewed the most recent developments
in the detection of reaction products in heavy-ion reaction studies.
The focus of the research conducted in the physics division of the HMI
is on heavy-ion-induced reactions using the accelerator VICKSI.

The symposium was supported by the HMI and organized by B. Gebauer,
D. Hilscher, H. Lettau, K. Maier, and W. von Oertzen of its physics
division.

The meeting was overshadowed by the untimely death on September 30,
1982 of Horst Lettau, an organizer of this symposium and one of the
most active physicists in the construction of detectors for heavy ions
at the HMI. The contribution by H.G. Bohlen commemorates his work. We
have lost a devoted physicist and a highly esteemed friend.

For the organizing committee

W. von Oertzen

Berlin, December 1982

TABLE OF CONTENTS

Introductory

Session I. Gas Detectors

Session II. Magnetic Spectrometers

Session III. Multidetector Systems

Session IV. General and Future Developments

Hans GEIGER als Wissenschaftler und Lehrer

von Otto Haxel, Heidelberg.

In diesem Hörsaal hat Hans GEIGER vom WS 36/37 bis etwa 43/44 in seiner Experimentalphysik-Vorlesung die Studenten der Natur- und Ingenieurwissenschaften begeistert und darüber hinaus in seinen mit großer Resonanz aufgenommenen Abendvorträgen viele Berliner Bürger mit den Wundern der modernen, genauer damals modernen Physik vertraut gemacht und sie fasziniert. Geiger verstand es, seine eigene Begeisterung für die naturwissenschaftliche Denkweise und insbesondere für die Physik auf seine Zuhörer zu übertragen und sie mitzureißen. Er konnte mit einfachen aber immer klaren Sätzen und anschaulichen Vergleichen auch komplexe Vorgänge einleuchtend darstellen. Er war ein begnadeter Redner. Diese Rednergabe war wohl ein Familienerbe, denn vor 100 Jahren (30. September 1882) wurde er als Sohn eines Gymnasiallehrers für alte Sprachen geboren. Der Geburtsort war Neustadt a.d. Weinstr. in der Rheinpfalz, die damals zum Königreich Bayern gehörte. Der Vater, offensichtlich sehr sprachbegabt wurde 2 Jahre später in die Landeshauptstadt München versetzt und schließlich auf einen Lehrstuhl für Indogermanische Sprachen nach Erlangen berufen. So wuchs Geiger in einem Gelehrtenhaushalt in München und Erlangen auf.

Nach dem Maturum und der obligaten einjährigen Militärdienstzeit begann Hans Geiger 1902 das Studium in Erlangen, -ein Semester auch in München- und trat 1904 in das Institut von Eilhard WIEDEMANN in Erlangen ein, um mit seiner Promotionsarbeit zu beginnen, nachdem ihm klar geworden war, daß sein Interessengebiet die Physik und nicht wie er zu Beginn des Studiums geglaubt hat, die Mathematik sei. 1906 promovierte er mit einer Arbeit über Strahlungs-, Temperatur- und Potentialmessungen in Entladungsröhren bei starken Strömen. Die Physik der Gasentladungen spielte damals eine Rolle wie heute die Hochenergiephysik, denn mit den Gasentladungsröhren konnten Kathodenstrahlen und Kanalstrahlen erzeugt werden, über die man Aufschlüsse über Existenz und Aufbau der Atome zu erhalten hoffte. Es war die Zeit als Philip LENARD den Nobelpreis für seine Arbeiten über Kathodenstrahlen und deren Durchgang duch Materie erhielt (1905) und seine ersten Überlegungen zu Atommodellen anstellte. Er hatte bereits erkannt, daß die Atome in ihrem Inneren leer sind und nur Kraftfelder enthalten. Es war aber auch noch die Zeit, in der ernsthaft die Meinungen von Wilhelm OSTWALD und Ernst MACH diskutiert wurden, die den Standpunkt vertraten, daß die Atomhypothese zwar für viele Gebiete der

Physik und Chemie eine fruchtbare Arbeitshypothese sei, daß aber die wirkliche Existenz der Atome noch lange nicht erwiesen sei.

Zurück zu Geiger. Sein Doktorvater Wiedemann hatte ihm eine Stelle bei Arthur SCHUSTER in Manchester/England, vermittelt, der sich durch seine e/m Bestimmungen einen Namen gemacht hatte. Ein Jahr verbrachte Geiger so bei Schuster und war gerade dabei wieder zu Wiedemann nach Erlangen zurückzukehren, als RUTHERFORD im Institut erschien, um das Institut zu besichtigen, denn er sollte die Nachfolge Schusters in Manchester antreten. Geiger zeigte ihm das Institut und beeindruckte Rutherford offensichtlich so nachhaltig, daß dieser ihm das Angebot machte, bei ihm weiter zu arbeiten. Auch Geiger war von der Persönlichkeit Rutherfords so angetan, daß er seine Heimreisepläne aufgab. Geiger wurde von Rutherford zunächst als Verwaltungsassistent für Bibliothek und Inventar angestellt. Doch erkannte er, daß Geiger noch andere Qualitäten besaß und machte ihn zu einem seiner engsten experimentellen Mitarbeiter. Geiger blieb 5 Jahre bei Rutherford in Manchester. Diese sollten für beide die arbeits- und erfolgreichsten Jahre ihrer Leben werden. In dieser Zeit entstand u.a. das Rutherfordsche Atommodell, bei dem Geiger wesentliche Geburtshelferdienste leisten konnte.

Geigers Erfahrungen in der Gasentladungsphysik kamen sogleich bei der Entwicklung einer elektrischen Zählmethode für Alpha-Strahlen zum Tragen. Die genaue Ermittlung der Zahl der von einer gegebenen Menge Radium ausgehenden Alpha-Strahlen erlaubte über die Halbwertszeit die direkte Messung der Loschmidt'schen Zahl, die damals nur sehr ungenau und indirekt über Sedimentationsgewichte und die Braun'sche Bewegung erfasst werden konnte. Ihre möglichst genaue Ausmessung war somit eine der fundamentalen Aufgaben der Experimentalphysik. Im Jahre 1908, d.h. schon im ersten Jahr der Zusammenarbeit wurde so das, was wir heute den Proportionalzähler nennen, geboren. Diese elektrische Zählmethode war von großer Bedeutung, denn die damalige Standardmessmethode war die Szintillationszählung, die mit gewissen subjektiven Fehlern behaftet war, denn die einzelnen Szintillationslichtblitze lagen in ihrer Helligkeit nicht allzuweit über der Reizschwelle des Auges. Daher konnten einigermaßen verlässliche Messungen nur innerhalb eines engen Intensitätsbereiches, etwa 20 - 40 Szintillation pro Minute, und nur bei gut dunkel adaptiertem Auge und ausgeruhtem Beobachter erzielt werden.

Heute, rückblickend wundern wir uns, daß diese elektrische Zählmethode nach ihrem erstmaligen Gebrauch wieder in der Versenkung verschwand und die Szintillationsmethode noch mehr als ein Jahrzehnt weiterhin das Feld beherrschte. Der Grund ist leicht einzusehen, denn so elektrisch war die Zählmethode gar nicht, verglichen mit dem was wir heute darunter verstehen. Es gab 1908 noch keine Verstärkerröhren, keinen Oszillographen und keine mechanischen Zählwerke. Die Impulse des Proportionalzählers mußten nicht minder mühsam mit dem Auge an auf höchste Empfindlichkeit getrimmten Fadenelektrometern abgelesen werden. Die photographische Registrierung war zwar möglich, aber nur unter großem Aufwand, denn eine Filmindustrie gab es auch noch nicht. Wollte man ein bewegliches, lichtempfindliches Band haben, so war man auf Photopapier angewiesen, das damals aber keineswegs handelsüblich war.

In der kurzen Zeitspanne von 1907 bis Ende 1912, als Geiger das Rutherfordsche Laboratorium verließ, um in Berlin-Charlottenburg, an der Physikalisch-Technischen Reichsanstalt die Leitung des Laboratoriums für Radioaktivität zu übernehmen, entstanden rund 30 Publikationen, darunter nicht wenige, die die Entwicklung der Physik ganz entscheidend beeinflußt haben. Auf diese will ich mich im folgenden beschränken.

Geiger hat mit der Szintillationsmethode, teils zusammen mit I.M. NUTTALL, eine große Zahl von Reichweiten verschiedener Alpha-Strahler ausgemessen. Er erkannte den systematischen Zusammenhang zwischen Zerfallskonstante des Strahlers und der Reichweite seiner Alpha-Strahlen, ein Ergebnis, das in die Physik als Geiger-Nuttall-Beziehung eingegangen ist, die für GAMOW zum Ausgangspunkt seiner Theorie des Alpha-Zerfalls geworden ist. Die Geigersche Reichweitenformel, nämlich die Aussage, daß die Reichweite der Alpha-Strahlen der dritten Potenz ihrer Geschwindigkeit proportional ist, erlaubte die Umrechnung von Reichweiten in Energien. Geiger selbst stand dieser Formel, die in der Literatur seinen Namen trägt, eher kritisch gegenüber, denn als ihr Urheber wußte er genau, daß sie nur eine Faustformel für den Laboratoriumsgebrauch war, die man nur benutzen sollte, solange es nichts besseres gab.

Die Gesetze der Statistik, angewandt auf den radioaktiven Zerfall, erweckten Geigers Interesse schon in den ersten Jahren seiner Tätigkeit bei Rutherford. Er untersuchte die statistischen Variationen der Zeitintervalle und deren Übereinstimmung mit den Gesetzen des Zufalls. Weiterhin galt sein Interesse der Kleinwinkelstreuung der Alpha-Strahlen, das was wir heute die Vielfachstreuung nennen. Auch hier interessierte ihn die Verteilung der Streuwinkel und deren Übereinstimmung mit den Gesetzen des Zufalls. Daher mußte es ihm sofort auffallen, als ihm

MARSDEN berichtete, daß gestreute Alpha-Strahlen gelegentlich auch unter großen Winkeln beobachtet werden konnten. Geiger und Marsden untersuchten diese Großwinkelstreuung, wir nennen sie heute Einzelstreuung, systematisch und berichteten ihre Ergebnisse Rutherford, der die Beobachtungen aber nicht glauben wollte und sie durch eine unerkannte Kontamination erklären zu müssen glaubte. Geiger und Marsden variierten schließlich das Material der Streusubstanz und konnten einen systematischen Zusammenhang zwischen dem Atomgewicht der Streusubstanz und der Zahl der getreuten Alpha-Strahlen herausarbeiten. Erst jetzt konnte Rutherford das für ihn unglaubwürdige Ergebnis als physikalische Realität hinnehmen. Die Voraussetzung für die Entdeckung des Atomkerns, für das Rutherfordsche Atommodell, war gegeben. Erst nachdem Geiger und Marsden die von der Rutherfordschen Streuformel geforderte Winkelabhängigkeit nachgewiesen hatten, wagte Rutherford seine neue Idee zu publizieren. Niels BOHR kam nach England und auch nach Manchester, wo ihn die Rutherfordschen Vorstellungen faszinierten. So entstand das Rutherford-Bohrsche Atommodell, und damit die Basis unserer heutigen Physik.

Bald danach verließ Geiger England, um an der Physikalisch-Technischen-Reichsanstalt (PTR) seine Stelle anzutreten.
Er hatte hier das Laboratorium für Radioaktivität einzurichten und die nötigen Eich- und Prüfmethoden zu erstellen, denn auch in Deutschland wurden radioaktive Substanzen, vor allem in der Medizin, im zunehmenden Maße in Gebrauch genommen. Der erst 30 Jahre alte Geiger fand hier die Unterstützung des noch viel jüngeren Walter BOTHE, mit dem er an der PTR mehr als ein Jahrzehnt erfolgreich wissenschaftlich zusammenarbeitete und mit dem ihn zeitlebens eine enge Freundschaft verband. Trotz der reichlichen Eichtätigkeit blieb immer noch Zeit für wissenschaftliche Arbeiten. Im Jahr 1913 entstand der Geigersche Spitzenzähler, mit dem nicht nur Alpha-Strahlen, sondern auch die wesentlich weniger ionisierenden Beta-Strahlen nachgewiesen werden konnten. Geiger nannte ihn Spitzenzähler, da das zählempfindliche Volumen durch eine feine Spitze, die einer Platte gegenübersteht, gegeben ist. Beim Zählen von ß-Strahlen wird der Spitzenzähler im Auslösebereich, wie wir ihn beim Geiger-Müller-Zählrohr kennen, beschrieben. Durch das kleine Zählvolumen von nur einigen mm³ wird der Nulleffekt entsprechend niedrig gehalten, was im Jahre 1913 für einen Zähler unbedingt erforderlich war, denn elektrische Registriervorrichtungen für hohe Impulszahlen standen noch nicht zur Verfügung.
Der Krieg von 1914-18 unterbrach Geigers wissenschaftliche Tätigkeit.

Nur CHADWICK arbeitete noch in der PTR. Er wurde als Engländer vom
Kriegsausbruch überrascht. Als Internierter konnte er - wenn ich eine
Erzählung von Geiger noch richtig in Erinnerung habe - in der PTR ar-
beiten und entdeckte dabei das kontinuierliche Betaspektrum. So gute
Sitten gab es einst. Eine Bestätigung dieser Geschichte konnte ich in
der Literatur nicht finden, aber immerhin fand ich in der Enzyklopädia
Britannica den Hinweis, daß Chadwick, der später das Neutron entdeckte
und Nobelpreisträger wurde, ein Schüler von Hans Geiger gewesen sei.

Erst nach dem Kriege 1919 konnte Geiger seine Arbeiten an der PTR
wieder aufnehmen, zusammen mit Bothe, der aus russischer Kriegsgefan-
genschaft zurückgekommen war. Sie arbeiteten über die Streuung von
Beta-Strahlen.

Auch eine kurze Zusammenarbeit mit EINSTEIN ergab sich in Zusammenhang
mit dessen Lichtquantenhypothese. Sie fand aber keinen Niederschlag in
einer Publikation von Seiten Geigers, da ein zunächst vermuteter Effekt
nicht beobachtet werden konnte, in Übereinstimmung mit späteren aus-
führlicheren Überlegungen von Einstein; wohl aber bedankte sich Einstein bei Geiger und Bothe für deren Bemühungen um die Aufklärung in
seiner diesbezüglichen Publikation.

Eine für die Entwicklung der Physik sehr wichtige Arbeit aus dieser
Zeit darf nicht übergangen werden. Im Jahre 1923 war der Comptoneffekt
entdeckt worden, den wir heute als schlagendsten Beweis für die Lichtquantenhypothese ansehen. Niels Bohr, H.A. KRAMERS und I.C. SLATER
vertraten in einer Arbeit über die Quantentheorie der Strahlung die
Meinung, daß der Stoß des Lichtquantes gegen das Elektron und die Bil-
dung eines energieärmeren Lichtquantes nicht als kausal bedingter Ele-
mentarakt aufgefasst werden dürfe und daß der Energie- und Impulssatz
nicht im Einzelprozess sondern nur gemittelt über eine Vielzahl von
Einzelprozessen gelten würde. Diese Fragestellung rief Geiger und
Bothe auf den Plan, die beim radioaktiven Zerfallsgeschehen sehr oft der
Frage der Abgrenzung von Zufall und Kausalität gegenüberstanden. Mit
zwei Spitzenzählern wiesen Geiger und Bothe sowohl das gestreute Rönt-
genquant wie auch das gestreute Elektron nach. Durch Koinzidenzmessun-
gen konnten sie zeigen, daß das Streuquant und das Elektron gleichzei-
tig d.h. innerhalb der damals noch bescheidenen Koinzidenz-Auflösungs-
zeit auftraten. Damit war gezeigt, daß auch im atomaren Elementarpro-
zess Impuls und Energiesatz gelten. Damit wurde, wie Max von LAUE in
seinem Akademienachruf auf Geiger sagt, die Physik vor einem Irrweg

bewahrt; außerdem wurde sie um eine neue experimentelle Möglichkeit, die Koinzidenzmethode, bereichert.

Im Jahre 1925 begann Geigers Hochschullaufbahn, als er einen Ruf an die Universität Kiel annahm. Er hätte schon früher, nämlich 1912 eine außerordentliche Professur an der Universität Tübingen annehmen können, bevorzugte aber damals die Anstellung bei der PTR. Die Umstellung auf das Universitätsleben fiel nicht ganz leicht, aber die Vorlesung, die sogenannte Große Vorlesung für alle Naturwissenschaftler und Mediziner bereitete ihm großen Spaß. Er besaß die seltene Gabe, sich in die Denkweise anderer hinein versetzen zu können und seine eigene Begeisterung auf seine Zuhörer zu übertragen, so daß seine Vorlesungen, und ebenso seine öffentlichen Vorträge, für die Hörer zum Erlebnis wurden.

Aber auch die Forschung kam zu ihrem Recht. In die Kieler Zeit fällt die Entdeckung des Geiger-Müllerschen Zählrohrs (1928), die Geiger auch außerhalb der Physik zu einem berühmten Mann gemacht hat.

Ich habe Geiger als junger Assistent einmal gefragt, warum es denn 20 Jahre gedauert habe, bis aus dem Proportionalzählrohr das Auslösezählrohr geworden war. Ich fragte, ob vielleicht einer der Gründe war, daß er in Kiel erstmals ein vollkommen unkontaminiertes Laboratorium zur Verfügung hatte. Geiger beschrieb die Situation so: Die radioaktive Kontamination der früheren Laboratorien hatte nur eine untergeordnete Rolle gespielt. Die Behauptung, in Manchester hätten nachts die Türklinken geleuchtet, sei eine ganz üble Verleumdung gewesen, denn jeder Mitarbeiter habe ja schon im eigenen Interesse auf peinlichste Sauberkeit wertgelegt. Naturgemäß seien in jedem Zimmer ein paar Präparate herumgestanden, aber bei Bedarf hätte man überall wo er bisher arbeitete, Räume mit normalem Null-Effekt finden können. Im übrigen hätten er und Rutherford schon 1908 gewusst, daß wenn sie die Spannung des Zählrohrs erhöhten, daß dann ihr Zähler bzw. das angeschlossene Elektrometer eine nicht deutbare Zappelei zeigte. Diese wurde jedoch als unkontrollierbare und nicht weiter interessierende Reaktion der positiven Ionen an der Kathode gedeutet. Als er Beta-Strahlen nachweisen wollte, habe er zum Spitzenzähler gegriffen und dieser habe seinen Zweck bestens erfüllt. Im übrigen habe er der Entwicklung der Zählmethoden, die nur Mittel zum Zweck seien, nie mehr Interesse entgegengebracht als etwa der Entwicklung von Ionisationskammern und Elektrometern. Erst in Kiel habe er sich wieder der alten Frage, was die positiven Ionen an der Kathode verursachen, zugewandt und einen seiner Doktoranden, nämlich

Walter Müller, mit der Aufgabe betraut, die Wirkung von positiven Ionen
auf den Zündvorgang einer Entladung zu untersuchen, in einer Anordnung,
die eben dem Zählrohr entsprach. Walter MÜLLER war ein Erfindertyp, mit
der inzwischen entstandenen Verstärkertechnik vertraut, der die Impulse
hörbar machen konnte. Jetzt war es möglich, auch eine höhere Impulszahl
wahrzunehmen. Müller beobachtete, daß die Impulszahl zurückging, wenn
er sich zwischen das Zählrohr und ein im benachbarten Zimmer befindli-
ches Präparat stellte. Geiger konnte zunächst noch nicht an die unwahr-
scheinliche Empfindlichkeit des Zählrohrs glauben, denn auf dem Szintil-
lationsschirm gaben Elektronen keinen Beitrag, höchstens eine schwache
Aufhellung bei ganz großen Intensitäten. In der Ionisationskammer rühr-
te der nach Wegnahme aller Strahler noch verbleibende Nulleffekt im we-
sentlichen von den wenigen aus der Wand austretenden Alpha-Strahlen her.
Demgegenüber war der Beitrag der Beta- und Gamma-Strahlen vernachlässig-
bar. Im Zählrohr sollten die Verhältnisse plötzlich umgekehrt sein!
Daß auch einzelne Elektronenpaare Impulse auslösen können, war auch für
Geiger eine Überraschung. Als Geiger Müllers Beobachtung erfuhr, wurde
sofort eine Absorptionsmessung gemacht, die den für Gamma-Strahlen er-
warteten Messwert ergab. Kalium, dessen natürliche Radioaktivität in
der Ionisationskammer nur mühsam zu finden ist, gab einen riesigen
Effekt. Das Zählrohr war geboren. Die nicht absorbierbaren Impulse wurden als solche der kosmischen Ultrastrahlung identifiziert.

Geiger erkannte sofort, daß das Zählrohr für den Nachweis der Höhen-
strahlen wegen seiner großen empfindlichen Fläche das geeignetste Instrument war. Eine große Zahl von Mitarbeitern in Tübingen und später
in Berlin, arbeiteten unter seiner direkten Leitung auf diesem Gebiet.
Noch in den Kriegsjahren, waren zumindest in den Ferienmonaten, der
große Hörsaal mit vielen Zählrohren ausgelegt, um das Phänomen der
großen Luftschauer zu untersuchen.

Die Technik der Zählrohr-gesteuerten Wilsonkammer wurde entwickelt. Die
Wilsonkammern selbst wurden immer größer um auch Schauer erfassen zu
können. Als die ersten Schwierigkeiten überwunden waren, zeigte es sich,
daß die Bilder umso schöner und klarer wurden, je größer die Kammer gebaut wurde. Ich habe noch manche Stunde im Zimmer von SCHÜTT und
ZÜHLKE , sowie mit DEUTSCHMANN, der die größte Kammer gebaut hatte,
in schöner Erinnerung.

Die große Durchdringungsfähigkeit der kosmischen Höhenstrahlung war das
ungelöste Problem. Aus Protonen konnte die durchdringende Komponente

nicht bestehen, denn Protonen werden durch Kernstöße absorbiert. Elektronen, das wußte man auch, wurden durch Bremsschaltung gebremst. Eine Super γ -Strahlung, an die zuerst geglaubt wurde, widersprach dem Koinzidenzversuch von Bothe und Kolhörster. Die Mesonen - des Rätsels Lösung - waren erst im Kommen.

Der Hochenergiephysiker von heute stellt mit Bewunderung fest, daß Geiger bereits vor 40 Jahren die Grundlagen zu den Fragestellungen und experimentellen Aufbauten gelegt hat, die heute das Gesicht der Hochenergie- bzw. der Elementarteilchenforschung prägen.

Leider hat Geigers Tod diesen Arbeiten ein frühes Ende gesetzt. Eine rheumatische Erkrankung fesselte ihn ans Bett und schwächte seinen Körper so, daß er die harten Berliner Nachkriegsmonate nicht überleben konnte. Er starb am 24. September 1945.

Man würde den Leistungen Geigers nicht gerecht werden, wenn man nicht auch seine literarische Tätigkeit hervorhebt. Er hat zusammen mit R. SCHEEL das 24-bändige Handbuch der Physik (1926-1933) herausgegeben, eine beachtliche Zahl gerade der bedeutendsten Bände allein. Seit 1936 hat er bis 1944 die Zeitschrift für Physik herausgegeben und durch kluge Auswahl der Arbeiten zu hohem Ansehen gebracht.

Durch die Erfindung des Geiger-Müllerschen Zählrohrs   - Geiger hat immer großen Wert darauf gelegt, daß der Name seines Schülers Müller mitgenannt wird - ist Geiger ein populärer Mann geworden. Der tickende Geigerzähler wurde eine Art Symbol des Atomzeitalters.

Auch wissenschaftliche Ehrungen wurden ihm zuteil, so erhielt er 1938 die Hughes-Medaille der Royal Society und im gleichen Jahr die Duddel-Medal der Physical Society London. Die größte Auszeichnung, wie sie seinen Schülern Chadwick und Bothe zufiel, der Nobel-Preis, ist ihm versagt geblieben. Vielleicht wäre es auch anders gekommen, wenn ihn der Tod nicht so früh aus fruchtbarer Arbeit gerissen hätte. Jedenfalls wäre er dann einer der populärsten Nobelpreisträger geworden.

H. Geiger im Hörsaal der TU Berlin, Ende
der dreißiger Jahre, umgeben von Studenten.

EMISSION AND ABSORPTION OF PHOTONS IN GASEOUS DETECTORS

G. Charpak

CERN, Geneva, Switzerland

ABSTRACT

In this paper some of the progresses due to a better understanding of the emission
and absorption of photons in gaseous detectors are discussed. The possibility of imaging photons from 4 to 10 eV offers many possible applications, discussed in this paper,
for X-ray imaging, high-energy calorimetry, particle identification, etc.

While ultraviolet photons have been known for a long time to play a major role in
the mechanism of the Geiger counter, more recent developments are due to a deeper understanding and control of emission and absorption of photons in gaseous detectors.
This is illustrated by the operation of wire chambers in the limited Geiger or streamer
mode, gas scintillation proportional counters, proportional photo-ionization scintillation counters, multistep avalanche chambers, and various types of detectors for photon imaging.

At a symposium held in conjunction with the commemoration of the 100th anniversary
of the birth of H. Geiger, I thought it appropriate to discuss the photon emission or
absorption in gaseous detectors, and this for several reasons. It so happens that my
group has been actively engaged, at CERN, in some developments based on the exploitation of these phenomena[1-3]. Also my personal interest in particle detectors grew
from the construction of Geiger counters in my first steps as an experimentalist:
looking in the dark at the wires illuminated by the excited ion sheath in Geiger
counters, I started the study of light pulses in proportional counters[4].

At the present time many groups have invented new devices based on the understanding or control of the various electromagnetic radiation phenomena connected with
the collisions of electrons in gases and I would like, in this contribution, to review
briefly some of these advances in detector techniques.

1. THE EMISSION OF PHOTONS IN GASEOUS IONIZATION DETECTORS

When electrons are drifting in gases under the influence of electric fields they
can experience a variety of inelastic collisions, leading to the emission of light.
In many cases, with mixtures commonly chosen in gaseous detectors for some empirical
reasons, a precise prediction is impossible because of the scarcity of experimental
data. The situation is understood in noble gases and in noble gas mixtures, and data
are available for some mixtures of noble gases with quenching agents such as N_2, CH_4,

or CO_2. Nevertheless, some simple guide-lines have proved to be effective in control-ling the operation of counters even with complicated mixtures.

1.1 Noble gases

Various states contribute to the emission of photons when the atoms of noble gases are excited by electrons accelerated by electric fields. The spectra are strongly de-pendent on the gas pressure and on the excitation modes: discharges, multiplication, drift without ionizing collisions. I will select only a few topics in this broad field, referring the reader to a detailed review[5] on the subject.

In discharges, besides the characteristic atomic lines, the spectra can exhibit a broad continuum with the following three main domains, as exemplified[6] in Fig. 1: the resonance peak corresponding to the fast allowed transitions from the lowest group of excited states, the "first continuum" close to this peak, and the "second continuum" at lower energy. The resonance peak is dominant at low pressures and disappears at higher pressures, as is visible from the spectra exhibited by discharges in argon at 50 and 400 Torr (Fig. 1) and the spectra determined under conditions of electron mul-tiplication near a wire[7] (Fig. 2) up to pressures of 20 atm. Under conditions where

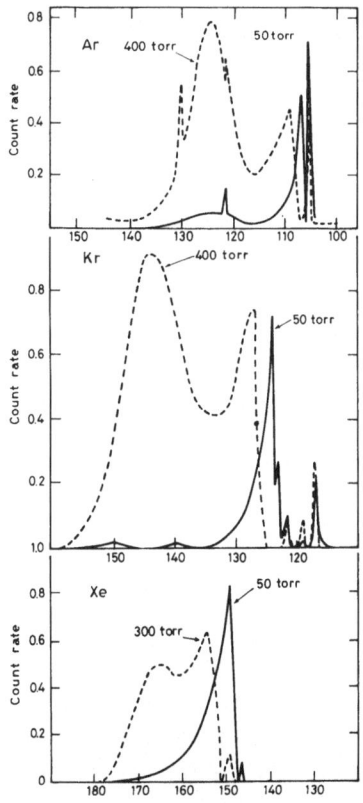

Fig. 1 Samples of the emission spectra of various noble gases (from Ref. 6). Discharge excitation.

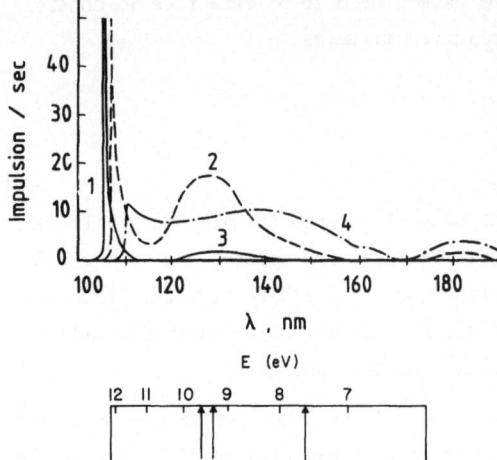

Fig. 2 Emission spectrum from argon excited by an avalanche. Wire counter: 100 μm wire diameter, gain 10^2. Pressure: P = 0.1 (curve 1), 1 (curve 2), 10 (curve 3), and 25 (curve 4) atm. 5.9 keV X-rays (from Ref. 7).

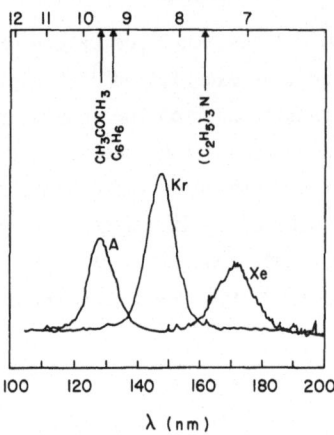

Fig. 3 VUV spectrum produced in noble gases by electrons of energies below ionization thresholds. The electrons drift in the pure noble gases at moderate electric fields, below the electron multiplication region (from Ref. 8).

no multiplication occurs, only the second continuum is visible as shown for pure noble gases[8] in Fig. 3. This corresponds, as we will see, to the operating conditions in the "gas scintillation proportional counters" (GSPCs); in this case the intensity of photon emission is nearly equal to the potential energy lost by an electron drifting in the field, divided by the average energy of the photons. This has been proved only in Xe, but probably gives the right order of magnitude for the other noble gases. It is at present widely accepted that the VUV photons in the "second continuum" are mainly due to the de-excitation of vibrationally relaxed excited molecular states

$$R_2^* \rightarrow 2R + h\nu \tag{1}$$

produced in three-body collisions

$$R + R + R \rightarrow R_2^* + R . \tag{2}$$

The time-structure properties of the emission of light are governed by the formation-time of reaction (2) and the decay-time of R_2^*. From what is known of the collision cross-section the formation-time, which is pressure dependent, is less than 100 ns at 1000 Torr. The decay-time of R_2^* ranges from 4 to 6 ns for the $^1\Sigma_u^+$ states, and reaches much longer times for the $^3\Sigma_u^+$ states, namely 3200 ns for Ar, 1700 ns for Kr, and 90 ns for Xe.

For mixtures of noble gases the spectra are radically modified by collisional de-excitation of the excited states of the host gas by guest atoms with lower energy states. This is illustrated[9] by Fig. 4, where it can be seen that at levels of Xe admixture to Ar inferior to 1% most of the photons correspond to atomic or molecular levels of Xe. The admixture of N_2 plays an important role in some detectors. The collisional transfer of the excitation energy of Ar to levels decaying rapidly with the emission of blue light in the 400 nm range is used as an efficient wavelength shifter, as illustrated[9] in Fig. 5.

Figure 6 shows the effect on the second continuum of small admixtures of CH_4 or CO_2 to Ar. One should notice that the effects are not linear. In Ar at 1 atm, the addition of the first 10^{-2} Torr has the same effect as the next 25×10^{-2} Torr. This shows that the different excited levels contributing to this continuum are not equally affected by the quencher. In detectors where mainly the fast components are active in the amplification mechanism they are certainly less affected by the quencher and we can thus understand why effects of VUV photons are still observed at much higher con-centrations than those considered in these static studies.

In addition it must be stated that there exists sufficient evidence, at present, that when a multiplication process occurs in gases it can involve higher excitation than that considered in the studies so far mentioned. There exists a copious emission

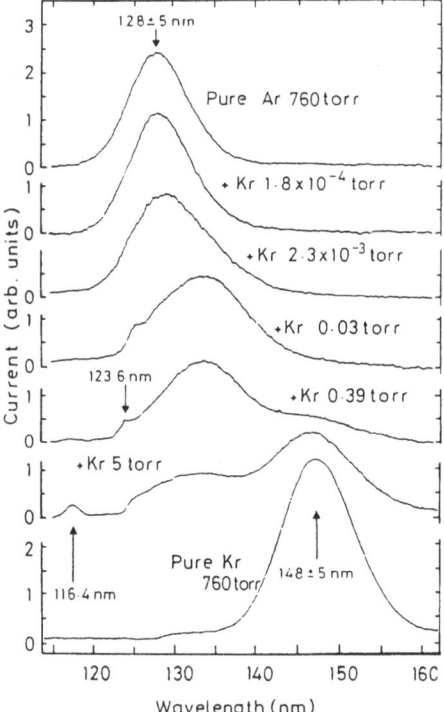

Fig. 4 VUV spectrum in mixtures of noble gases. The conditions are the same as in Fig. 3 (from Ref. 9).

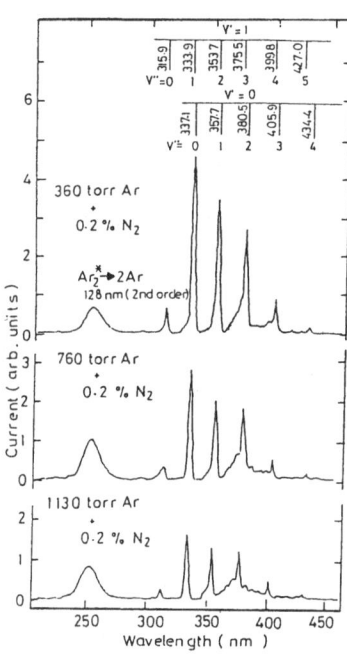

Fig. 5 Photon spectrum in mixtures of Ar and N_2 (from Ref. 9).

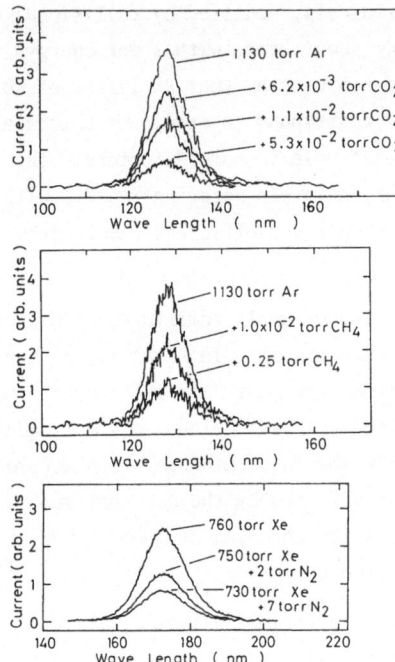

Fig. 6 Photon spectrum in Ar with the admixture of quenching agents (from Ref. 9).

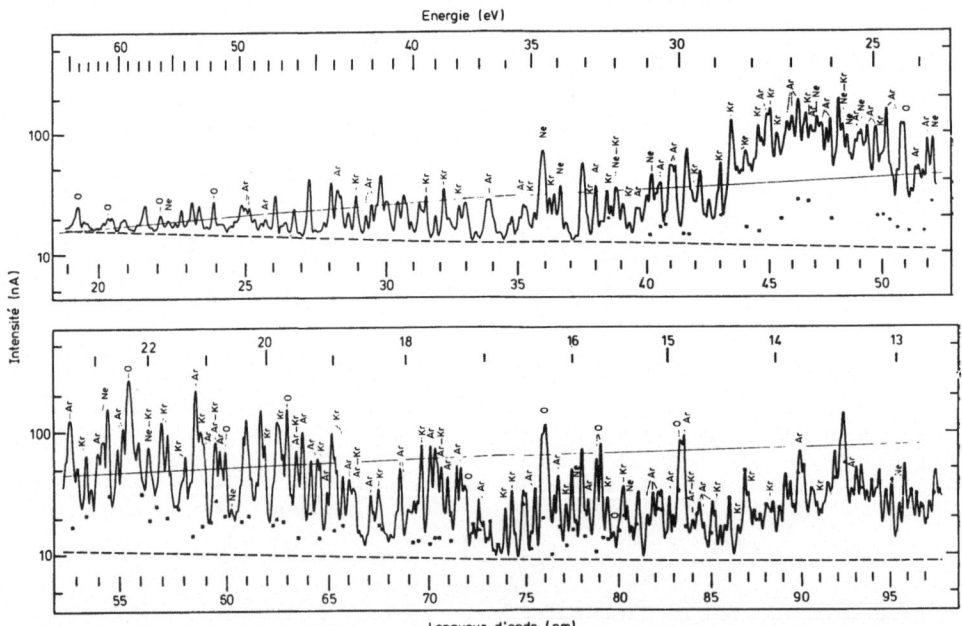

Fig. 7 Spectrum of photons excited in a mixture of noble gases by a discharge. The considerable variety of lines shows that when the excitation is produced by high-temperature electrons the emission is not restricted to a few limited lines or bands (from Ref. 10).

of VUV photons not displayed on the above-mentioned spectra. This is evident from the operation of some counters in the "limited streamer mode" or some Geiger counters if we accept the generally adopted scheme of avalanche propagation by VUV photons. It is illustrated by the wealth of lines observed in a mixture of noble gases excited by a discharge[10] as seen in Fig. 7.

2. THE DETECTION OF PHOTONS BY GASEOUS DETECTORS

We consider only photons in the VUV or lower energy range. The new developments of gaseous detectors for photon imaging are in many respects connected with the photon-emitting processes in gases.

To detect a photon one has to absorb it with a gaseous component with a low enough ionization potential, or on a photocathode. The first approach was pioneered by Ypsilantis and Séguinot[11] and has been developed actively for Čerenkov light imaging. Figure 8 shows the photo-ionization and absorption properties of some vapours, together with the transparency of some optical windows.

Metallic cathodes have too low a photo-ionization yield to be of practical interest. I will not mention here alkali photocathodes, whose combination with gaseous detectors has been and still is the object of active studies[12]. I will only mention the liquid photocathode[13] recently invented by Anderson, which consists of a thin layer of tetrakis(dimethylamine)ethylene (TMAE -- a tetraaminoethylene) condensed on the cathode of a low-pressure multiwire chamber. Table 1 shows the properties of the various tetraaminoethylenes in the gaseous or liquid form. With ionization potentials of only 3.5 eV a considerable range of applications is at hand.

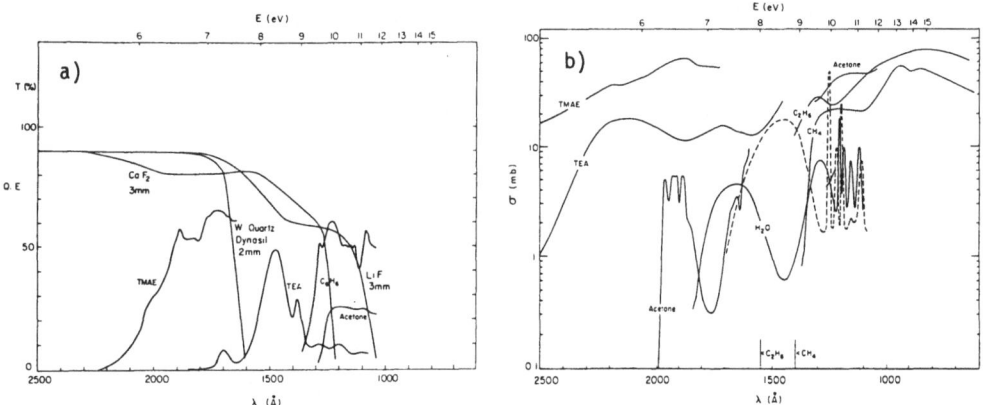

Fig. 8 a) Quantum efficiency of various vapours and transparencies of some windows. TEA: triethylamine. TMAE: tetrakis(dimethylamine)ethylene. b) Absorption cross-sections of some vapours and gases.

Table 1

Measured values of ionization potentials for several
tetraaminoethylenes dissolved in trimethylsilane

Compound	I_g (gas)		E_{th} (liquid)	
	nm	eV	nm	eV
TMAE	231	5.36	350	3.54
TMB1	229	5.41	340	3.65
TMAB	221	5.60	324	3.38
TMPD	200	6.20	281	4.40

2.1 The Geiger counter

Geiger-Müller counters were at the peak of their importance in the 1930's. They
were perfectly adapted to the existing level of low-sensitivity counting equipment and
the difficulty of constructing stable high-voltage supplies.

In the original counter working with pure noble gases the discharge in the counter was propagated mainly by secondary effects of the ions and the photons on the
cathode, releasing new generations of avalanches, with some tricks to prevent continuous discharges, such as high resistors in series or reversing voltages.

The addition of quenching vapours by Trost[14] was a serious progress which led to
the self-quenched counters. With 10^8 ions/cm the counter delivered a pulse of nearly
constant value, corresponding to the surrounding of the central wire by the ion sheath.
These slow-moving ions produce a dead-time paralysing the counter for several milli-
seconds, in general. The discharge along the wire propagates at a speed of about
3 cm/µs. The inconvenient length of the dead-time and the growing availability of
cheap sensitive electronics has led to the abandonment of the Geiger counter in most
cases.

However, now and then interest is revived in this type of operation for projects
concerned with large volumes of detectors with, for instance, 10,000 or 100,000 detecting tubes. The limitation of the discharge along the wire can be obtained by various
mechanical or electrical means, thus reducing the dead region to a limited portion of
the wire.

The possibility of exploiting the propagation velocity along a wire in order to
localize the original avalanche has also been considered. This method, first proposed
by Lauterjung and Gruhle[15], was studied for multiwire structures by myself and
Sauli[16]. It was indeed found that it is possible to choose the gas and the geometrical parameters in such a way that no propagation is observed from wire to wire in a
multiwire chamber, but only along a single wire, with extremely easy detection of the

arrival of the streamer at the end of the wire, by a pick-up electrode. The propagation velocity is in the range of 5 cm/µs and is very easy to measure. Intrinsic fluctuations limit the accuracy to about 3 mm (FWHM) for 20 cm length. This is sufficient for some applications. For large detectors the dead-time is incompatible, in most cases, with the background noise.

2.2 Limited Geiger or streamer mode

Now and then it has been observed that it is possible to obtain, in wire counters, saturated pulses as in Geiger counters, but limited in time duration, in other words not connected to a propagation along the whole length of the wire[17]. In the early study, at CERN, of the properties of multiwire chambers such operating conditions were mentioned[18].

The so-called magic gas, Ar + isobutane + freon was for a long time a favourite in many detectors because of the advantages of large, nearly saturated, pulses[19]. However, no clear understanding of the mechanism was available until two independent investigations, which showed, indirectly, that the discharge mode with the magic gas could be characterized as a limited streamer mode, or a string of avalanches not reaching the cathode[20,21]. While these studies were of academic nature, an observation made at Saclay and CERN raised a renewal of interest for practical reasons. It was observed[22] that with a proper choice of geometric parameters, fast high-current saturated pulses could be obtained, very similar to Geiger pulses, but short in time, showing that the propagation, if any, along the wire, was very limited. Photographs of the light-emitting sheath of ions around the wire confirmed this interpretation as a limited Geiger mode[23,24]. This was highly interesting at a time when large detectors for proton lifetime measurements were envisaged and when the rate limitations due to the localized dead-time region were of no importance. This was extensively exploited in the plastic tubes developed by Iarocci and collaborators[25] for a large-size proton-decay detector. It was followed by a series of systematic studies by many groups clarifying the condition for a limited Geiger or a limited streamer mode.

The study of Alexeev et al.[23,24] showed very clearly, with photographs integrating many events, the correctness of the interpretation and the various conditions for the obtainment of these modes of operation; see Fig. 9.

The study of Atac et al.[26], with an image intensifier showing individual events, permitted refinement of this study and confirmed the interpretation; see Fig. 10. The gas most suited to limited streamer operation is, according to these last authors, Ar + CH_6. They found that the dead-time region can be limited to 3 mm along the wire and are considering using this fact for counting of electrons in large-size calorimeters.

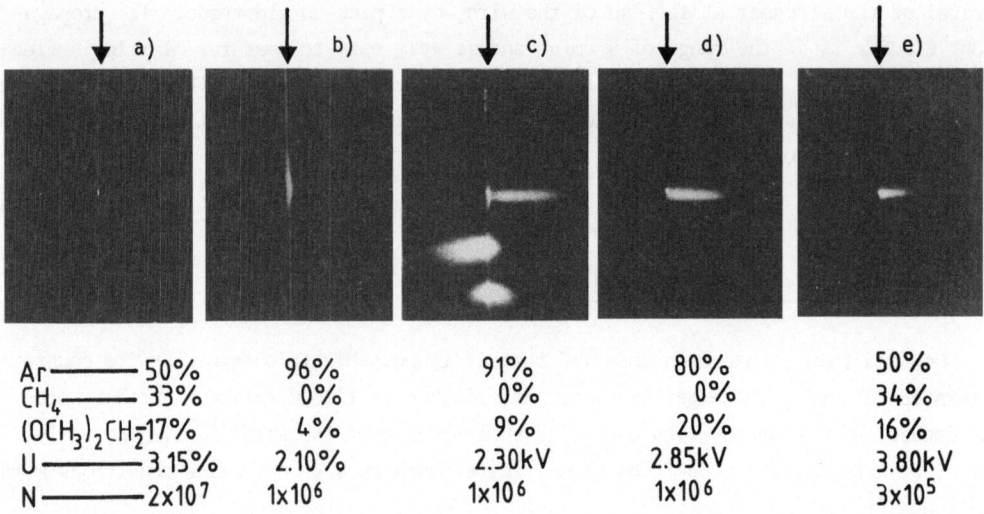

a)	b)	c)	d)	e)

Ar————— 50%	96%	91%	80%	50%
CH₄————— 33%	0%	0%	0%	34%
(OCH₃)₂CH₂–17%	4%	9%	20%	16%
U —————3.15%	2.10%	2.30kV	2.85kV	3.80kV
N —————2x10⁷	1x10⁶	1x10⁶	1x10⁶	3x10⁵

Fig. 9 Limited discharge modes. The propagation of a discharge in a detector can be limited along the wires or in cathode-anode space according to the geometry and the gases. The pictures show the distribution of avalanches under various conditions, integrated over many events (from Ref. 24).

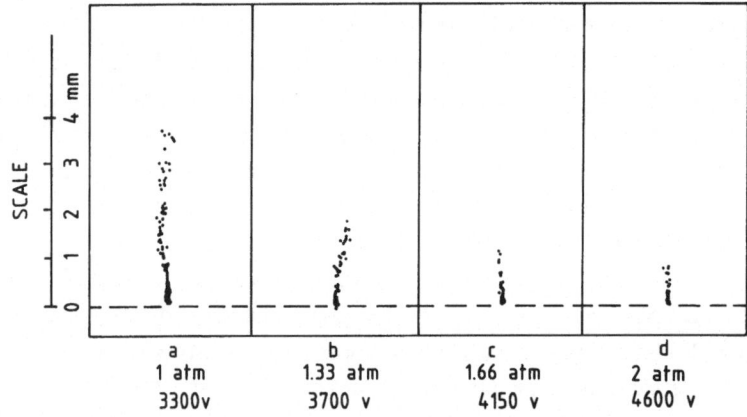

a	b	c	d
1 atm	1.33 atm	1.66 atm	2 atm
3300v	3700 v	4150 v	4600 v

Fig. 10 Self-quenching streamers at various pressures with 50% argon, 50% ethane, and 100 μm anode wire (from Ref. 25). Single events photographed with an image intensifier.

3. THE GAS SCINTILLATION PROPORTIONAL COUNTER (GSPC)

The direct exploitation of the VUV photons followed the work of Policarpo[5,27] and is a technique adopted in some fields since it gives energy resolution for medium energy X-rays nearly a factor of two better than with ordinary wire chambers (7.5%, FWHM at 5.9 keV) with even better results than solid-state detectors in the energy range close to 0.1 keV.

In this detector the X-photons are absorbed in a space filled with a noble gas, and the ionization electrons drift, through a grid, to a space of a few millimetres, where a constant electric field is adjusted to produce VUV light without electron multiplication. This VUV light is converted to visible light by a thin layer of wavelength shifter deposited on the quartz window closing the chamber. This visible light is detected by photomultipliers.

The localization of the position of the initial ionization electrons can be obtained by the proper weighting of the light pulses from several photomultipliers[28]. These counters have found applications in some space experiments[29] and have also raised interest for some medical applications as was shown by the work of Nguyen et al.[28], where, with xenon pressures of 10 atm, a sizeable efficiency is obtained up to the region of 100 keV with a much better energy resolution than scintillation counters.

One may wonder, however, whether this type of counter is not superseded now, in all cases where localization is required, by the detector described in the next section.

4. THE "PHOTO-IONIZATION PROPORTIONAL SCINTILLATION" (PIPS) CHAMBER

It is possible to detect the VUV photons produced by ionization electrons drifting in an electric field, in a proportional wire chamber, by choosing a window transparent to the VUV photons and a gas filling with an admixture of a photo-ionizable vapour[11], see Fig. 11. This was proposed by Policarpo[30] and gave rise to a rapid development. The first detector based on this principle was built with a LiF window and triethylamine (TEA) vapour, whose threshold is around 7.5 eV [2]. The energy was close to 10% (FWHM) for 5.9 keV X-rays.

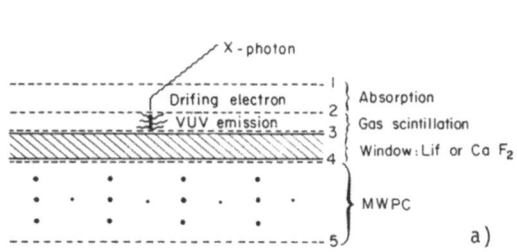 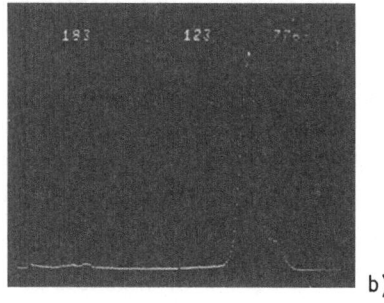

Fig. 11 Photo-ionization proportional scintillation counter (PIPS). a) Schematic view. The X-rays are absorbed in a pure noble gas. The drifting ionization electrons produce VUV light in a scintillation gap. The VUV photons traverse a window and are detected and localized in a wire chamber with a photo-ionizable vapour (TEA, TMAE, acetone, etc.). b) Energy resolution: 5.9 and 6.4 keV from a ^{55}Fe source: 7.8% FWHM (D. Anderson). Gas filling: 400 Torr Ar + 100 Torr isobutane + 0.3 Torr TMAE. CaF$_2$ window. Gas scintillation proportional counter with 760 Torr xenon.

Anderson[31] showed that TMAE vapour is quite advantageous for these applications. The threshold is 5.4 eV and permits the use of quartz windows. In addition the quantum efficiency is higher than 50%, i.e. better than any known solid photocathode. He obtained 7.3% (FWHM) for 5.9 keV X-rays and an imaging detector based on this principle has already been successfully tested in flight for a satellite observatory.

Although a localization accuracy of about 1 mm has been obtained in this detector[32], one should keep in mind that the very low vapour pressure of TMAE (0.3 Torr at 20°) is responsible for a large mean free path for VUV photon absorption, namely 1.8 cm at 20 °C and about 1.2 mm at 30 °C, and that a very broad distribution of several centimetres width is obtained with such a device. It is the centre of gravity of the distribution which is determined with accuracy. However, for imaging of simultaneous photons in an image, this could be a serious drawback. Triethylamine, on the contrary, permits higher partial pressures, typically 25 Torr at 20° and the mean free path is only 1.5 mm. For some applications this may be an essential feature. The coupling of a scintillation xenon chamber with photo-ionizable multiwire chambers is also giving rise to active research for medical applications, where imaging detectors with a high-energy resolution is desirable in some specific cases.

5. THE SINGLE-PHOTON VUV IMAGING

The imaging of single photons in the VUV, UV, or visible range obviously plays a considerable role in many domains of science. In high-energy physics the possibility of identifying charged particles of given momentum by their Čerenkov angle in various radiators is of primordial importance. Since the first suggestion[11] of using photo-ionizable vapours in multiwire chambers for this purpose, rapid progress has been accomplished.

Multistep structures[33] have proved to be excellent tools for the detection and amplification of single electrons produced by the ionization of vapours such as TEA or TMAE. They permit higher amplifications than single-step wire chambers by reducing substantially, for a given level of amplification, the secondary electrons liberated by the photons or the ions produced in multiplicative avalanches.

A chamber with a surface of 80 × 40 cm² is being operated in an experiment at Fermilab. It gives a two-dimensional accuracy of about 400 μm (r.m.s.). Figure 12 shows VUV images obtained with such a chamber. For cases where a considerable multiplicity of photons has to be handled simultaneously, a drift chamber making use of the large mean free path of VUV photons in TMAE (1.2 cm at 30 °C) offers a very good solution. The two-dimensional distribution is then transformed into a three-dimensional distribution. Photons very close to each other being absorbed at different depths can

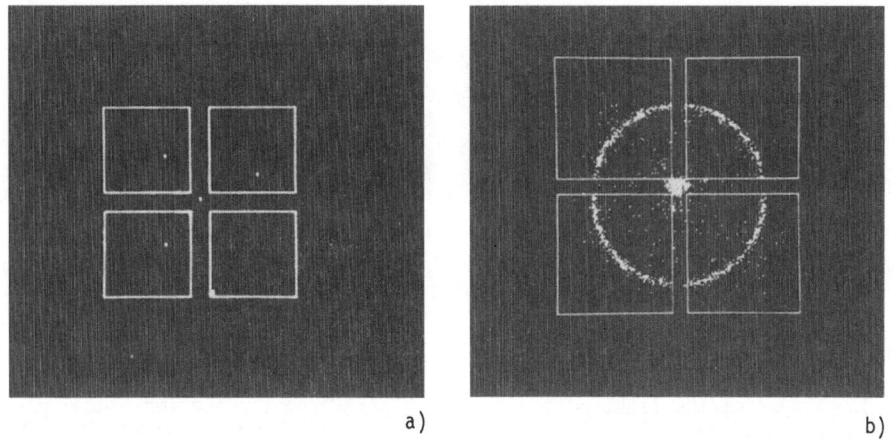

a)   b)

Fig. 12 Images of Čerenkov light in a multistep avalanche chamber (from Ref. 31).
First step: parallel gap, second step: MWPC, gas mixture Ar + CH$_4$ + TEA. Total gain
5×10^7. a) Single event; b) 200 events. Čerenkov ring radius = 6.8 cm. Beam of
200 GeV pions.

be more easily detected without ambiguity. Čerenkov rings containing about 20 detected
photons are easily recorded. The gas used in the drift chamber is pure methane with
TMAE and, possibly, a small admixture of isobutane[34]).

The main difficulty of the approach seems to be the copious emission of VUV photons by avalanches in methane, giving rise to parasitic photoelectrons. The multistep
chamber offers interesting possibilities in the suppression or reduction of these parasitic effects. The transfer space from the first amplifying gap to the second amplifying structure can be made long enough to absorb the parasitic photons before they
can reach the conversion space where the VUV photons to be imaged are absorbed. However, multistep structures are not compatible with all the gas mixtures. We have so
far observed that helium + isobutane + TMAE works satisfactorily with a multistep
structure and should be a good candidate. In this case a multistep structure of a
large depth, say 5 cm, could easily permit the separation of photons absorbed at different depths and thus resolve large multiplicities in the Čerenkov rings into much
smaller multiplicities within a narrow range of drift times.

THE IMAGING OF PHOTONS BELOW 1 keV BY ELECTRON COUNTING

The imaging of 1 keV electrons illustrates the possible advantages of using the
photons emitted by avalanches in gaseous detectors instead of the induced charge
pulses.

The detection of particles by the flash of light produced by the avalanches near
a wire has been exploited in the scintillation drift chamber[1,3,35,36]). The advantage
was the much smaller effect of space-charge, since the detection is possible with a

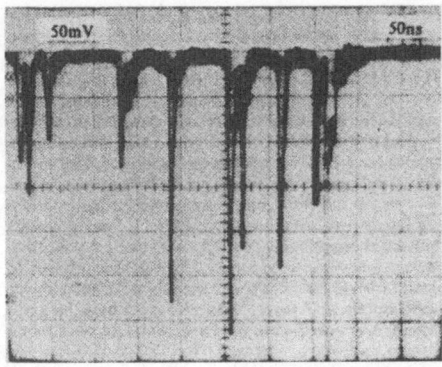

Fig. 13 Traces of the light pulses for a C K X-ray event. The individual peaks correspond to avalanches initiated by single electrons. The gas used was Ar + CH_4 (4%) + + CO_2 (5%) + N_2 (7.5%) with a drift-region field strength of $\sim$ 10 V cm^{-1} and a gas gain of 10^6 (from Ref. 17).

smaller amplification resulting in much higher possible rates. However, the intrinsic duration of the light pulses limited the particle separation to nearly the same value as with the use of the charge-induced pulses. Siegmund et al.[37] showed that, by a proper mixture of N_2 and quenching agents with Ar, it is possible to reduce the pulse width of the light flashes produced by avalanches in a gap limited by parallel grids to 2 ns only, almost a factor of ten better than with the induced pulses. By absorbing the X-rays in a conversion space with a low electric field transferring electrons to the amplifying gap it is possible to separate, with a photomultiplier, the pulses produced by individual electrons (Fig. 13). The energy resolution is then limited only by the fluctuations in the gain.

Up to 1 keV this method seems promising and gives the ultimate possible resolution with a gaseous detector. We have shown, at CERN[38], that the UV light produced in such a mixture, which is essentially due to the addition of N_2, can be conveniently converted, with almost 100% yield, by a plastic wavelength shifter, thus permitting very flexible geometrical structures.

7. THE DETECTION OF PHOTONS WITH A LIQUID PHOTOCATHODE
IN A MULTIWIRE CHAMBER

Anderson[13] has investigated the possibility of detecting photons with liquid TMAE condensed on the cathode of a multiwire chamber. Table 1 shows that liquid tetraaminoethylenes permit reaching a photo-ionization threshold as low as 3.5 eV. The first tests with condensed TMAE showed that a quantum efficiency of 1% was achieved with a threshold at about 4.2 eV. Interesting applications appeared immediately from the observation that a BaF_2 scintillator emits photons detectable by this photocathode.

An efficiency of about 50% was achieved with 0.6 MeV γ-rays and this may offer interesting applications for positron cameras. At energies in the GeV range the energy resolution can be quite good; and this first happy marriage of a high-density scintillator, of only 2.2 cm radiation length, with a multiwire chamber offers great prospects for photon calorimeters, since the spatial development of a shower can be obtained together with the total energy loss.

3. CONCLUSIONS

Developments of recent years show that the gaseous detectors of the multiwire type, or with amplification between parallel grids, permit a considerable extension of their applications, owing to the progresses made in the detection of single photons of energies ranging from 4 to 10 eV.

The detection of the VUV photons emitted by scintillation gaseous converters permits an improved energy resolution in imaging X-ray detectors.

The detection of VUV photons from Čerenkov light permits a serious progress in particle identification.

The detection of the scintillation light from BaF_2 crystals permits the development of a new class of calorimeters.

REFERENCES

1) G. Charpak, S. Majewski and F. Sauli, Nucl. Instrum. Methods 126 (1975) 381.
2) G. Charpak, P. Policarpo and F. Sauli, IEEE Trans. Nucl. Sci. NS-27 (1980) 212.
3) G. Charpak, S. Majewski and F. Sauli, IEEE Trans. Nucl. Sci. NS-23 (1976) 202.
4) G. Charpak and G.A. Renard, J. Phys. Radium 17 (1956) 585.
5) A.J.P.L. Policarpo, Physica Scripta 23 (1981) 539.
6) G.S. Hurst and A. Klots, Adv. in Rad. Chem. 5 (1976) 1.
7) V.D. Peskov, J. Applied Spectroscopy (USSR) 30 (1979) 860.
8) M. Suzuki and S. Kubota, Nucl. Instrum. Methods 164 (1979) 197.
9) T. Takahashi, S. Himi, M. Suzuki, Jian-zhi Ruan (Gen) and S. Kubota, Emission spectra from Ar-Xe, Ar-Kr, Ar-N_2, Ar-CH_4, Ar-CO_2 and Xe-N_2 gas scintillation proportional counter, Rikkyo University (Tokyo, Japan) Report RUP 81-16, Dec. 1981.

10) S. Girard, G. Lévêque and J. Robin, J. Phys. E: Sci. Instrum. 12 (1979) 719.

11) J. Séguinot and T. Ypsilantis, Nucl. Instrum. Methods 142 (1977) 377.

12) F. Sauli, Rediscovering the gaseous photodiode, preprint CERN-EP/82-26 (1982), submitted to Nucl. Instrum. Methods.

13) D. Anderson, Extraction of electrons from a liquid photocathode into a low pressure wire chamber, preprint CERN-EP/82-100 (1982), submitted to Phys. Lett.

14) A. Trost, Phys. Z. 35 (1935) 801.

15) K.H. Lauterjung and W. Gruhle, A new direct localizing and measuring device for extended radiation, in 2nd United Nations Int. Conf. on the Peaceful Uses of Atomic Energy, Geneva, 1958.

16) G. Charpak and F. Sauli, Nucl. Instrum. Methods 96 (1971) 363.

17) H. Neuert, in Kernphysikalische Messverfahren (G. Braun, Karlsruhe, 1966).

18) G. Charpak, D. Rahm and H. Steiner, Nucl. Instrum. Methods 80 (1970) 13.

19) R. Bouclier, G. Charpak, Z. Dimčovski, H.G. Fischer, F. Sauli, G. Coignet and G. Flügge, Nucl. Instrum. Methods 88 (1970) 149.

20) G. Charpak, G. Petersen, A. Policarpo and F. Sauli, IEEE Trans. Nucl. Sci. NS-25 (1978) 122.

21) J. Fischer, H. Okuno and A.H. Walenta, Nucl. Instrum. Methods 151 (1978) 451.

22) A. Breskin, A. Diamant-Berger, G. Marel, G. Tarlé, R. Turlay, G. Charpak and F. Sauli, Nucl. Instrum. Methods 123 (1975) 225.

23) G.D. Alekseev, N.A. Kalinina, V.V. Karpukhin, D.M. Khazins and V.V. Kruglov, Nucl. Instrum. Methods 153 (1978) 157.

24) G.D. Aleskeev, D.M. Khazins and V.V. Kruglov, Self-quenching streamer discharge in a wire chamber, Dubna report D13-12027 (1978).

25) G. Battistoni, E. Iarocci, M.M. Massai, G. Nicoletti and L. Trasatti, Nucl. Instrum. Methods 152 (1978) 423.

26) M. Atac, A.V. Tollestrup and D. Potter, IEEE Trans. Nucl. Sci. NS-29 (1982) 388.

27) A.J.P.L. Policarpo, Space Sci. Instrum. 3 (1977) 77.

28) H. Nguyen Ngoc, J. Jeanjean, H. Itoh and G. Charpak, Nucl. Instrum. Methods 172 (1980) 603.

29) J. Davelaar, G. Manza, A. Peacwick, B.G. Taylor and J.A.M. Bleeker, IEEE Trans. Nucl. Sci. NS-28 (1980) 196.

30) A. Policarpo, Nucl. Instrum. Methods 153 (1978) 389.

31) D. Anderson, Nucl. Instrum. Methods 178 (1980) 125.

32) C.J. Hailey, W.H.M. Ku and M.H. Vartanian, IEEE Trans. Nucl. Sci. NS-29 (1982) 138.

33) G. Coutrakon, M. Cribier, J.R. Hubbard, Ph. Mangeot, J. Mullie, J. Tichit, R. Bouclier, A. Breskin, G. Charpak, G. Million, A. Peisert, J.C. Santiard, F. Sauli, C.N. Brown, D. Finley, H. Glass, J. Kirz and R.L. MacCarthy, Proc. Nuclear Science Symposium, San Francisco, 1981 (ed. G.F. Knoll), IEEE Trans. Nucl. Sci. NS-29 (1982) 323.

34) E. Barrelet, T. Ekelöf, B. Lund-Jensen, J. Séguinot, J. Tocqueville, M. Urban and T. Ypsilantis, A two-dimensional, single-photoelectron drift detector for Cherenkov ring imaging, preprint CERN-EP/82-09 (1982), submitted to Nucl. Instrum. Methods.

35) M. Simon and T. Baum, A scintillation drift chamber with 14 cm drift path, preprint University of Siegen S1 82-6 (1982).

36) V.I. Baskakov, V.K. Chernjatin, B.A. Dolgoshein, Y.N. Lebedenko, A.S. Romanjuk, V.H. Fedorov, I.L. Gravilenko, S.P. Konovalov, S.N. Majburov, S.V. Muravjev, V.P. Postovetov, A.P. Shmeleva and P.S. Vasiljev, Nucl. Instrum. Methods 158 (1979) 129.

37) O.H. Siegmund, J.C. Culham, I.M. Mason and P.W. Sanford, Nature 295 (1982) 678.

38) D. Anderson and G. Charpak, Some advances in the use of the light produced by electron avalanches in gaseous detectors, preprint CERN-EP/82-05 (1982), submitted to Nucl. Instrum. Methods.

GAS FILLED HEAVY ION DETECTORS

H. Stelzer
GSI, Darmstadt

I. Historical Remarks and Introduction

Gas-filled detectors are, beside the anorganic scintillator, the oldest instruments to detect ionizing radiation. An ionization chamber was already used in the first experiments with cosmic rays in the beginning of this century. A major drawback of this instrument was the missing amplification device for the weak signal of an ion-chamber. It was not possible to record the pulse of a single ionizing event. In 1908 Rutherford and Hans Geiger published a paper (Rutherford 08) in which they described a device which was able to detect the single pulses of alpha particles. This first gas-filled detector which was able to record the pulses of individual ionizing events was in fact, in modern terminology, a proportional counter in cylindrical geometry. The internal gas-amplification was high enough that the single pulses of this counter could be directly recorded by a galvanometer. It was only in 1928 when Hans Geiger and W. Müller (Geiger 28) invented the nowadays called Geiger-Müller counting tube with its much higher internal gas gain. Here the gas multiplication around the thin wire of the tube is so high that even minimum ionizing particles like beta particles could be recorded without any further amplification. The pulse-height of such a Geiger-Müller counter is, as it is well-known, independent of the primary ionization, which is often a disadvantage. Only with the upcoming of vacuum tube amplifiers in the 1930's the advantage of a proportional counter or an ionization chamber, namely their proportionality between primary ionization and output pulse-height, could be fully exploited and came more and more into use. Nowadays, the Geiger-Müller counting tube is mainly of historical relevance, because the above mentioned non-proportionality between primary ionization and output signal and its long dead time in the order of a ms, but nevertheless, they are still in use for some special applications (Jones 81). One should point out that Geiger fully realized the importance of a high electric field in the gas amplification process and that such high fields could easily be produced around thin wires or sharp points (Geiger point counter (Geiger 12)). One may safely state that all the developments of gas-filled detectors with internal gas amplification is based on the pioneering work of Hans Geiger.

Another mile-stone in the evolution of gas-filled nuclear particle detectors is Charpak's invention of the multi-wire proportional chamber in 1968 (Charpak 68). He realized that each of thin parallel wires, spaced 2 mm apart and placed between two electrodes acting as cathodes, behaves like a proportional counter, essentially independent from each other. In this way large-area position sensitive detectors, especially needed in High Energy Physics, could easily be constructed. Within few years, nearly all electronic counter experiments in High Energy Physics, which used up to then mainly optical, acoustical or magneto-strictive spark-chambers as a position-sensing device, changed over to this new type of detector or to its close relative, the drift-chamber. It offered a much higher count-rate capability, shorter dead- and memory time and a direct data-transfer into the computer, in contrast to optical data recording, just to mention a few of the numerous advantages of a MWPC. The rapidly growing use of multi-wire chambers is, of course, closely related to the development of modern solid state electronics in the 1960's.

The instrumentation in nuclear and heavy ion physics was dominated by solid state detectors (silicium- and germanium diodes), since their upcoming in the 1960's. To the experimentalist they offered unique advantages compared to the up to then mainly used proportional counters, ionization chambers and scintillation counters: an excellent energy-resolution, much better than obtained up to then, a linear response in the charged particle's energy, a high stopping power, which resulted in small size detectors, the possibility to produce very thin energy-loss detectors, the operation of the detector directly in the vacuum of the scattering chamber, and, especially important for those physicists who want to do more serious things than detector development, the easy availability off-the-shelf. A historical review of the development of semiconductor detectors is given by McKenzie (1979), who, together with D.A. Bromley, was one of the pioneers of this new detector technology.

With the beginning of the 1970's a interest in the physics of phenomena related to very heavy ions arose. New accelerators, like the UNILAC at the GSI at Darmstadt provided beams of projectiles as heavy as uranium. Now, in the spectroscopy of very heavily ionizing radiation, some serious drawbacks of the up to then so extremely successful semiconductor detectors became apparent. I will just mention them, without going into details: the pulse-height defect, the plasma time jitter, the degradation of the detector performance after exposure to intense radiation and their small size, which made it extremely costly to build large solid-angle detection systems. For these reasons, in many laboratories a new interest in gas-filled detectors arose. In the following I will try to illustrate the research which has been undertaken during the past decade on gas-filled counters for heavily ionizing radiation, by describing some specific examples.

II. Gas-filled Ionization Chambers

II.1 General Remarks

In an ionization chamber the electrons which have been deliberated by a nuclear charged particle in the sensitive gas-volume along the track of the particle are measured. The electric field, which separates the positive and negative charge-carriers, is either parallel or perpendicular to the particle's trajectory. It turns out that the number of produced electrons, to a high degree of accuracy, depends linearly on the energy the particle has lost in the gas-volume of the chamber. This holds true for a wide range of projectile charges and velocities. The linear response of a gas-filled ionization chamber means that there is a fixed mean energy required to create one electron-ion pair. This energy, usually denoted by W, amounts to about 30 eV for the commonly used counting gases and is about a factor of ten smaller for silicon detectors. The fact that W is practically independent of the particle's energy (at least above some hundred keV) comes about because the competition between ionization and excitation-processes is rather independent of energy and the amount of energy going into kinetic energy of gas atoms is negligible Only at low energies at some hundred keV, the fraction of the energy-loss of the particle due to non-ionizing collisions increases, producing a corresponding increase in W, since there is less energy available for ionization processes.

However, one should bear in mind that the experimental evidence for the independence of W of the charge of the heavy-ion projectile is rather scarce (ICRU 79). The best thing then, and what is normally done, is to calibrate the ion-chamber with projectiles with known Z and energy. An energy-resolution as low as 0.7 % FWHM is routinely achieved, thus proving that the number of free electrons produced by a charged particle in a gas is an excellent means to determine the energy the particle has lost in the sensitive volume of the ionization chamber.

Basically two important parameters of a charged particle can be determined in an ion-chamber: 1. the total energy, when the particle is stopped in the chamber. 2. the nuclear charge Z, by measuring the energy ΔE the particle has lost on a certain path x and its total energy E. Very often gas-filled ion-chambers, especially small ones, are simply used as ΔE-counters, whereas the total energy is measured by silicon detectors mounted directly in the ion-chamber. Additionally, a gas-filled ionization chamber can be made position-sensitive by various methods, which I will discuss later. This feature is of special importance for large-size ion-chambers.

The main advantages of gas-filled ionization chambers, compared to solid-state detector telescopes, may be summarized as follows:
1. large size detectors can be built. In this way a broad range of the angular distribution of the emitted particles can be measured with one set-up. Furthermore the large solid-angle saves precious beam-time.
2. The response of an ion-chamber is essentially linear in the deposited energy. A pronounced pulse-height defect like in solid-state detectors is not observed, but recombination effects are obviously present. A heavy ion with a given energy yields a some percent higher signal, when the gas-pressure in the chamber is reduced by about a factor of two (but still high enough to stop the particle). This effect can not be explained by an electro-negative impurity.
3. no radiation damage occurs in the usually used counting gases like methane, iso-butane or tetrafluor-methane.
4. the dynamic range (i.e. the range of Z-values and energies which can be measured simultaneously) is much broader than of a solid-state detector telescope. Furthermore, the effective thickness (in mg/cm²) of the detector can easily be adopted to the experimental requirements by simply changing the gas pressure.

All these points listed above are of special importance and relevance in heavy ion physics. The demands of experiments with heavy ions have triggered most of the recent developments of gas-filled ionization chambers.

II.2 A Modern Ion-chamber

In the following I will discuss, as an example of the state-of-the-art of modern ionization chambers, a chamber which is currently be assembled by our group (Gobbi 81) at GSI. An artist's view of the

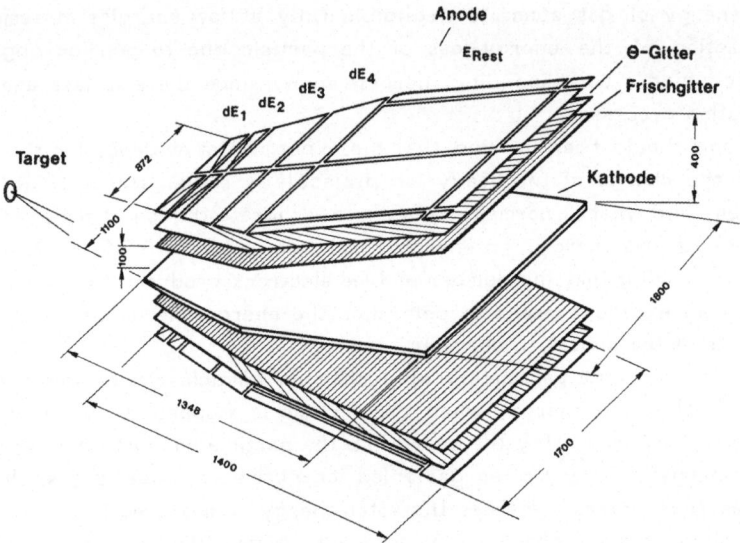

Fig. II.1 Perspective view of the Mammoth Twin ion-chamber.

Mammoth Twin ion-chamber is shown in Fig. II.1. The geometrical dimensions of the detector have been determined by two criteria: the demands of the experiments to be performed with this instrument and, on the other hand, the technical feasibility of this device.

This ion-chamber will be used in a new kinematic coincidence set-up for experiments with heavy ions with up to 20 MeV/u kinetic energy. In deep inelastic collisions, which will be further investigated with this apparatus, the projectile-like reaction products have nearly the projectile velocity, when the collision occured with a low Q-value. To stop heavy ions with this energy in hydrocarbon gases, one needs ~30 mg/cm² (for ^{238}U) and ~80 mg/cm² (for ^{20}Ne) (Huber 80). The active depth of the chamber has been chosen to 125 cm, which corresponds to a thickness of 9 mg/cm² per 100 mbar methane or 33 mg/cm² per 100 mbar iso-butane. With an operating gas-pressure of some hundred mbar, even the light and fast reaction products are stopped in the ion-chamber. Of course, a higher gas-pressure would reduce the necessary depth of the chamber, but there are several reasons against it:

- low energy particles would stop directly behind the entrance window and there always slight inhomogeneities of the electric field are present, which may cause incomplete electron collection by the anode.
- a higher gas pressure implies a thicker entrance window which causes more energy - and angle - straggling and a stronger support structure for the entrance window, which yields dead zones.
- it is still an open question wether at a higher gas pressure recombination effects do occur, at least for the heaviest ions. Again, a lower operating gas pressure is advisable.

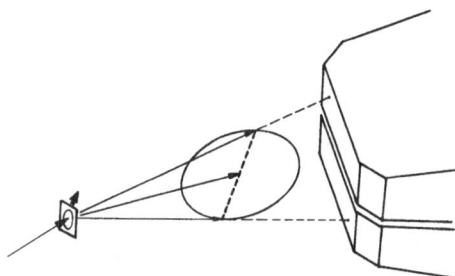

Fig. II.2 Sketch of the kinematics of a splitting process.

The general lay-out of the chamber is based on an earlier design (Sann 75). It is again a twin-chamber, but now the lower and upper half (fig. II.2) are completely independent detectors. The chamber body is made of ordinary structural steel, the interior is electroplated with nickel.

A massive iron-plate in the medium plane of the detector separates the gas-volume of the two ion-chambers and is strong enough to operate the upper and lower half at completely different gas-pressures. In this way particles with largely different energies can be measured simultaneously and the dynamic range is greatly increased. Such operating conditions are necessary, when one wants to investigate heavy ion reactions, where a high-energy forward emitted projectile-like fragment undergoes splitting. The resulting secondary fragments emitted in forward and backward direction within the moving frame of the primary particle are then observed at forward angles in the laboratory frame with significantly different energies, a kinematic situation which is sketched in fig. II.2. The relative angle of the two splitting fragments ranges from 0 to 60°. The entrance window of this twin ionization chamber is 800 mm wide (in the plane) and 2 x 100 mm high (out of plane). This dimension has been dictated by the technical feasability of the window.

Located 700 mm away from the target, the detector has an angular acceptance of ±30° (in the plane) which just fits to the above described splitting process. The distance of 700 mm of the entrance window to the target is dictated by the need of a reasonable long flight-path for time-of-flight measurements. The time stop-signal is provided by a large-area, but nevertheless very thin parallel-plate-avalanche counter, which is installed directly in front of the chamber. The massive plate in the medium plane (fig. II.3), which separates the upper and lower chamber, carries on both sides the cathode planes. The structure, which is ± 10° inclined to the medium plane, carries the anode plates, together with the Frisch-grid and the θ-grid.

The measurement of the energy the nuclear particle has lost across a given path (the width of the dE-plates) together with its total energy (equal to the sum of the signals on all anode plates) yields information about its nuclear charge. An optimum Z-resolution is obtained, when the ratio of $\Delta E : E_{total}$ is as high as .75 (Sistemich 76). The large dynamic range of the nuclear charges of the products from a heavy ion reaction requires a subdivision of the anode in several independent electrodes, which makes it possible to form the optimum $\Delta E : E_{total}$ ratio over the whole energy- and nuclear charge range. These considerations led us to the design of the anode shown in fig. II.1. There are four plates for the energy-loss measurement of the particle across the dE plates, which are 60, 60, 200 and 200 mm resp. wide. The last big plate determines the residual energy E_{Rest}. All the anode plates are split in two halves to lower the capacitance.

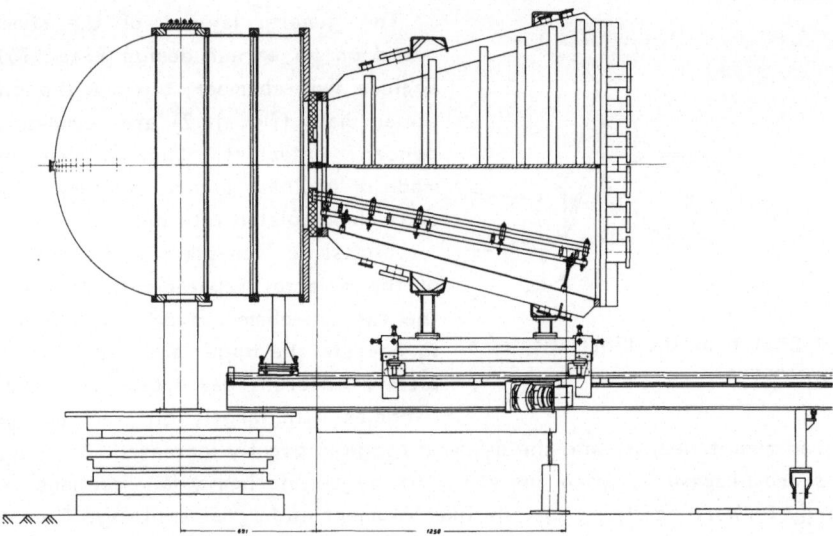

Fig. II.3 Side-view of the Mammoth ion-chamber with the scattering chamber.

The distance between the Frisch-grid and the anode is 40 mm. Between these two electrodes the so-called θ-grid is mounted. This grid consists of 50 μ wires which run parallel to the track of a particle coming from the target. The wire-spacing at the entrance window is 2 mm and at the chamber end 6 mm. This electrode determines the in-plane scattering angle. When the primary electrons drift to the anode, they induce a signal on the nearest wire when they pass through the θ-grid. All the wires are soldered to a discrete L-C-delay line and the wire which carries the signal is determined by measuring the time-difference between the signals at the two ends of the delay-line. With smaller chambers of this type, a position resolution of about 2 mm has been achieved by this method. The out-of-plane coordinate is measured via the time-difference between the passage of the charged particle through the chamber and the anode signal, since this signal only appears when the primary electrons have passed the Frisch-grid. This method to measure the Y-coordinate works of course only if there is a constant drift-velocity and hence a constant reduced field-strength E/p throughout the active chamber-volume. In earlier, smaller chambers, the time-difference between the cathode signal, which is prompt with the passage of the particle, and the anode signal was used to extract the Y-coordinate. But here in this big chamber the cathode signal will have a too slow rise-time to get out a reasonable timing information. Instead, the already mentioned parallel-plate avalanche counter placed in front of the ion-chamber will provide the timing. Another important aspect of this avalanche counter is to provide the fast trigger signal that a particle has hit the ion-chamber. All the signals of this chamber are rather slow, in the μs range, and, when using the ion-chamber in coincidence with other fast detectors, for example parallel-plate avalanche counters as recoil detectors, one would have to delay all the fast signals by some μs.

The cathode is on negative potential, the Frisch-grid is on ground, and the anode and the θ-grid are on positive potential. These last two are chosen such that no electrons are lost on the two grids. In the cathode Frisch-grid volume one needs a homogeneous reduced electric field strength of E/p = 1 V/(cm • Torr), when the chamber is operated with pure methane. At this field methane has a maximum drift velocity of 10 cm/μs. With one constant cathode potential all along the depth of the chamber, the electric field would fall off as 1/Y, Y being the cathode Frisch-grid distance, which varies between 100 and 400 mm. This strongly varying field not only influences the drift-velocity, a effect which deteriorates the position determination in the Y-coordinate, but has furthermore the desastrous effect that the primary electrons are not properly collected by the different dE anode plates.

The electric field and the potential distribution has been calculated with a computer program (Stelzer 82). A quite homogenoeus electric field with a correct collection of the primary electrons onto the different anode-plates can be achieved when the cathode is subdivided into segments. Each segment is on a potential according to its mean distance to the Frisch-grid. At the entrance, the first cathode segment is 4 mm wide, and each of the following is 4 mm wider than the preceeding one. Thus one needs about 23 different segments. Such a structure can easily be realized with printed-circuit material.

The chamber will be ready for the first test runs till the end of this year, the first experiments with this big detector are planned in spring 1983.

II.3 A new way to determine the in-plane scattering angle in an ion-chamber

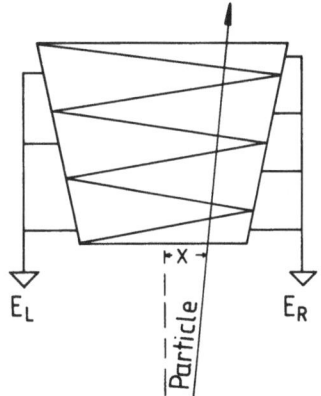

Fig. II.4 Position determination by the triangle method (Rosner 81).

Recently a new position measurement for ionization chambers has been reported by a group from the MPI Heidelberg (Rosner 1981). They cut the first ΔE-anode plate in triangles and connected the left and the right triangles of this saw-bladed structure to two amplifiers (fig. II.4). The energy-loss of a particle divides between left and right triangles according to the impact position x of the particle, and this impact position is given by the simple relation:

$$x = (E_R - E_L)/(E_R + E_L)$$

They report good linearity and obtained a position resolution of 0.3 mm with 132 MeV ^{32}S ions.

Our group (Gobbi 81) tried this method in our 120 cm deep ionization chamber and compared it with the position obtained with the above described θ-grid. This θ-grid method gives a sufficiently good position information (Δ x ~ 2 mm), but for short tracks, when the particle stops already after some cm, the resolution gets worse or the θ-signal is even missing. With the new method we did not achieve the good results reported by Rosner et al. We observed a 5% cross-talk, measured with a pulser between adjacent segments, which gets a big correction, when the signals on the left and right side are very different for large x . Furthermore, the specific ionization of the particle increases along the path across this saw-bladed structure. In principle one can correct for this effect, but this makes the analysis more complicated. Finally, one should notice that this method needs twice the amount of electronics. Normally we use the θ-grid to determine the in-plane scattering angle, but for events with no θ-signal we apply the triangle method.

II.4 Bragg-Curve Spectroscopy

The most simple way to extract the information from an ionization chamber is to col-lect just all the electrons a nuclear charged particle has deliberated in the active gas-volume of the chamber on one anode plate. In this way the total energy the particle has spent in the chamber is measured. In addition to the total energy information about the nuclear charge of the particle can be obtained by determining the portion of energy the particle has lost over a given path-length x. This can be achieved by subdividing the anode into two or more seperately read-out plates, as it is done in the big chamber described in the preceding chapter. Even more detailed information about the nature of the particle and its physical properties can be obtained by determining the spatial dis-tribution of the electrons produced along the path of the particle in the gas, i.e. if one measures the Bragg-curve. This Bragg-curve spectroscopy (BCS) has first been pro-posed and tested by C. Gruhn (Gruhn 79 and 82). Another BCS-detector with very interesting figures-of-merits has been built by a group of the TU Munich (Schiessl 82).

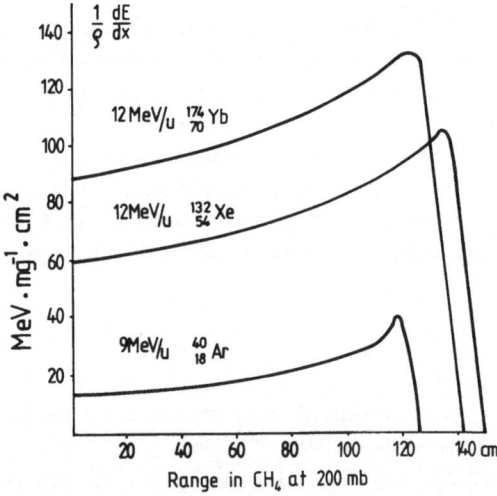

Fig. II.5 Bragg-curves of three different ions in methane at 200 mb.

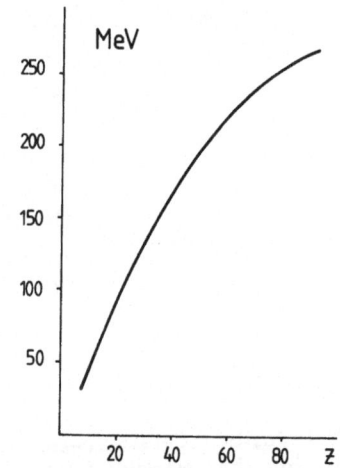

Fig. II.6 Energy-loss of a heavy ion with nuclear charge Z on its last 20 cm in methane at 200 mb.

Let me first try to point out why the Bragg-curve spectroscopy is a promising meth-od to determine the energy and the identity of a heavy ion. In fig. II.5 are plotted the Bragg-curves of three different heavy ions, of Argon, Xenon and Ytterbium at 9 MeV/u and 12 MeV/u resp. The specific energy-loss is plotted versus the path length of the heavy ion in methane at 200 mb. I have chosen this representation because, firstly, methane is our favourite ion-chamber gas due to its high drift-velocity and secondly, to illustrate the geometrical size of detectors needed to measure heavy ions at these energies. The enhancement of the Bragg-peak at the end of the path of the particle ranges from about 3 for the "light" heavy ion down to about 1.5 for the "heavy" heavy ion. The dE/dx-values in this figure and the following have been taken from the Northcliff-Schilling tables (Northcliff 70). What is actually measured is not the peak of the dE/dx-curve, but the energy the heavy ion has lost on its last portion of its range. By looking at fig. II.5, when one stays with methane at 200 mb as chamber gas, one sees that one should measure the energy the particle has lost on its last 20 cm. The energy a heavy ion with nuclear charge Z looses on its last 20 cm in methane at 200 mb is plotted in fig. II.6. The slope of this curve ranges from 4.8 MeV per nuclear charge at low Z-values down to around 1.6 MeV per nuclear charge at high Z-values. This means one needs an energy-resolution in the Bragg-peak of about 0.5 % to resolve single Z-values at high Z, which is certainly not easy to achieve.

How has such a BCS-detector been realized by the two groups? Both ionization chambers have the electric field lines parallel to the particle track. The collection of the electrons along the particle track has the inherent disadvantage that now the charge carriers are in close neighbourhood for a much longer time than in a configuration where the electric field is perpendicular to the track. Recombination effects may lead, at least for very heavily ionizing particles, to a loss of electrons. A schematic view of the detector of the Munich group is shown in fig. II.7.

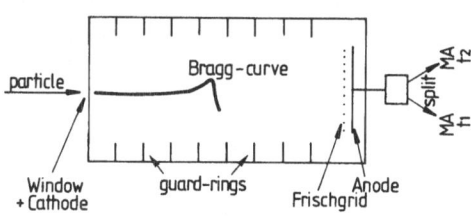

Fig. II.7 BCS-detector of the Munich group (Schiessl 82).

The entrance window serves as the cathode, which is on ground potential. The electrons the particle has produced in the gas drift to the anode and are sensed by the amplifier when they have passed the Frisch-grid. The chamber is 16 cm deep. The separation of the ionization in the Bragg-peak from the total ionization is done by a simple trick: the preamplifier output signal is split and fed into two main-amplifiers with different shaping times t_1 and t_2. T_1 is about the collection time of the electrons from the Bragg-maximum and the signal from this main-amplifier is thus proportional to the ionization around the Bragg-peak, and, as we have seen from fig. II.6, is therefore a unique function of the nuclear charge. The shaping time t_2 of the second amplifier is chosen to be longer than the total collection time of the electrons and is thus proportional to the total energy.

Both groups report a Z-resolution of about 1.2 % at Z = 16 (Schiessl 82) and at Z = 26 resp. (Gruhn 82). This figure-of-merit is about the same as the one achieved with 'conventional' ion-chambers, where the electrons are collected perpendicular to the track and the nuclear charge is determined by a $\Delta E-E_{Rest}$ measurement on two anode plates. This method works best, as already mentioned, when ΔE and E_{Rest} have about the same magnitude, and under this optimum condition, a Z-resolution of 1.1% at Z = 92 has been achieved (Sann 80). But, since the length of the anode plates is fixed, this optimum condition is only fulfilled in a rather limited energy- and Z-range. By principle of operation, a BCS - detector always takes the optimum part of the ionization track as the Z-signal and thus should have a much bigger dynamic range with good Z-resolution.

One way to take advantage of the Bragg-peak measurement in an ion-chamber with the electric field lines perpendicular to the track would be to divide the anode in 10 cm wide plates. The two or three last anode plates which show a signal will be added together and taken as the Bragg-peak signal (cf. fig. II.5,6). Combined with the measurement of the total energy by summing up all the anode plates, this should yield an optimum Z-resolution over a wide dynamic range. A chamber with 140 cm depth would need 14 preamplifiers, mainamplifiers and ADC channels; an amount of electronics still tolerable.

III. Low-Pressure Proportional Counters

III.1 Parallel-Plate-Avalanche Counters

Since their upcoming in the instrumentation technique for heavy ion experiments in 1975 (Hempel 75, Stelzer 76), parallel-plate avalanche counters (PPAC) have found wide-spread use predominantly as timing detectors for heavily ionizing radiation. Its simple design and operation, its excellent timing results (a time-resolution of 120 ps (FWHM) is obtained with a small size detector, for large-area detectors a time-resolution of 200-300 ps FWHM is routinely achieved under experimental conditions) and its reliability are one of the main attractive features of this gas proportional counter. Very soon a PPAC could be made position-sensitive by inserting a grid of wires between the anode and cathode and by dividing the cathode into stripes (Breskin 77, Eyal 78, Just 78, Harrach 79).

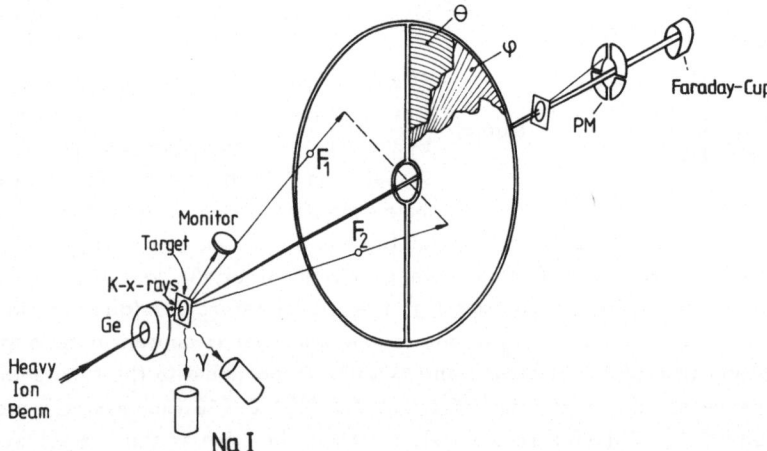

Fig. III.1 Experimental set-up with a ring-counter (Bock 82).

As an example of the great flexibility one has in the design of a PPAC I would like to mention the ring-counters (Gaukler 77) which have been built in the last years in our detector laboratory (Bock 82, Lieb 82). The cathode made of printed-circuit material has a etched ring pattern; the rings are read out by means of a tapped delay-line and thus the scattering angle θ of the particle is directly determined. The azimuthal angle ᶲ is obtained by inserting a grid of wires, with the wires in radial direction, between anode and cathode; if only a rough ᶲ-determination is needed, the subdivision of the anode-foil into azimuthal segments will do the job. The counter has a hole in the middle to let the beam pass through. A typical experimental configuration with such a ring-counter is shown in fig. III.1 (Bock 82). The two reaction-products from a binary reaction of the beam in the target are detected in the ring-counter, which measures their scattering angles and their time-of-flight-difference. These quantities completely determine the kinematics of the binary reaction. The sodium-iodide crystals around the target measure the accompaning ᵧ-rays. The counters denoted by PM monitor the beam position.

A problem specifically met in heavy ion physics is the wide dynamic range of the specific ionization dE/dx of the particles to be detected. Sometimes particles have to be registered which differ in their energy-deposit in the gas of the PPAC by more than a factor of 100. In an experimental environment with a high flux of very heavily ionizing radiation the high-voltage applied to the PPAC has to be lowered by about 10 or 20 volts to avoid sparking, and 100% detection efficieny for light particles, e.g. α-parti-

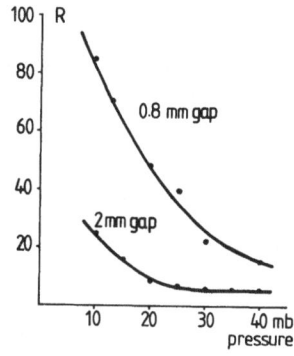

Fig. III.2 Quenching in a PPAC (cf. text).

cles, is no longer achieved. We found out a simple trick to improve the dynamic range of the detection efficiency of a PPAC by simply increasing the operating pressure from the usual 10 mb to about 30 mb. Obviously, at this higher pressure a self-quenching effect in the gas-multiplication process occurs. In fig. III.2 is plotted the pulse-height ratio R for fission fragments and α-particles from a ^{252}Cf-source versus the operating gas-pressure. The small-size PPAC with a 0.8 mm anode-cathode gap shows at a low pressure a pulse-height ratio of about 90, a value one expects from dE/dx tables (Northcliff 70) and levels off to about 10 at > 40 mb pressure. The large-size detector with a 2 mm gap exhibits already at low pressure a strong quenching effect. A ratio of ~ 30 has also been observed by other groups (Hempel 75, Just 78). Above 25 mb the pulse-height ratio of fission fragments and alpha-particles is independent from the gas-pressure and is about 6. Our group (Gobbi 81) has already often made use of this feature of a PPAC. "Light" heavy ions are efficiently suppressed by operating the PPAC with a low pressure and a reduced high-voltage, whereas at a higher pressure and increased HV all the light ions are registered with full efficiency.

A French group (Urban 81) investigated the response of a PPAC with a 5 mm gap at 100 mb i-butane to minimum-ionizing particles, which have a specific ionization a factor of 1000 smaller than α-particles at 5 MeV. In a planar electrode configuration it is not possible to maintain at 100 mb pressure the high electric field one needs to multiply sufficiently high the few primary electrons the particle has left behind in the detector. The maximum attainable reduced field strength is 80 V/(cm • mb), about a factor of five to six smaller than normally reached in low-pressure counters. The efficiency is thus only 80%, the time-resolution is 3.7 ns (FWHM).

As long as we use PPAC's in our laboratory, there is a discussion about the question whether the very low energy secondary electrons emitted by the heavy ion traversing the cathode foil contribute to the signal. These electrons which are produced directly at the cathode are most important for a timing measurement, since they have always a constant drift-time and by far the biggest gas-multiplication across the whole anode-cathode gap (according to the Townsend-law of the gas-multiplication process). The yield of secondary electrons depends on the specific energy-loss of the heavy ion and ranges typically between some ten and some hundred in the forward direction (Pferdekämper 77). The yield of primary electrons due to ionization in 1 mm i-butane at 10 mb is roughly about a factor of 20 higher, thus outnumbering completely the secondary electron emission process. One could think of evaporating the cathode with a substance, like lithium-fluoride, to enhance the secondary electron yield. Tests with such a counter are planned in our laboratory in the near future.

III.1 The Double-grid Avalanche Counter

Following the idea of Charpak's multi-step-avalanche chamber, where the overall multiplication process is subdivided into two steps (Charpak 78, Breskin 79), one of us (Lynen 80) suggested to use this principle of operation in a very low pressure heavy ion detector. The schematic view of such a double-grid-avalanche counter (DGAC) is shown in fig. III.3. The detector consists of four electrodes, denoted by S, K, A and T. S and T are thin mylar foils (1.5 µm) or stretched polypropylen foils (0.6 µm) uniformly evaporated either with silver or gold.

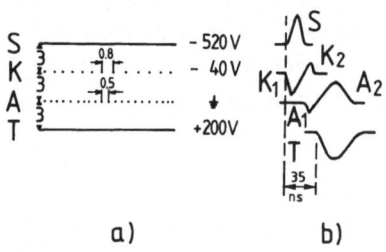

a)

b)

Fig. III.3 Schematic view of a double-grid avalanche counter.
a) electrode configuration.
b) shape of the observed signals.

The operating gas-pressure is 2 mb i-butane. The electrodes K and A are orthogonal wire grids of 20 µm gold-plated tungsten wires. The wire-spacing is 0.8 mm on the K-electrode and 0.5 mm on the A-plane. The spacing between adjacent electrodes is 3 mm. Typical potentials applied to the different electrodes are indicated in fig. III.3 as well. Fig. III.4 shows a plot of the electric field lines in such a configuration (Stelzer 82). One distinguishes three regions with quite a different field strength. In the high field region between S and K a reduced field-strength $E/p = 800$ V/(cm $\bullet$ mb) is reached, nearly twice as high as normally

achieved in PPAC's. In this gap the main amplification process occurs. Most of the secondary electrons are caught by the wires of the K-grid and there a fast (rise-time 3ns) negative signal (labelled K_1 in fig. III.3b) appears, which induces on S a fast positive signal. Close in the vicinity around a wire of the K-plane, in the very high field region, most probably a second amplification occurs, as already pointed out by Breskin (1982). The rest of the electrons drift into the low-field region K-A with $E/p = 150$ V/(cm $\bullet$ mb), where only little gas-amplification occurs. Practically all the electrons pass through the A-grid and reach the medium-field region A- T ($E/p = 300$ V/(cm $\bullet$ mb)), where again multiplication takes place. A rather slow negative signal with a rise-time ~ 12 ns appears on T, which induces a positive signal A_2 on the A-grid. The signal-component A_1 most probably comes from electrons caught by the A-plane, which induces the K_2-component on K. The signal on T is 35ns later than the signal on K and S. This time-difference can be used to estimate the drift-velocity of electrons in iso-butane at these high E/p values. The distance between K and T is 6 mm, hence the drift-velocity is 17 cm/µs, about a factor of three higher than at low E/p values (Bohrmann 76).

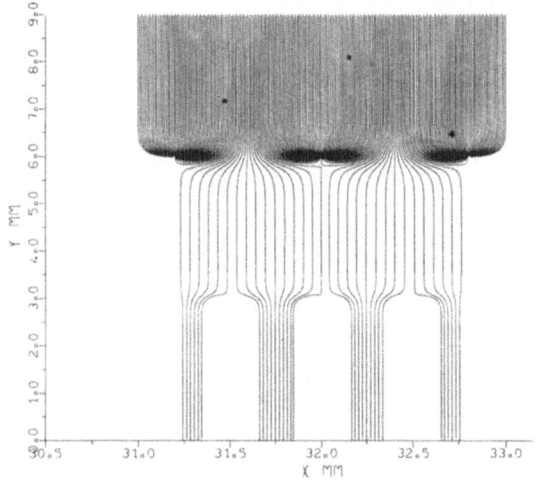

Fig. III.4 Electric field lines in a DGAC. Electrode 'S' is located at Y=9 mm. Potentials as in Fig. III.3.

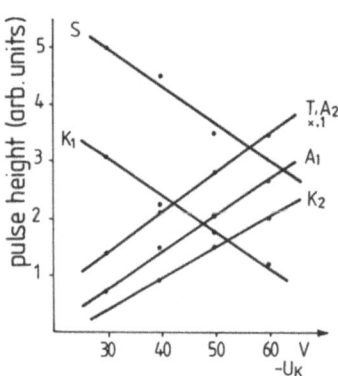

Fig. III.5 The signals of a DGAC as a function of U_K.

This qualitative picture of the gas-amplification process and of the development of the signals on the different electrodes has been gained by studying the pulse-height and the pulse-form of all the signals as a function of the potentials applied to the four electrodes. Fig. III.5 shows the pulse-height of the signals as a function of the potential U_K applied to the K-grid. This plot clearly shows the mutual dependence and correlations of all the signals. A more negative potential on K decreases the field between S and K, but increases the field in the second amplification gap.

All the signals are in the order of some mV and, after a fast, low-noise amplifier, they all show an excellent signal-to-noise ratio. The signal from the S-electrode is taken as a timing-signal. The measured timing peak of the light fragments of a [252]Cf source has a width of 400 ps (FWHM). Taking into account the intrinsic velocity-distribution of these fragments, their energy-loss-straggling in the foils of the preceding start-counter and the contribution of the start-counter, one calculates a time-resolution of the DGAC (active area 20 x 30 cm²) of 200 ps (FWHM). By looking at the field-line distribution of such a counter (fig. III.4) one should expect a strong variation of the driftpath-length and hence of the drift-time of the electron to the nearest wire of the K-grid, depending on the impact point of the nuclear particle on the counter. But due to the transversal diffusion of the electron cloud on their way to the K-grid, these driftpath differences are smeared out. The transversal diffusion of the electrons during their drift through the 3 mm S - K gap can be estimated as follows:

1. at atmospheric pressure, at low E/p values, the transversal diffusion is (Jean-Marie 79): δx = 235 µm per 10 mm drift

2. the diffusion scales with 1/SQRT(p) at constant E/p (Farr 78)

Neglecting the E/p dependence, one gets at p = 2 mb:

ϑ x = 0.235 • SQRT(500) • 0.3 mm FWHM

ϑ x ~ 1.60 mm FWHM.

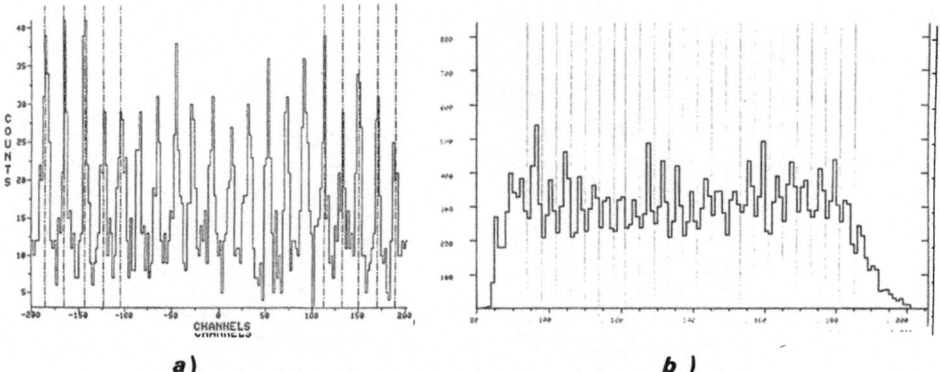

a)   b)

Fig. III.6 a) Position spectrum of a DGAC (illiminated roughly homogeneous with elasti-
tically scattered 5.9 MeV/u Xe-ions) with a 4 mm tap structure.
b) the same for a DGAC with 2 mm taps.

The grids K and A are used for the position determination in the two coordinates.
The wires are grouped in 4 mm wide bins and soldered together to a tapped

Fig. III.8 Installation of DGAC's in a big
scattering chamber (Gobbi 81).

Fig. III.7 Experimental arrangement of
10 DGAC's around the target (Lynen
82).

Fotos: A. Zschau, GSI

delay-line. As delay-line we use either 3 ns per tap long Lemo-cables or chips. The
delay-line is read out at both ends, a method which allows a rough interpolation
between two taps. Furthermore, since the sum of the two signals has to be a constant,
namely the total length of the delay-line, multi-particle hits on the detector can be

clearly distinguished. Fig. III.6a shows a position spectrum of 5.9 MeV/u Xe-ions, elastically scattered on Au; the 4 mm wide tap structure is clearly resolved. This shows that the diffusion of the electron cloud is less than 4 mm. Fig. III.6b shows the position spectrum of another detector which has a 2 mm wide tap structure; here the tap structure is much less resolved, which indicates a transversal diffusion in the order of 2 mm, in fair agreement with the above estimated value.

The signal from the T-electrode (cf. fig.III.3) is well suited for an energy-loss measurement. The resolution is in the order of 25%.

Due to the low operating pressure of 2 mb, even large-size (20 x 30 cm²) DGAC's can be built which have only 10 mm wide frames. These frames are permanently glued together and are thus gas-tight and do not need an extra housing to contain the counting gas. This has the very important advantage that these counters can be installed very close together, thus minimizing dead spaces between them. A further advantage of the low pressure is that the S and T-electrode can serve as well as the gas-windows, which considerably reduces the total thickness of the counter, without deterioating too much the time- and energy-loss resolutions.

These double-grid avalanche counters have been built by our group since 1981 for two large experimental set-ups:
1) for experiments at the SC at CERN with 84 MeV/u carbon-ions, 10 DGAC's in trapezoidal form (active area 220 cm²), were arranged around the target to detect the light and heavy fragments (Lynen 82). The dynamic range is wide enough to register with full efficiency light ions A > 10 in the presence of heavy target-residues. Fig. III.7 shows the experimental arrangement.
2) Twelve 20 x 30 cm² counters of this type have been built for a new kinematic coincidence set-up for experiments at 20 MeV/u at the upgraded UNILAC (Gobbi 81). Fig. III.8 shows the installation of the counters inside the scattering-chamber.

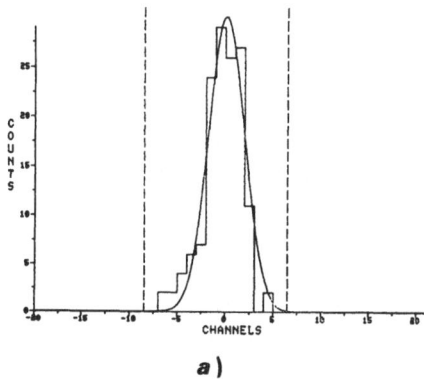 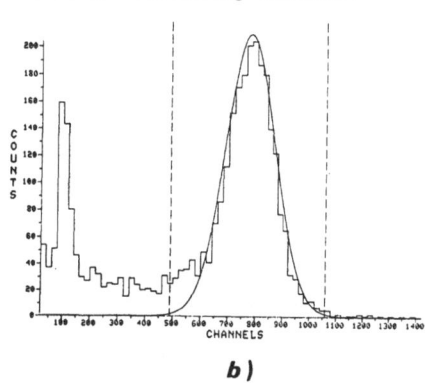

a)   b)

Fig. III.9 a) Time resolution of a DGAC (cf. text). (1 ch = 100 ps). b) Pulse-height spectrum of elastically scattered Xe-ions in a DGAC.

These briefly mentioned set-ups have already been used sucessfully in several experiments. Fig. III.9a shows the performance of the timing of a DGAC. In this figure the difference between calculated and measured time is plotted. The calculated time has been obtained from the scattering angles of elastically scattered Xe on Au at 5.9 MeV/u. The time spectrum shown includes the contribution from the uncertainty in the angle-determination (due to the 4 mm position-resolution in the DGAC's and the 4 mm wide beam spot).

Fig. III.9b shows the pulse-height spectrum in a DGAC of Xe-ions from the above reaction. Events below the dashed line at channel 500 are background events.

Quite recently a small-size DGAC (5 x 7 cm²) has been built and tested (Keller 82), which differs in some aspects from the above described ones:

1) The wire distance in the K- and A-plane is 1 mm.
2) The A-T gap has been increased from 3 to 6 mm.
3) Two neighbouring wires are soldered together to a tap of a delay-line, i.e. the tap width has been decreased from 4 mm to 2mm.

Let me briefly describe the main differences in the performance of this new designed DGAC compared to the old one:

1. Despite the increased wire distance of 1 mm (instead of 0.8 mm) in the K-plane, the counter still has an excellent time-resolution of <200 ps (FWHM), as determined with a ^{252}Cf-source and a 1.4 MeV/u Kr-beam. This effect can be understood by remembering the estimated transversal diffusion of ~1.6 mm of the electrons (see above) across the 3 mm wide gap. Fig. III.10a shows a time-spectrum

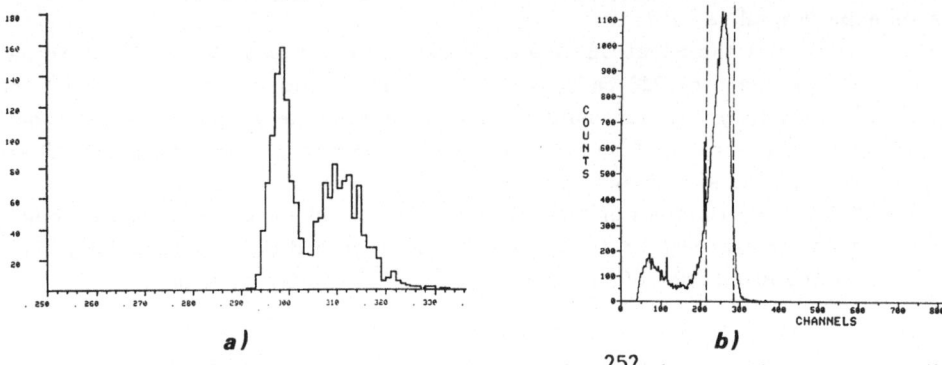

a)   *b)*

Fig. III.10 a) TOF-spectrum of the fragments of a ^{252}Cf-source between a small-size PPAC and a DGAC (1 ch = 100 ps).

b) Pulse-height spectrum of 1.4 MeV/u Kr-ions. The resolution is ~20 %. (The spikes at channels 120 and 220 are pulser signals.)

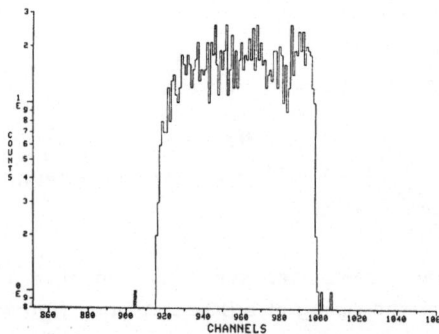

Fig. III.11 Position spectrum of 1.4 MeV/u Kr-ion in a DGAC (cf. text).

of the light and heavy fragments of a ^{252}Cf-source. The observed width of 400 ps (FWHM) includes the velocity- and energy-loss straggling in the start-counter and the contribution of the start-counter (a small-size PPAC) itself. The total thickness of the counter is <300 µg/cm² (the entrance window, the S-electrode and the T-electrode; all are stretched polypropylene foils with some silver-coating). It would be highly desirable to reduce the thickness furthermore. One way to do that would be to replace the S-foil by a narrow grid of wires. But now it is doubtful wether the good timing properties are still maintained. The primary electrons produced directly at S (acting as cathode, cf. fig. III.3), which are most important for the timing properties of the detector, experience an inhomogenous electric field, since at the beginning of the drift to the anode, there is no smearing-out effect due to a diffusion process.

2) Due to the bigger A-T gap, the signals on A and T are now completely decoupled. The signal on A is now negative, caused by electrons caught by the wires and is no longer a positive signal induced by T. This offers the possibility to adjust the pulse-height on T (for example, to match the pulse-height to the range of the charge-sensitive ADC which measures the energy-loss) without affecting the signal on A. Fig. III.10b shows the pulse-height spectrum of 1.4 MeV/u Kr-ions.

3) The effect of the reduced tap-width of 2 mm on the position determination has already been discussed (fig. III.6). Fig. III.11 shows the position-spectrum measured on the K-grid when the detector is illuminated with 1.4 MeV/u Kr-ions through a circular mask with 5.5 mm diameter. In the Y-coordinate a narrow cut has been applied. From the sharp slope at the edges of the mask a resolution of o.25 mm (FWHM) is calculated. On the Y-coordinate, the A-grid (cf. fig. III.3), the position resolution is a factor of two worse. This finding can be explained by the longer drift-path of the electrons through the detector and hence a bigger transversal diffusion. On this detector commercially available delay-line chips have been used. Two types (BELFUSE and RHOMBUS) have been tested and compared. They showed similar performance. The linearity of these delay-lines is good enough for a position-resolution of about 2 mm, but certainly not adequate for a resolution of <1 mm.

IV. Summary and Outlook

I have tried to illustrate the present status of gas-filled heavy-ion dedectors. Despite their venerable age, gaseous detectors are in a still ongoing stage of development, continously modified and adapted to new experimental requirements. Heavy ion experiments require detection systems which are able to record simultaneously in only one detector several parameters, like time-of-arrival, position or energy, of particles covering a broad range of specific ionization. Gas-filled detectors have proven to be well suited to meet these demands.

In the near future, the interest in heavy-ion physics will focus on higher energies and the demands for detectors will change. On the one hand, very thin detectors, like the one used up to now, will be needed to register particles from the target-fragmentation region, but the fast forward-going products with their much less specific ionization and their high multiplicity will require detection systems which resemble more and more the ones used in High Energy Physics.

I thank all my colleagues for the many helpful discussions during the preparation of this manuscript, especially Dr. A. Gobbi and Prof. U. Lynen.

Literature

Bock 82: R. Bock, Y.T. Chu, M. Dakowski, A. Gobbi, E. Grosse, A. Olmi, H. Sann, D. Schwalm, U. Lynen, W.F.J. Müller, S. Bjornholm, H. Esbensen, W. Wölfli, E. Morenzoni, accepted for publication in Nucl. Phys.

Bohrmann 76: S. Bohrmann, Diplomarbeit, Universität Heidelberg, 1976, unpublished

Breskin 77: A. Breskin, and N. Zwang, Nucl. Instr. & Meth. 146(1977)461

Breskin 79: A. Breskin, G. Charpak, S. Majewski, G. Melchart, G. Petersen, and F. Sauli, Nucl. Instr. & Meth. 161(1979)19

Breskin 82: A. Breskin, Nucl. Instr. & Meth. 196(1982)11

Charpak 68: G. Charpak, R. Bouclier, T. Bressani, J. Favier, and C. Zupancic, Nucl. Instr. & Meth. 62(1968)262

Charpak 78: G. Charpak, and F. Sauli, Phys. Lett. 78B(1978)523

Eyal 78: Y. Eyal, and H. Stelzer, Nucl. Instr. & Meth. 155(1978)157

Farr 78: W. Farr, J. Heintze, K.H. Hellenbrand, and A.H. Walenta, Nucl. Instr. & Meth. 154(1978)175

Gaukler 77: G. Gaukler, H. Schmidt-Böcking, R. Schuch, R. Schule, H.J.Specht, and I. Tserruya, Nucl. Instr. & Meth. 141(1977)115

Geiger 12: H. Geiger and E. Rutherford, Phil. Mag. 24(1912)618

Geiger 28: H. Geiger and W. Müller, Phys. Z. 29(1928)839 and Phys. Z. 30(1929)489

Gobbi 81: A. Gobbi, G. Augustinski, R. Bock, H. Daues, S. Gralla, K.D.Hildenbrand, M. Ludwig, W.F.J. Müller, A. Olmi, M. Petrovici, W. Quick, H.Sann, H. Stelzer, J. Toke, GSI Scientific Report 1981

Gruhn 79: C.R. Gruhn, in: Proc. Symposium on Heavy Ion Physics from 1O to 200 MeV/u, BNL-51115(1979), p. 471

Gruhn 82: C.R. Gruhn, M. Binimi, R. Legrain, R. Loveman, W. Pang, M. Roach, D.K. Scott, A. Shotter, T.J. Symons, J. Wouters, M. Zisman, R. Devfries, Y.C. Peng, and W. Sondheim, Nucl. Instr. & Meth. 196(1982)33

Harrach 79: D.v. Harrach, and H.J.Specht, Nucl. Instr. & Meth. 164(1979)477

Hempel 75: G. Hempel, F. Hopkins, and G. Schatz, Nucl. Instr. & Meth. 131(1975)445

Huber 80: F. Huber, A. Fleury, R. Bimbot, D. Gardes, Ann. Phys. Fr. Vol. 5(1980)

ICRU 79: International Commission on Radiation Units and Measurement, ICRU Report 31(1979)

Jean-Marie 79: B. Jean-Marie, V. Lepeltier, and D. L'Hote, Nucl. Instr. & Meth. 159(1979)213

Jones 81: A.R. Jones and R.M. Holford, Nucl. Instr. & Meth. 189(1981)503

Just 78: M. Just, D. Habs, V. Metag, and H.J. Specht, Nucl. Instr. & Meth. 148(1975)283

Keller 82: J.G. Keller, K.-H. Schmidt, Ch. Sahm, and H. Stelzer, to be published in Nucl. Instr. & Meth.

Lieb 82: K.P. Lieb, S. Brüssermann, H. Emling, E. Grosse, J.Stachel, and P. Sona, to be published

Lynen 80: U. Lynen, private communication

Lynen 82: U. Lynen, R. Bock, Y.T. Chu, P. Doll, R. Glasow, A. Gobbi, K.D. Hildenbrand, H. Ho, W. Kühn, H. Löhner, W.F.J. Müller, A. Olmi, D. Pelte, H. Sann, R. Santo, H. Stelzer, U. Winkler, invited talk at the Int. Conf. on Selected Aspects of Heavy Ion Reactions, Saclay, May 3-7, 1982, Nucl. Phys. A387(1982)129

McKenzie 79: McKenzie, Nucl. Instr. & Meth. 162(1979)49 and references therein

Northcliff 70: L.C. Northcliff, and R.F. Schilling, Nucl. Data Tables A7(1970)

Pferdekämper 77: K.E. Pferdekämper, and H.G. Clerc, Z. Physik A280(1977)155

Rosner 81: G. Rosner, B. Heck, J. Pochodzalla, G. Hlawatsch, B. Kolb, and A. Miczaika, Nucl. Instr. & Meth. 188(1981)561

Rutherford 08: E. Rutherford and H. Geiger: Proc. Roy. Soc. Lond., Ser. A81(1908)141

Sann 75: H. Sann, H. Damjantschitsch, D. Hebbard, J. Junge, D. Pelte, B. Povh, D. Schwalm and D.B. Tran Thoai, Nucl. Instr. & Meth. 124(1975)509

Sann 80: H. Sann, private communication

Schiessl 82: Ch. Schiessl, W. Wagner, K. Hartel, P. Kienle, H.J. Körner, W. Mayer and K.E. Rehm, Nucl. Instr. & Meth. 192(1982)291

Sistemich 76: K. Sistemich, P.Armbruster, J.P. Boucquet, Ch. Chauvin, and Y. Glaize, Nucl. Instr. & Meth. 133(1976)163

Stelzer 76: H. Stelzer, Nucl. Instr. & Meth. 133(1976)409

Stelzer 82: this programm is based on: H. Buchholz, Elektrische und magnetische Potentialfelder, Springer-Verlag, Berlin 1957. It has originally been written to calculate field distributions in drift-chambers and multiwire proportional chambers (cf. K. Bethge (ed.) Experimental Methods in Heavy Ion Physics, p. 167. Lecture Notes in Physics, Vol. 83, Springer-Verlag 1978).

Urban 81: M. Urban, W.R. Graves, and C. Heil, Nucl. Instr. & Meth. 188(1981)47

NEW TRENDS IN LOW-PRESSURE GASEOUS DETECTORS

A. Breskin*
Department of Physics
Weizmann Institute of Science
Rehovot, Israel

ABSTRACT

Multiwire proportional chambers (MWPCs) operated at very low gas pressures are shown to be an efficient timing and imaging tool for heavily ionizing particles. Their properties are described. Amplification in steps is shown to be feasible at low pressure, providing higher gains and gating possibilities. Both techniques can be extended to single electron detection at gains $>10^7$ with possible applications to photon detectors. A new type of a 4π Tracking Range and Energy Chamber, TREC, for heavy ions is presented.

1. Introduction

Since the early and glorious days of nuclear physics and detector physics, and both domains have always been closely related, the basic structure of a gaseous detector didn't really change. It has always been defined as a volume of gas contained between an appropriate set of two electrodes having between them an electric field. Nowadays we certainly know more about the phenomena occurring between the anode and the cathode during the amplification process; we are better equipped to study them. But even so, almost eight decades after the discovery of gaseous amplification by Geiger and Rutherford[1], there are still many obscure phenomena to be studied and clarified concerning the mechanisms that govern the development of an avalanche.

While photons and photon-mediated avalanches are at the leading front of particle physics and X-ray instrumentation (see contribution of G. Charpak to these proceedings). nuclear instrumentalists dealing with heavily ionizing particles, being somewhat more conservative, are still taking advantage of the good old electrons in single or multiple avalanche processes.

A large variety of modern and rather sophisticated detection techniques, often differing by their operation mechanism or operation conditions, is not any more purely dedicated to a particular discipline; it is amazing to note how they change fields of application. High energy physicists are using nowadays Geiger tubes, solid-state detectors and even low pressure techniques. On the other hand, nuclear physicists are adopting drift chambers, MWPCs, streamer chambers and even time projection chambers.

In the present paper I would like to draw your attention and discuss some new approaches to the detection of heavily ionizing particles and single electrons.

---

*The Hattie H. Heineman Research Fellow

We shall see that MWPCs operated at very low gas pressures (0.3-3 torr) present some outstanding timing and imaging properties: time resolution 100 ps (fwhm), position resolution 100 μm (fwhm), and seem to be an ideal tool for high rate heavy ion detection. I will discuss the possibility and the mechanism of amplification in steps, even at low gas pressures, having as a main feature a considerable increase in gain, and the possibility of gating the detector on selected events. Applications of both techniques to single electron detection will be discussed in the text. Finally it will present a new instrument developed recently for some "exotic" fission studies: the TREC, a 4π Tracking Range and Energy Chamber that may find applications in nuclear physics far beyond its original design.

2. Low-pressure MWPCs

The guide-line for developing high accuracy time and position detectors for heavily ionizing particles was the feeling that the present gaseous detectors, widely employed in this field, have limitations. Drift chambers combined with parallel plate avalanche counters (PPAC)[2,3] and position sensitive PPACs[4,5] are the most widespread techniques for timing and imaging of heavy ions and even of low energy light particles[6]. Typical resolutions of those counters are of the order 0.2-0.4 ns and 0.3-1 mm (fwhm); the pressure range is of the order of 5-20 torr. In order to efficiently determine the kinematical parameters of reaction products with reasonable resolutions, large size systems are required, implying rather large surface detectors and thus raising a window thickness problem. It is difficult with present techniques to build large area thin transmission detectors often required as START detectors for time-of-flight experiments; on the other hand, the STOP detectors are limited in time resolution and are not always compatible with the excellent timing properties of pulsed beams or even of secondary emission START detectors[7].

MWPCs[8] operating at normal gas pressures were never considered as time measuring devices, for the simple reason that their time resolution is of several tens of ns. This is due to the long drift time of electrons, released in the sensitive volume, towards the sense wire where amplification occurs (fig.1a). First measurements at low pressures (3 storr), have shown that time resolutions of 2.5 ns (fwhm) could be reached with 5.5 MeV α-particles[9]. This result was simply attributed to a faster drift time at higher reduced fields. Some further investigations[10,11] have shown clearly that the amplification mechanism at low pressures is entirely different (fig.1b). The reduced electric field strength E/p in the constant field collection region reaches values of several hundreds of V/cm·torr which are very close to those reached in PPACs, thus indicating that some amplification already occurs at this region. The field in the vicinity of the wires is about two orders of magnitude higher as shown in fig. 2. Electrons released in the sensitive volume gain enough energy to start an avalanche in the constant field region; a second amplification step occurs when the

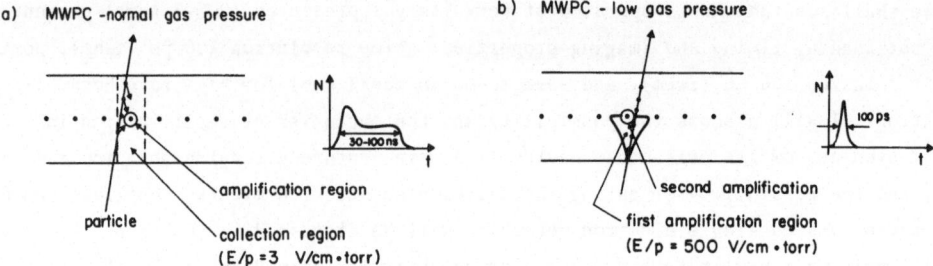

Fig. 1: Operation mechanism of MWPCs. a) At normal gas pressures electrons collected from the sensitive volume are amplified in the region of the wires. b) At low pressures electrons start an avalanche in the constant field region; a second amplification step occurs when the electron swarm reaches the wires.

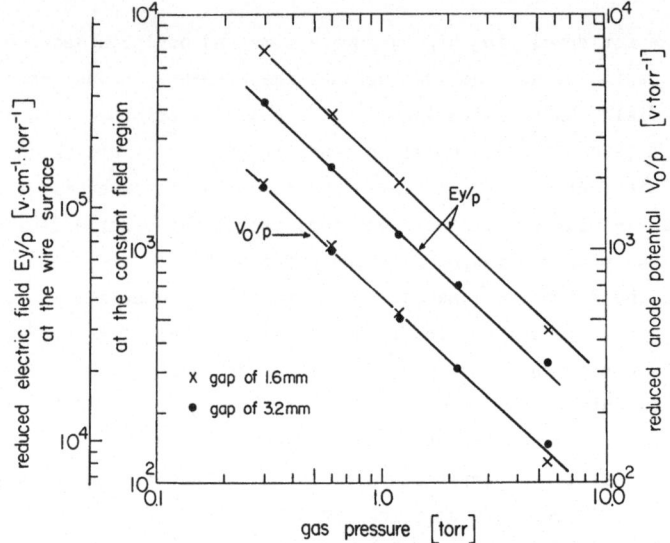

Fig. 2: Reduced anode potential V_0/p and reduced electric fields E/p at the constant field region and on the surface of the anode wires, as function of the pressure of isobutane. Data measured with 5.5 MeV α-particles.

electron swarm reaches the wires. The avalanche has an exponential growth, with a dominant contribution from the charges released in the vicinity of the cathodes. The time resolution will depend on the fluctuations in the position of these primary charges.

2.1 General Properties

The properties of low-pressure MWPCs have been investigated during the last few years with various gases, pressures, geometries and particles. I will summarize their most important properties and for more details refer the reader to previous works in

this field [10,11,12,13]. Most of the properties have been studied with detectors
having 10 µm anode wires, 1 mm apart and cathodes made of thin, 0.3-2.5 µm, aluminized
polypropylene or hostaphan foils. The gap between the anode wires and each of the
cathodes varied between 1.6-6.4 mm. The chambers have been operated in a pressure
range of 0.3-10 torr, mainly with isobutane in which the highest gains (10^5-10^6) and
the best time resolutions have been reached.

Let us summarize some general properties and features of this operation mode:
- At low pressures we do have a double step amplification process in MWPCs. The
  contribution of the wires to the total gain varies from a factor of 5 to 2000,
  according to the gap between the electrodes and to the pressure.
- At equal pressures and electric fields, and even with lower maximum potentials,
  MWPCs offer higher gains than PPACs.
- Due to the high values of E/p, the positive ions produced during the avalanche
  process are removed more than an order of magnitude faster than at normal gas
  pressures. It is shown in fig. 3 that the rise time of the ion component of
  the charge pulse is smaller than 1 µsec for a gap of 3.2 mm. This fast ion
  collection considerably reduces space-charge effects thus allowing a high rate
  operation, of the order of 10^5 c/s·mm^2. Fig. 4 shows only a small drop in
  pulse-height at these rates.
- The fast electronic component of the charge pulse (fig.3), essential for timing,
  is more pronounced in the MWPC. The rise time of current pulses is of the
  order of 2-3 ns.

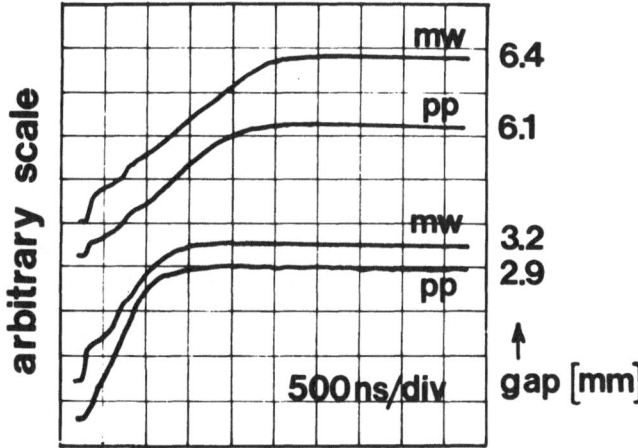

Fig. 3: Charge pulses from the MWPC and the PPAC of a comparable gap,
operated at 1.2 torr of isobutane and at the same reduced electric field
in the constant field region. Both types of counters have the same fast
ion collection times but the electronic component of the pulse, essential
for fast timing, is more pronounced in the MWPC.

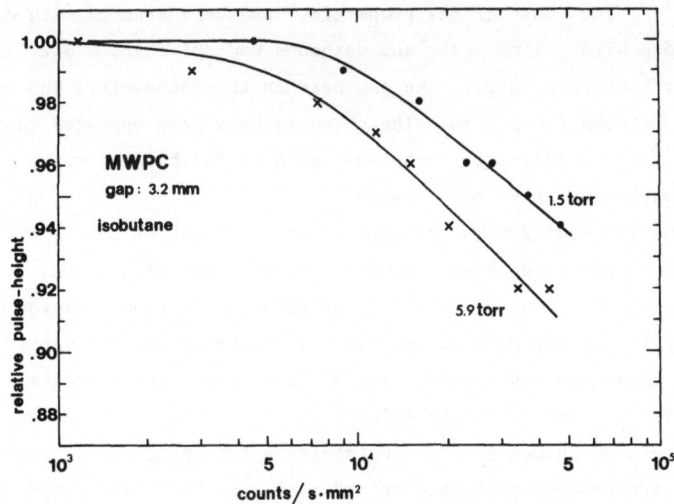

Fig. 4: Relative pulse heights from a MWPC as a function of
particles rate. Operation conditions: ^{16}O ions, gap=3.2 mm,
1.5 and 5.9 torr of isobutane.

2.2 Time Resolution

The time resolution was measured with a 50 mm^2 surface barrier detector placed
behind the MWPC, having 100 μm depletion width. The signals were processed with low-
noise, 50Ω input, fast preamplifiers [14] and constant-fraction timing discriminators
(ORTEC 934). The overall time resolution as function of pressure is plotted in Fig. 5
for a 40x40 mm^2 detector. The best results with 27 MeV ^{16}O ions, 140 ps (fwhm), have
been reached at 0.3-1.2 torr. If we estimate a reasonable contribution of 100 ps due
to the solid-state detector then the intrinsic resolution of the MWPC is of the same
order. An intrinsic resolution of 135 ps (fwhm) has been measured in the same way
with an 80x100 mm^2 detector at 1.6 torr and a 120 MeV ^{58}Ni beam. This result is not
corrected for any kinematical broadenings. A careful time scanning across the surface
of the detector has shown smooth behaviour with a maximum delay of 160 ps at the corners.
This delay is partially due to the pulse propagation time along the detector and can
be easily corrected by software. Better time resolutions may be achieved by extracting
the signals from individual wires. The time resolution was not affected by counting
rates as high as $5 \cdot 10^4$ c/s·mm^2. Presently we are investigating the resolution as function of the avalanche position between adjacent wires.

2.3 Position Resolution

When an avalanche occurs on an anode wire, a positive charge is induced on the
cathodes [15]. The position of the avalanche can be determined by measuring the centre
of gravity of the induced charge distribution using various methods, such as delay

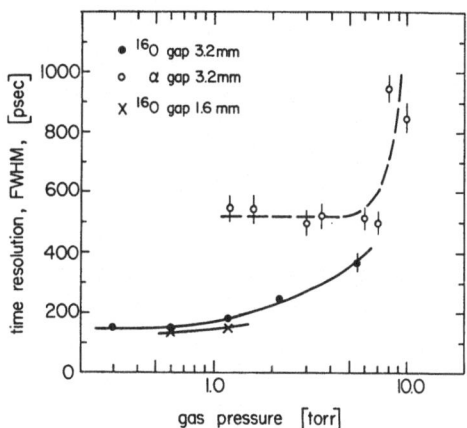

Fig. 5: Overall time resolution measured between the MWPC and a solid-state detector as function of gas pressure for 5.5 MeV α-particles and 27 MeV ^{16}O ions. If we estimate a 100 ps contribution of the SSD, then the intrinsic time resolution at the pressure range of 0.3-1.2 torr is 100 ps FWHM.

lines[16], charge division[17] or direct computation[18].

It has been shown[19] that at a normal gas pressure the avalanche position along a MWPC wire can be measured with an accuracy of about 80 µm (fwhm) and that inclined tracks allow an interpolation between anode wires when electrons are shared between two or more wires. A resolution of 350 µm (fwhm) has been measured between wires, 2 mm spaced. In the low-pressure operation mode, the avalanche already has a certain width before reaching the wires - mainly due to electron diffusion. We have shown[20] that the avalanche width is larger than our wire spacing, namely 1 mm, by the fact that we interpolate the position between wires with good accuracy even at normal incidence. Some other observations proving the large spread of the avalanche are the important number of adjacent wire hits (40-60%) reported by the authors of ref. 21 and the possibility of preamplification and transfer mechanism in a low-pressure pure hydrocarbonic gas, where a photon-mediated process is excluded[22]. This will be discussed in detail in section 3.

We have investigated the localization capabilities of low-pressure MWPCs[20], applying one of the simplest and most unexpensive methods: the delay-line read-out illustrated in fig. 6. We have built an 80x100 mm^2 MWPC having 10 µm anode wires, 1 mm apart. The gap between anode and cathode is 3.2 mm. The cathodes are made of 50 µg/cm^2 polypropylene foils on which 50 µg/cm^2 gold strips have been evaporated.

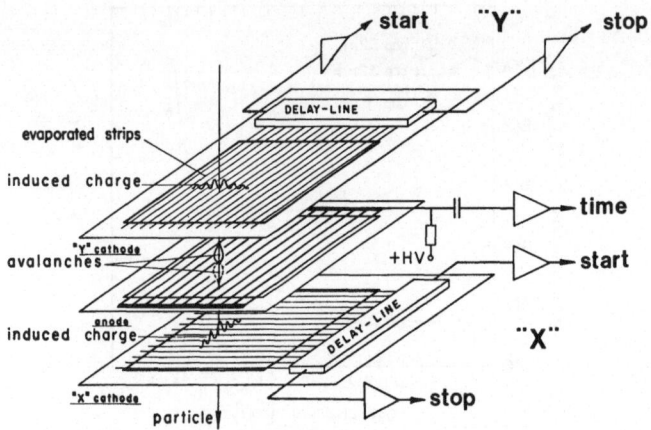

Fig. 6: The principle of the delay-line, induced charge position
read-out from MWPCs. The position on each coordinate is obtained by
measuring the time difference between the induced pulses propagating
towards the two ends of the delay-lines.

The "X"-coordinate strips, running orthogonal to the wires, and thus reading the positions of the avalanche along the wires, are 1 mm apart (center to center). The Y-coordinate strips, parallel to the wires, interpolating the position between the wires, are 2 mm apart. The distance between two adjacent strips is 0.2 mm. The strips are connected to the taps of integrated delay lines. The position of the avalanche, on each of the two coordinates, is obtained by measuring the time difference between the induced pulses propagating towards the two ends of the delay lines, as shown in fig.6. On the X-coordinate we are using 10 delay-lines of the type PE 20619 having a delay of 5ns between taps and a Z_0 of 200Ω. On the Y-coordinate we use 4 delay-lines, PE 28100, having a delay of 5ns between taps and a Z_0 of 50Ω. The delay-lines are matched to 50Ω cables via pulse transformers. Low noise, ground-gate FET preamplifiers[23] and constant-fraction timing discriminators are used to process the signals.

Position resolution was measured at 1, 1.6 and 3 torr of isobutane, with fission-fragments of ^{252}Cf. A collimator made of 100 µm slits was placed at a distance of 3 mm from the entrance cathode; their image on the anode is 150 µm. The distance between slits (center to center) is 2.1 mm in order to allow a full scanning between wires in 0.1 mm steps.

Despite the fact that the two coordinates are of an absolutely different nature, an avalanche along a wire and an avalanche between wires, and the strips being of different sizes, the position resolutions are practically the same. At a pressure of 1 torr the signals from the delay-lines are practically in the noise level and the resolution is of the order of 1 mm (fwhm). At 1.6 torr the situation is much better and the resolution is of the order of 300 μm (fwhm) on both coordinates. Fig. 7 shows the position distribution of the collimated source after 48 hours of continuous counting due to the limited rate of the source. Fig.8 shows the result at 3 torr for 12 hours of counting; the intrinsic resolution is of the order of 100 μm (fwhm).

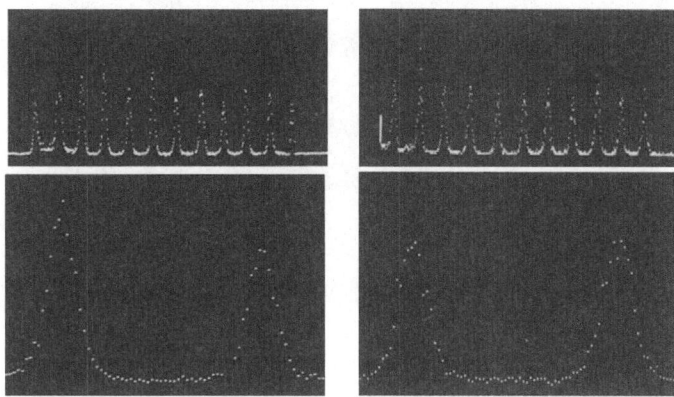

Fig. 7: The position distribution of the 150 μm collimated source. Distance between peaks - 2.1 mm. Left: X-coordinate right: Y-coordinate. p=1.6 torr of isobutane, HV=510 V. The fwhm is of the order of 330 μm. Intrinsic resolution: 300 μm (fwhm). Counting time: 48 hours.

I would like to emphasize that the delay-lines, having a limited frequency band-width, strongly damped the cathode-induced pulses. While the time signal of the anode is about two orders of magnitude above noise, even below 1 torr, the corrolated cathode induced signals, having originally about half of the anode charge each, are practically at the noise level after having passed through the delay-lines. We are investigating at present the direct computation of the center of gravity method[18] having the following advantages:

- it should give the ultimate accuracy determined by the spacial distribution of the charges;
- the measurement is local, in the sense that the induced pulses do not have to propagate along the whole length of the chamber.

We believe that this method will provide us with resolutions below 100 μm, even at very low pressures.

The spacial distribution of the cathode induced pulses in a counter having a gap of 3.2 mm have a width of the order of 2 mm (fwhm) as shown in fig. 9. Read-out strips having a width of 1 mm will be sufficient for the centroid computation method. The narrow distribution of induced pulses, typical to the avalanche mode in parallel-plate

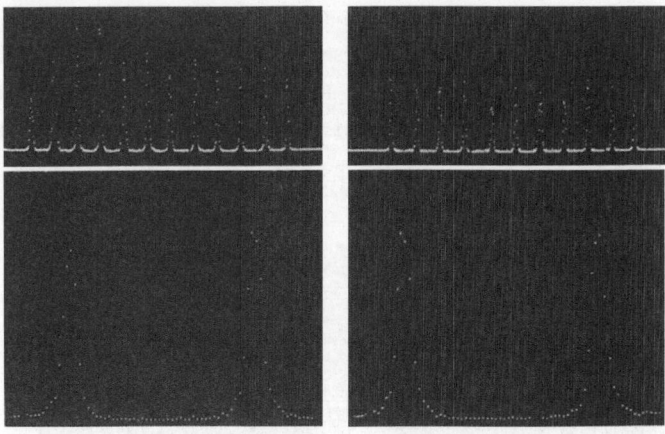

Fig. 8: Position distribution at p=3 torr, HV=560 V. The
fwhm is of 180 μm. Intrinsic resolution: 100 μm fwhm.
Counting time: 12 hours.

geometries, offers the possibility of efficiently detecting several simultaneous
particles with a good double track resolution. This can be achieved by correlating
direct and induced signals recorded from anode wires and cathode strips running in
various directions.

2.4 Applications to Heavy Ion Reactions

Several low-pressure MWPCs have been successfully operating in our laboratory
for more than two years as fast, highly transparent START detectors for TOF measurements in various heavy-ion experiments. A good example is shown in fig. 10 illustrating a versatile and highly efficient heavy-ion identification system[24] composed of
various low-pressure gaseous detectors measuring TOF, angular distribution (θ, ϕ)
energy loss (ΔE) and energy (E) of one or two fragments in coincidence. Many other
START detectors, having total thicknesses of 100-160 μg/cm^2 are employed in proton or
heavy ion induced fission experiments, presently performed in several laboratories[25].
These detectors are described in detail in ref. 13. Position sensitive MWPCs are
under construction to replace PPACs used as STOP detectors in these experiments.
Besides their attractive timing and imaging properties they are much less sensitive to
electrons emerging from the target, often limiting the gain of the counters and requiring rather high electronic thresholds. The difference in sensitivity to electrons is
shown in fig. 11 for a MWPC placed at 40 mm from a target, followed by a PPAC at a distance of 250 mm from the target. Both detectors have, after amplification, the same
total charge. The difference in the rise-times is also visible in the figure. The

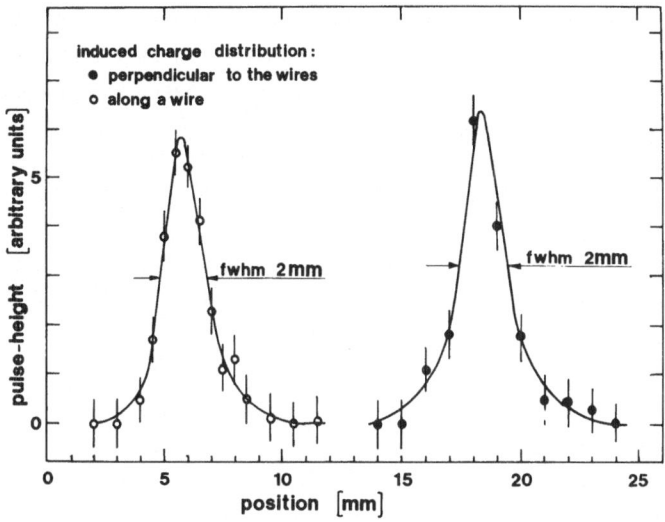

Fig. 9: Spatial distribution of the cathode induced charges in a MWPC with gap of 3.2 mm. On the left - the distribution along the wires. On the right - the distribution perpendicular to the wires. The distribution was measured with a collimated α-source, at a pressure range of 1.5-10 torr.

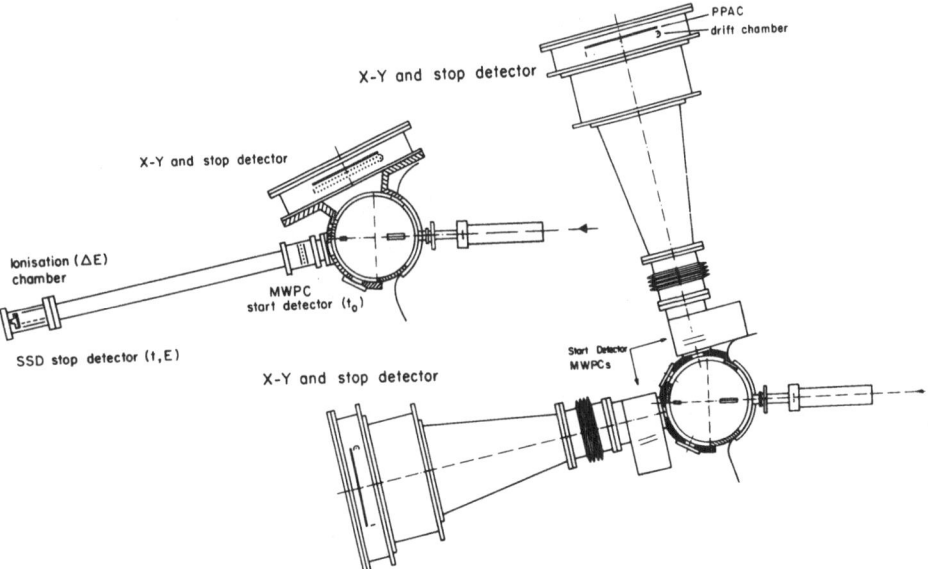

Fig. 10: A schematic view of the Weizmann Institute heavy ion identification system[24]. On the right the kinematical coincidence system, measuring velocity, angular distribution and roughly the energy loss of binary reaction products. On the left, a forward TOF arm, measuring velocity, energy loss and energy in fusion reactions; a combination of this arm with a large solid angle imaging detector is suited for deep inelastic collisions.

possibility of covering detection areas with very thin windows, of a few tens of $\mu g/cm^2$, is not only important for thickness reduction of transmission detectors but opens up the possibility to efficiently detect low energy heavy recoils and even heavy elements from isotope separators.

2.5 Single Electron Detection

MWPCs operated at 1-4 torr of isobutane can efficiently detect single electrons as shown in fig. 12[13]. A chamber was irradiated with a UV lamp and single photo-electrons extracted from the cathodes could be detected far above noise level at counter gains of about 10^7.

In the last few years there has been an increasing interest in single photo-electron detection and imaging in various types of photon detectors, mainly in the domain of high energy physics[26]. Gaseous counters with photoionizable gas mixtures[27] have proved to be efficient VUV photon detectors when used as Čerenkov ring imaging de-vices for the identification of relativistic particles[28]. Single photoelectrons, photoproduced in the gas, are detected in various types of detectors; the most efficient of them are multistep avalanche and MWPCs[28-30] and drift chambers[31]. Among the photo-ionizable vapors used so far, are triethylamine (ionization potential Ig=7.4 ev) and tetrakis (dimethylamine) ethylene, called TMAE[32] (Ig=5.4 ev). Both vapors have low vapor pressures at room temperatures and therefore one may imagine some low-pressure VUV photon detectors, even for Cerenkov ring imaging. In such a case the detector would not be sensitive to parasitic effects due to beam particles or to photon feed-back in the gas.

A new type of photon detector was recently introduced by D. Anderson[33]. It con-sists of a liquid photocathode made of TMAE, coupled to a low-pressure MWPC. The idea is to photoionize liquid TMAE, that might have an ionization potential as low as 3.7 ev in a liquid phase, and a quantum efficiency approaching 90-100%, and to extract and amplify the photo-electrons with a low-pressure MWPC. It has been shown that electron extraction is more efficient at lower gas pressures in isobutane and methane[34], and we have shown that counter gains are higher too. The development of this method is pursued by the author mainly in view of applications for calorimetry in high energy physics using BaF_2 scintillators.

Another low-pressure application to calorimetry is the gas-filled photodiode recently proposed as a cheaper, and perhaps more comfortable substitute for the photo-multiplier, for light read-out of scintillators[35]. Gas-filled photodiodes have the advantage, over vacuum photodiodes already used for this purpose[36], of being able to amplify the photo-electrons extracted from a solid photocathode, thus enabling the use of less expensive electronics. Photodiodes having a Bialkali photocathode, a parallel grid amplification structure, and filled with argon methane and isobutane at various pressures (0.1-250 torr) have recently been tested showing that gaseous gains of the order of 20-100 can be reached[37]. The problem of life-time of the photocathode,

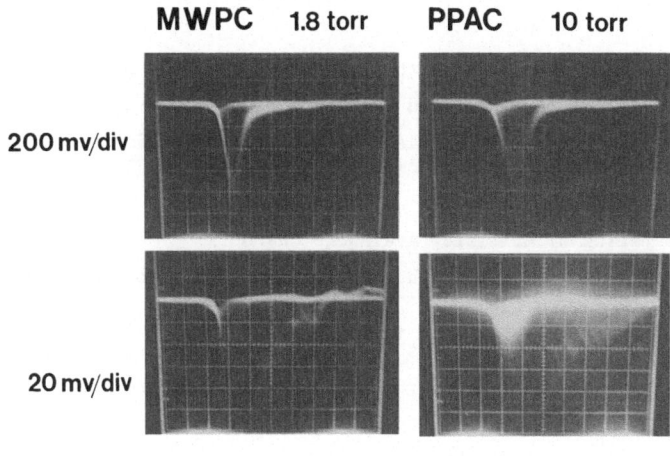

MWPC 1.8 torr   **PPAC** 10 torr

200 mv/div

20 mv/div

5 ns/div

Fig. 11: Time pulses of fission fragments from proton-induced fission at 475 MeV, recorded from a MWPC START detector followed by a PPAC. Both counters deliver the same total charge, but their sensitivities to parasitic electrons from target differ by orders of magnitude. Operation conditions: isobutane, MWPC: p=1.8 torr, HV=490 V, t_r=2.5 ns. PPAC: p=10 torr, HV=730 V, t_r= 5 ns.

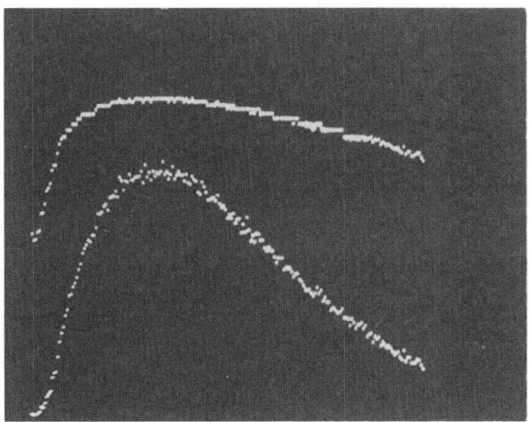

Fig. 12: Pulse-height distribution (normal and logarithmic scales) of single photo-electrons photoproduced on the cathode of a MWPC (L=2x3.2 mm, s=1 mm, d=10 μm). Isobutane, 4 torr, HV=750 V. The charge at the peak is of the order of 1pC (gain of the order of 10^7).

56

suffering from ion feedback and impurities of the filling gas is presently under investigation[38].

Two possible solutions may increase the gain of the low-pressure photodiode: the use of a MWPC instead of the parallel grid structure or the multistep avalanche chamber that, besides the high gain, allows also the gating possibility on selected events.

3   Low-pressure Operation of Multistep Avalanche Chambers

The preamplification of an initial charge and the transfer of the primary avalanche to a second amplification step, as shown in fig. 13, has been a subject of an intense study during the last few years[39-42]. Using pure electrostatic considerations it was proved that an efficient charge transfer from the preamplification region to the transfer region ($E_p \gg E_t$) can only occur if the avalanche has a considerable lateral spread[41]. It has to be of the same order or to exceed the wire spacing of electrode b. At normal pressures and with standard gas mixtures the electron diffusion, in a simple avalanche model, is about an order of magnitude too small to allow an efficient transfer. Large photon-mediated avalanches can be obtained in some binary gas mixtures, where the ionization potential of one of the molecules is lower than the energy of the excited state of the other atom or molecule. At low gas pressures electron diffusion leads to large avalanches and the multistep operation mode can take place even in organic gases like isobutane or methane[22].

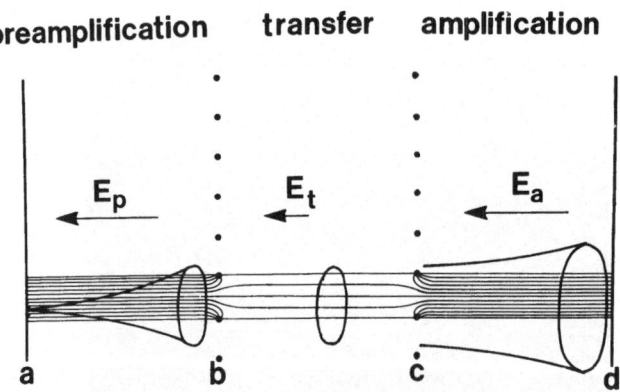

Fig. 13: The principle of amplification in steps. A charge injected into the high field preamplification region produces an avalanche. If the lateral spread of the avalanche is of the order or exceeds the wire spacing of electrode b the avalanche can be efficiently transferred into the second amplification stage. The transfer efficiency is roughly E_t/E_p.

A multistep operation has recently been performed with two parallel plate counters having as electrodes 93% transparent grids, and with a parallel plate counter followed by a MWPC[22]. The detectors operated at a pressure range of 2-20 torr. We
have found that higher transfer gains (ratio between the double-step gain and a single-
step gain) could be reached at the lower pressures and with lower transfer fields,
diffusing the avalanche before the amplification in the second step. Transfer gains
of 30 could be reached with α-particles at 2 torr of isobutane while the gain was an
order of magnitude lower at 10 torr. Very low total gains could be reached with
methane and the transfer gain was limited to a factor of two.

We have efficiently detected single electrons photo-produced by a UV lamp on the
first cathode of the multistep structure. A transfer gain of 30-100 could be reached
at pressures of 5-10 torr. The two single-electron pulse-height distributions shown
in fig. 14 differ by a factor of 30 in gain. The total charge at the peak of the
double-step distribution is of the order of 3pC, corresponding to a total gain of $2 \cdot 10^7$.

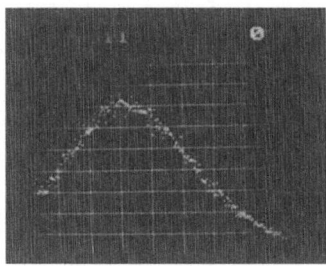 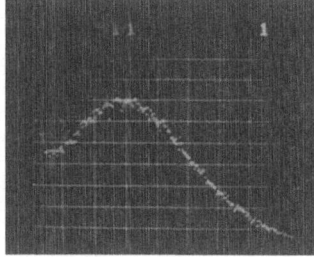

Fig. 14: Pulse-height distributions of single-photoelectrons photo-
produced on the first cathode of the detector. Left, a single gap
PPAC, charge at the peak: 0.1 pC. Right, a PPAC followed by a MWPC,
the charge at the peak is 3 pC. Isobutane, p=5 torr. The pulses at
the beginning of the double step distribution are not due to noise
but to photoelectrons produced at the second step.

This method seems to be particularly attractive for single-photoelectron detection,
for example in liquid photocathode chambers or in gas-filled photodiodes:
- higher gains can be reached;
- an electrical gate[42] can be applied in the transfer region in order to trigger
  the device only on selected events thus increasing the rate capability;
- a proper gate may stop back-moving positive ions, drastically lowering secondary
  cathode-induced defects[38];
- multielectrode structures may allow an unambiguous position read-out of multiple
  events[41].
A work along similar lines in this field, for applications with heavy ions, is presented
by H. Stelzer in these proceedings.

4   The TREC, a 4π Low-pressure Tracking Range and Energy Chamber

A new instrument recently developed for some "exotic" fission studes, related to
the search of heavy relics from the Big Bang [43], is the TREC: Tracking Range and Energy
Chamber [44]. It is a three-dimensional, position sensitive proportional detector of the
Time Projection Chamber (TPC) type [45,46]. The TREC differs from the original TPC, which
may be defined as an "electronic Bubble Chamber" used only for tracking purposes of minimum ionizing particles, by the fact that the heavy particles produced
inside its sensitive volume are fully stopped. The detector is designed to follow
tracks along their trajectories and to measure their direction, total range in the gas,
energy and most probably the specific ionization along the trajectory - over a solid
angle close to 4π.

The detector described here was originally developed for the study of the possible
anomalous spontaneous fission of ^{252}Cf that may reveal the existence of heavy (10-10^6
GeV) particles produced in the hot, dense, early Universe [47-49]. Such heavy particles
cannot be produced at accelerators but their presence inside a nucleus would drasti-
ally modify the kinematics of different nuclear processes, one of which is spontaneous
fission. The total energy released in the case of ^{252}Cf, 180 MeV, is normally shared
between two fission fragments, inversely proportional to their masses. If a fission
of nuclei with attached relics occurs, practically all the kinetic energy will be taken
by one fragment, leaving a clear signature of the presence of an abnormal mass.

The TREC is shown schematically in fig. 15.

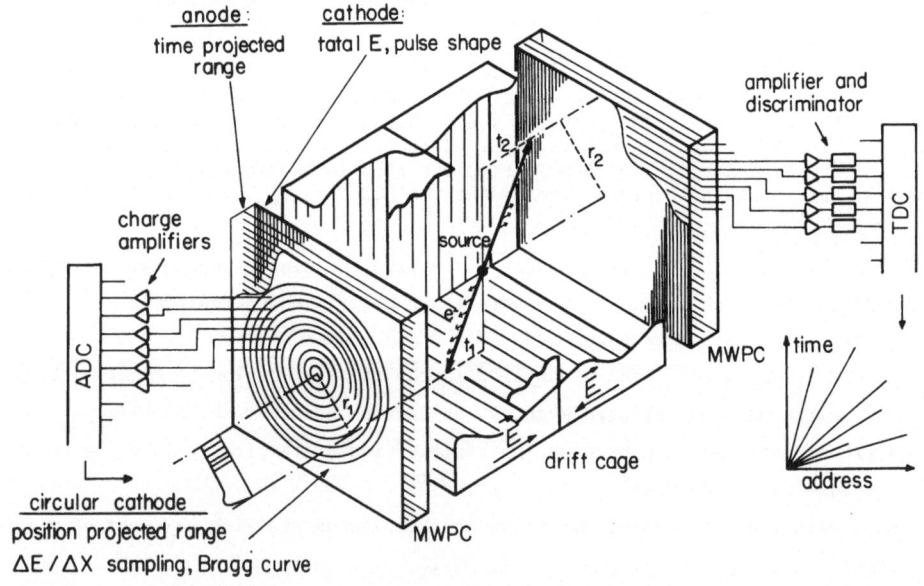

Fig. 15: A schematic view of the TREC principle and its electronics

It consists of a large drift volume 2x100 mm long, having two end-cap multiwire proportional chambers (200x200 mm^2). It is placed in a large vacuum-tight vessel and operates at low gas pressures of the order of 100-200 torr. A ^{252}Cf point-source is deposited on a thin foil support placed at the center of a conductive plate that separates
the counter into two independent and equal parts. The drift field is defined along the
square cage by 1 mm copper strips, 4 mm apart (center to center) on a G10 board. The
constant potential gradient is obtained by a chain of 1 MΩ resistors. A negative
potential is applied to the central plate and the first MWPC cathodes are grounded.

Fission fragments emerging from the source (3 mm in diamter) are fully stopped in
the drift volume, ionize the gas and give rise to electrons that drift towards the
MWPCs, where amplification occurs. A direct charge pulse is produced on the anode
wires (at positive potential) while pulses of an opposite polarity are induced on the
two cathodes.

The following parameters are measured:

- the total energy (E) of each fragment is determined from the charge impulse re-
  recorded on the first cathodes, made of 60 μm wires, 1.27 mm apart.
- The position-projected range (r) is recorded from the Bragg-like distribution
  of charge pulses induced on the rear cathodes. These cathodes are made of con-
  centric rings on a printed board, 1.1 mm wide, 1.27 mm apart (center to center).
  Pulses from consecutive strips are read-out through charge sensitive amplifiers[50] and a system of CAMAC ADCs[51]. The output of the ADCs versus the strip
  addresses represents a Bragg curve, the length of which is equal to the position-
  projected range.
- The time-projected range (t) is recorded from the arrival time of successive
  electrons produced along the track, to the anode wires of the MWPC. Time signals from individual wires (20 μm diameter, 1.27 mm apart) are amplified,
  shaped[52] and digitized with a system of CAMAC TDCs[53]. The TDCs output versus
  the wire address allows one to follow the track, wire by wire, till its origin and
  to measure its total time-projected range. The total range and the angle of
  emission are obtained from the measurement of both time and position projected
  ranges.

Two other parameters that may be of importance are the following:

- The rise-time of the fast current pulse induced on the first cathode is a direct
  measure of the emission angle[54].
- The ADC output from each ring is proportional to the specific ionization along
  the trajectory. Information on the Bragg curve of the type proposed by C. Gruhn
  (see contribution to these proceedings) may be obtained for a 2π angular range
  and may be used for particle identification.

The detector is already in use. Fig. 16 shows the energy distribution of 6.1 MeV
α-particles and fission fragments from the ^{252}Cf source. The detector operates with
an amplification factor of the order of 100 only. The distributions shown in fig. 16
are for the full angular range. There is certainly an angular correlation to the

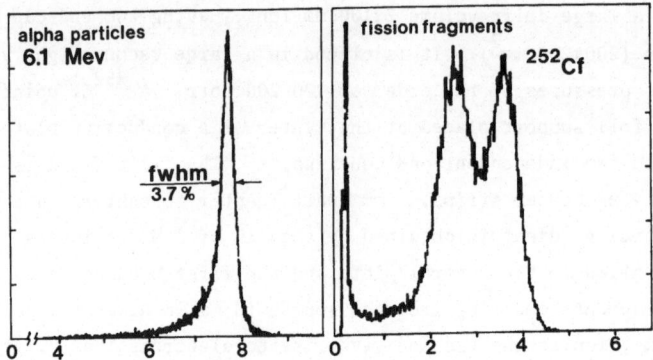

Fig. 16: Energy distribution from the TREC. Left: α-particles, p=150 torr, anode potential 2100 V, drift field 2.2 V/cm·torr. Right: fission-fragments, p=120 torr, anode potential 1600 V, drift field 2.2 V/cm·torr. The distributions are integrated over a solid angle of 2π

measured charge, not yet corrected for, due to space charge effects. Examples of tracks as recorded by the TDCs are shown in fig. 17. The sensitivity to the full angular range is well shown (vertical and horizontal tracks). It should be noted that the α-particles are not fully stopped at the pressure of 120 torr of isobutane. The ADC system for the position-projected range is presently being installed.

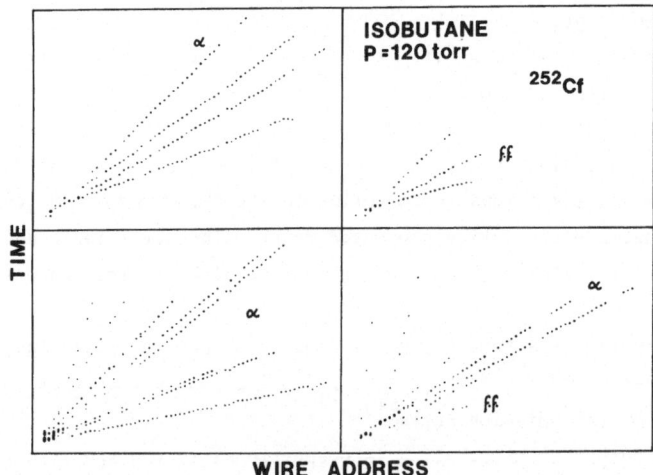

Fig. 17: Examples of tracks of α-particles and fission frag-ments emerging from the ^{252}Cf source. The α-particles are not stopped in the detector at the present pressure of 120 torr. The zero channels of the TDCs are not yet adjusted.

The TREC may find various applications far beyond its original design, and become a versatile and flexible tool for the study of various nuclear reactions. Being able to operate at variable pressures, the same instrument can be adapted to particular problems of detecting particles of different energies and ionizations. Among immediate applications is the detection of heavy recoils from a target placed at the center of the chamber. We are at present designing such a detector for future experiments with relativistic heavy ions, the main problem being the ionization of the gas by the direct beam of the order of 10^8 particles/s·cm^2, passing through the chamber. We expect this to be solved by a proper electric field gating[42] or by "killing" the central sensing part of the detectors[55]. Among other applications one may consider the detection of charged particles from neutron-induced reactions, etc.

5. Conclusions

The aim of this article was to draw the attention of the reader to some new approaches to the detection of heavily ionizing particles and, at the other extreme, single electrons using the same techniques.

New detectors are not only developed when there is a demand for a particular instrument but also sometimes in order to create such a demand... This philosophy has its proof in having successfully contributed to the progress in nuclear physics over the last few decades.

References

1) E. Rutherford and W. Geiger, Proc. Roy. Soc. A81 (1908) 141.
2) A. Breskin, I. Tserruya and N. Zwang, Nucl. Instr. and Meth. 148 (1976) 275.
3) I. Tserruya, A. Breskin, R. Chechik, E. Duering, S. Kaplanis, N. Trautner and N. Zwang, Nucl. Instr. and Meth. 196 (1982) 225.
4) A. Breskin and N. Zwang, Nucl. Instr. and Meth. 146 (1977) 461.
5) Y. Eyal and H. Stelzer, Nucl. Instr. and Meth. 155 (1978) 157.
6) A. Breskin, R. Chechik and N. Zwang, Ann. Israel Phys. Soc. 4 (1981) 254.
7) J. Girard and M. Bolore, Nucl. Instr. and Meth. 140 (1977) 279.
8) G. Charpak, Ann. Rev. Nucl. Sci. 20 (1970) 195.
9) F. Binon, V.V. Bobyr, P. Duteil, M. Guanere, L. Hugon, M. Spighel and J.P. Stroot, Nucl. Instr. and Meth. 94 (1971) 27.
10) A. Breskin, R. Chechik and N. Zwang, Nucl. Instr. and Meth. 165 (1979) 125.
11) A. Breskin, R. Chechik and N. Zwang, IEEE Trans. Nucl. Sci. NS-27 (1980) 133.
12) A. Breskin, Nucl. Instr. and Meth. 141 (1977) 505.
13) A. Breskin, Nucl. Instr. and Meth. 196 (1982) 11.
14) I.S. Sherman, R.G. Roddick and A.J. Metz, IEEE Trans. Nucl. Sci, NS-15 (1965) 500.
15) G. Charpak, D. Rahm and H. Steiner, Nucl. Instr. and Meth. 80 (1970) 13.
16) R. Grove, I. Ko, B. Koskovar and V. Perez-Mendez, Nucl. Instr. and Meth. 99 (1970) 381.
17) See refs. given by J.C. Alberi and V. Radeka, IEEE Trans. Nucl. Sci. NS-23 (1976) 25.
18) G. Charpak, A. Jeavons, F. Sauli and R. Stubbs, CERN 73-11 (1973).
19) G. Charpak, G. Petersen, A. Policarpo and F. Sauli, Nucl. Instr. and Meth. 148 (1978) 471.
20) A. Breskin, R. Chechik and N. Zwang, "High accuracy delay-line position sensing in low-pressure MWPCs", 1982, in preparation.
21) J. Stahler, G. Hemmer and G. Presser, Nucl. Instr. and Meth. 164 (1979) 305.

22) A. Breskin, G. Charpak and S. Majewski, "On the low-pressure operation of multi-step avalanche chambers", 1982, in preparation.

23) M. Birk, A. Breskin and N. Trautner, Nucl. Instr. and Meth. 137 (1976) 393.

24) I. Tserruya, A. Breskin, R. Chechik, E. Duering, S. Kaplanis, N. Trautner and N. Zwang, Nucl. Intr. and Meth. 196 (1982) 225.

25) Z. Fraenkel et al. Experiments performed at TRIUMF, LASL, ORNL, BNL. Unpublished results.

26) G. Charpak, "The role of photons in gaseous detectors". Contribution to these Proceedings.

27) J. Séguinot and T. Ypsilantis, Nucl. Instr. and Meth. 142 (1977) 377.

28) G. Coutrakon, M. Cribier, J.R. Hubbard, Ph. Mangeot, J. Mullie, J. Tichit, R. Bouclier, A. Breskin, G. Charpak, J. Million, A. Peisert, J.C. Santiard, F. Sauli, C.N. Brown, D. Finley, H. Glass, J. Kirz and R.L. McCarthy, IEEE Trans. Nucl. Sci. NS-29 (1982) 323.

29) J.R. Hubbard, G. Coutrakon, M. Cribier, Ph. Mangeot, H. Martin, J. Mullie, S. Pallanque and J. Pelle, Nucl. Instr. and Meth. 176 (1980) 293.

30) R. Bouclier, G. Charpak, A Cattai, J. Million, A. Peisert, J.C. Santiard, F. Sauli, G. Coutrakon, J.R. Hubbard, Ph. Mangeot, J. Mullie, J. Tichit, H. Glass, J. Kirz and R. McCarthy, CERN-EP/82-83, submitted to Nucl. Instr. and Meth.

31) E. Barrelet, T. Ekelöf, B. Lund-Jensen, J. Séguinot, J. Tocqueville, M. Urban and T. Ypsilantis, CERN-EP/82-09, submitted to Nucl. Instr. and Meth.

32) D. Anderson, IEEE Trans. Nucl. Sci. NS-28 (1981) 842.

33) D. Anderson, "Extraction of electrons from a liquid photocathode into a low-pressure wire chamber", CERN-EP/82-100, submitted to Phys. Lett.

34) D. Anderson, Private communication.

35) F. Sauli, "Rediscovering the gas-filled photodiode", CERN-EP/82-36, submitted to Nucl. Instr. and Meth.
    - D. Miller, private communication.

36) W. Kononenko, B.W. Robinson, W. Selove and G.E. Theodosion, Nucl. Instr. and Meth. 186 (1981) 585.

37) G. Charpak et al., "The gas photodiode as a possible large area photon detector". Contribution to the IEEE Nucl. Sci. Symposium, Washington, October 1982.

38) S. Majewski, private communication.

39) G. Charpak, G. Melchart, G. Petersen, F. Sauli, E. Bourdinaud, P. Blumenfeld, C. Duchazeaubeneix, A. Garin, S. Majewski and R. Walczak, CERN 78-05 (1978).

40) G. Charpak and F. Sauli, Phys. Lett. 78B (1978) 523.

41) A. Breskin, G. Charpak, S. Majewski, G. Melchart, G. Petersen and F. Sauli, Nucl. Instr. and Meth. 161 (1979) 19.

42. A. Breskin, G. Charpak, S. Majewski, G. Mulchart, A. Peisert, F. Sauli, F. Mathy and G. Petersen, Nucl. Instr. and Meth. 178 (1980) 11.

43. G. Barbiellini, A. Breskin, R. Chechik, G. Hermann, T. Johansson, S. Polikanov and N. Trautmann, "On the search for exotic heavy relics from the Big Bang". 1982, in preparation.

44. A. Breskin, T. Johansson, S. Polikanov and J.C. Santiard, "The TREC, a 4π, low-pressure Tracking Range and Energy Chamber", 1982, in preparation.

45. W.W. Allison, C.B. Brooks, J.N. Bunch, J.H. Cobb, J.B. Lloyd and R.W. Plenning, Nucl. Instr. and Meth. 119 (1974) 499.

46. D. Nygren, "Proposal for PEP facility based on the time projection chamber", (PEP4, Dec. 1976).

47. P.H. Frampton and S.L. Glashow, Phys. Rev. Lett. 99 (1980) 1981.

48. J. Ellis, T.K. Gaisser, G. Steigman, Nucl. Phys. B177 (1981) 427.

49. R.N. Cahn and S.L. Glashow, Science 213 (1981) 607.

50. Developed by J.C. Santiard - CERN.

51. LeCroy, ADC data acquisition system No.2280.

52. J.C. Santiard, CERN-EP/80-04 (1980).

53. LeCroy, time digitizing system no.4290.

54. A. Breskin, G. Charpak and F. Sauli, Nucl. Instr. and Meth. 136 (1976) 497.

55. S. Majewski and F. Sauli, CERN-EP/75-14 (1975).

RESPONSE OF GAS SCINTILLATORS TO HEAVY CHARGED PARTICLES[*]

M. Mutterer, P. Grimm, H. Heckwolf[+], J. Pannicke[++], W. Spreng[+++] and
J.P. Theobald

Institut für Kernphysik, Technische Hochschule Darmstadt,
D-6100 Darmstadt, FR Germany

Basic features of scintillation detectors based on the luminescence of noble gases
and relevant gas mixtures are traced in the light of known radio luminescence processes. The application of this detection regime to heavy ions is outlined and experimental results are reported.

1. Introduction

An energetic charged particle passing a gaseous detector medium looses energy predominantly by ionization and excitation processes. In a rare gas which is the main
component in many commonly used detector gases (e.g. Ar-CH$_4$) approximately one third
of this energy is dissipated by non-ionizing excitation, an important amount being
reemitted by radiation. However, nearly all gaseous detectors developed throughout
their long history have been considered as mere charge devices, omitting the piece
of information carried away by the photons.

The intense research work performed on gas-ionization detectors of different types
during the past 20 years has made them nearly unique tools in wide research fields
like elementary-particle physics and nuclear physics with heavy ions[1,2]. Their
great success is mainly due to their insensibility to radiation damage and their
capability to be adjustable to many requirements of very complex present-day experiments. The physical processes underlying the drift and diffusion of charges,
charge multiplication and avalanche built-up etc. are well understood and controlled
today[3].

Compared to gas-ionization detectors, the development of gas scintillators based on
the luminescence of noble gases was relatively slow. An early period of active gas
scintillator research in the fifties[4] was obviously interrupted by the advent of
solid-state detectors. For many years then, only a few groups followed this line of
research development and it is mainly due to a group at the University of Coimbra,
Portugal, that the basic principles of gas scintillation detectors are quite well
known today. Especially their active research work on a scintillation mode where a
moderate electric field applied to the gas volume yields intense secondary light
emission through electroluminescence, the so-called scintillation proportional counters[5,6], has led to X-ray detectors of an energy resolution unmatched by any wire

[*] Supported in parts by Gesellschaft für Schwerionenforschung, Darmstadt
and Deutsche Forschungsgemeinschaft, Bonn-Bad Godesberg.
present addresses: [+]Univ. Frankfurt, [++]ILL, Grenoble, [+++]GSI, Darmstadt

chamber, with important applications[7],[8], e.g. in X-ray astronomy. For applications in other fields, esp. nuclear physics, the potentialities of gas scintillators are not yet exploited or even fully considered. Since the early days of gas scintillator research a large progress has been made in the understanding of gas luminescence by other lines of research than detector development, mainly fundamental atomic physics, and the physics of gas discharges and laser excitation.

Starting a few years ago, we have focussed our interest on the application of gas scintillators to heavy ions. The aim of this work was not to find optimal detector designs for particular heavy ion experiments but rather to work out the potentialities of these detector regimes.

This paper summarizes our work, without going into much details which have been described in previous papers[9-14]. At first, we shortly summarize what is known about light emission processes in rare gases, to an extent needed for an interpretation of the experimented findings. Radio luminescence has recently been exhaustively reviewed by M. Salete S.C.P. Leite[15] and by A. Policarpo[16], with a closer relation to detector applications.

2. Gas Scintillation Mechanisms

A schematic diagram of the deexcitation modes in a rare gas following excitation by charged particle impact[15] is shown in Fig. 1. The primarily excited atomic states can due to the large level spacing in noble gase atoms be assembled to three groups: lowest excited states X^*, higher excited states X^{**}, and ionized levels X^+. All these groups can form by collisions with groundstate atoms diatomic molecular states, some of them being weakly bound excimer states (X_2^*, X_2^{**} and X_2^+) of some stability. In a gas under pressure the formation of dimers competes with radiative atomic deexcitation, reducing the natural life time of the atomic states and contributing molecular decays to the observed radioluminescence. Emission spectra of all rare gases (except He) exhibit sharp lines in the vacuum-ultraviolet (VUV) wavelength region of their first excited states, i.e. resonance radiation, only at pressures below ~ 50 torr. At higher pressure, continuous spectra from the decay of the excimer states dominate, shifted to lower wavelength but lying still in the VUV, below 2000 Å, for all noble gases

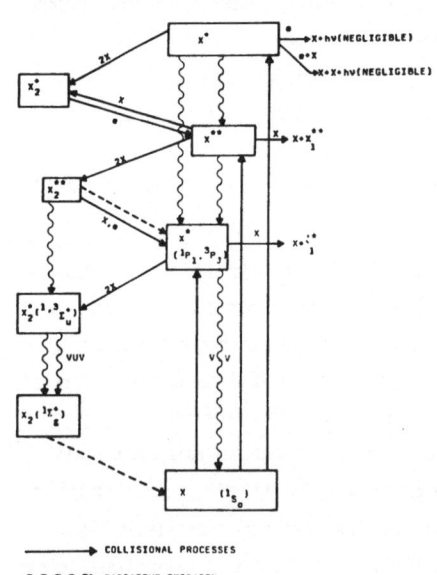

------→ COLLISIONAL PROCESSES

〜〜〜〜 RADIATIVE EMIS.ION

------→ DISSOCIATION PROCESSES

Fig. 1: Deexcitation processes in rare gases following charged particle impact (Ref. 15).

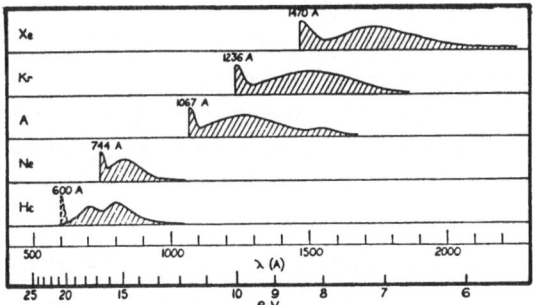

Fig. 2: Schematic representation of rare gas continua (Ref. 19).

(Fig. 2). It is well established that the VUV continua originate from the decay of the lowest molecular states $^1\Sigma_u^+$ and $^3\Sigma_u^+$ formed essentially by three-body collisions by the first excited atomic levels $^3P_1$ and $^3P_2$[17-19].

$$X^*(^3P_1, \ ^3P_2) + 2^1S_0 \rightarrow X_2^*(^1\Sigma_u^+, ^3\Sigma_u^+) + ^1S_0 \tag{1}$$

$$X_2^*(^1\Sigma_u^+, \ ^3\Sigma_u^+) \rightarrow X_2 \ (^1\Sigma_g^+) + h\nu$$

$$X_2(^1\Sigma_g^+) \rightarrow 2 \ ^1S_0$$

The so-called first continuum in the vicinity of the resonance lines is ascribed to transitions from high vibrational levels in $^1\Sigma_u^+$ and $^3\Sigma_u^+$ to the repulsive ground-state $^1\Sigma_g^+$ and the lower energetic second continuum is related to the same transitions after the molecular states have been vibrationally relaxed. The second continuum dominates at higher pressure ($p \geq 400$ torr) where relaxation is fast. The lowest molecular singlet states $^1\Sigma_u^+$ desintegrate in the range of ns ($\tau_s = 4.2$ ns for Ar and 6.2 ns for Xe) whereas the triplet states $^3\Sigma_u^+$ are much slower

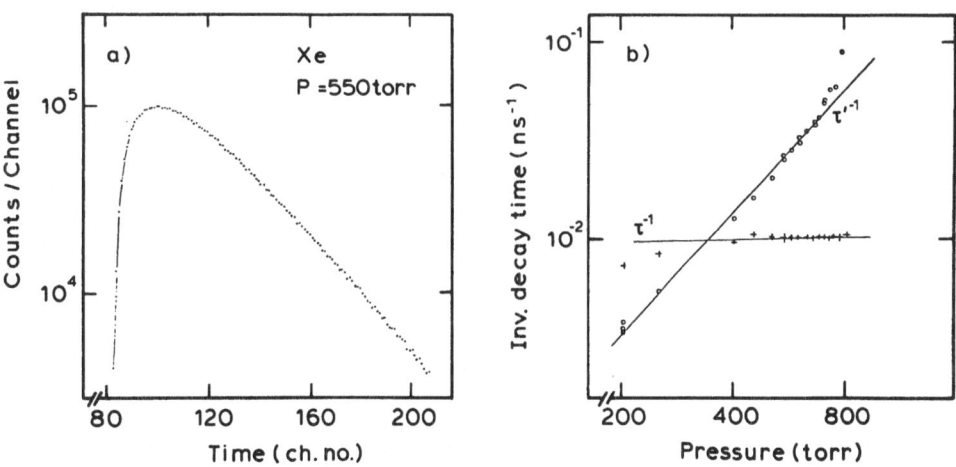

Fig. 3: Time spectrum (a) and decay times (6) of the second continuum decay in Xe (data taken from Ref. 17)

(τ_t = 3.2 μs for Ar and 100 ns for Xe). Scintillation decay via the second continuum according to Eq. (1) exhibits the $1/p^2$ dependence of the collisional formation process of molecular states followed by their characteristic decay (Fig. 3).

At pressures around 200 torr, the argon spectrum shows a weak third continuum, not observed in other noble gases, at somewhat longer wavelengths (~ 2200 Å) attributed to an emission from the group of higher excited molecular states Ar_2^{**}. This continuum disappears at higher pressure where dissociative depopulation of Ar_2^{**} to atomic states Ar^* becomes dominant, followed again by process (1). Radiative decays between highly excited states are generally weak and no detailed information on them is presently available, except that they contribute weak fast components to radio luminescence in the visible or near infrared[20].

Ionized atomic levels X^+ which are created by charged particle impact roughly twice as frequent as neutral excited states also form molecular states X_2^+ with deeper potential minima (esp. a $^2\Sigma_u^+$ state), with a rate about ten times faster than the low excitation states described above. They decay mainly by dissociative recombination with thermalized electrons to lower lying atomic levels in X^{**}, cascading down by collisions and feeding finally again the VUV continuum according to Eq. (1). The intensity and time structure of recombination luminescence depends on the ion and electron density distributions, the electron thermalization and diffusion processes and the recombination[21,22]. They are thus functions of the velocity and atomic charge of the particles and strongly depend on gas composition and pressure. For weak ionizing β and α particles and a gas pressure below ~ 1 bar, no recombination luminescence is observed at a time scale of some μs used in detector applications, whereas in the highly ionized tracks of heavy ions or fission fragments recombinations are frequent, with the consequences that here scintillation light output increases at high ionization density.

Collisional energy transfer processes are also very important in binary mixtures of noble gases and mixtures of noble gases with molecular gases, like N_2. At pressures of several hundert torr, a small amount of a heavier rare gas (0.001 to 0.1 % of Xe in Ar or Kr and of Kr in Ar) leads to the appearance of resonance atomic and continuous molecular emission of the admixture gas, together with the disappearence of the emission continua of the host gas[23,24]. This effect has the consequence that small Xe admixtures to Ar or Kr shift the original spectrum to longer wavelengths, better matching the spectral responce of photon sensors. As mixtures of rare gases with molecular gases, mainly Ar-N_2 was studied so far. The measured spectra suggest a highly efficient transfer of atomic energy both from the Ar^* and Ar_2^* levels to the second positive band in N_2[24-25]. This line spectrum in the range from 2000 to 4000 Å fits to the response of standard quartz-window photo tubes.

3. Fundamental Properties of Gaseous Scintillation Counters

In the light of the given qualitative picture of rare gas luminescence some basic
features of scintillation detectors based on noble gases and relevant gas mixtures
can be traced:

a) Since the first excited states being the main sources of luminescence are very
   energetic in rare gases (~ 2/3 of the ionization potentials), scintillation
   efficiency is high. For the luminescence yield defined by the energy radiated
   per unit energy lost in the medium by the ionized particle, values up to ~ 30 %
   have been reported[26].

b) The emitted molecular radiation cannot excite atomic states in the same gas.
   This implies a high transparency, allowing large-volume or high-pressure devices of uniform luminosity.

c) The deactivation of excited atomic or molecular species reproduces initial ground-
   state atoms both in pure gases and mixtures, maintaining the original gas composition even at a high absorbed dose. Thus, gas scintillators are expected to
   be of long-term stability, free from radiation damage.

d) Ionized species contributing to light emission at high ionization densities play
   an important role only in the vicinity of the highly ionized column along a particles track were recombinations are frequent. The main sources of luminescence,
   however, are quite insensitive to spatial charges once the charge carriers have
   been separated by diffusion or by weak external fields. As a consequence, gas
   scintillators are expected to allow high counting rates, free from space charge
   limitations. Also diffusion of electrons near the entrance foil, affecting the
   performance of gas-ionization counters, will not influence scintillation counters,
   allowing compact designs with short absorption lengths. The additional source of
   luminescence due to column recombinations can be used with profit in detectors,
   to identify heavy particles by applying pulse-shape discrimination techniques.

e) The dominant role of collisions in the deactivation process implies a strong
   dependence of the scintillation decay times on pressure, and on the dopant concentration in the case of gas mixtures. So, gas pressure is an important parameter which must be considered carefully in projecting a scintillator for a particular experiment.

f) The importance of collisional processes makes scintillation performance also very
   sensitive to spurious contaminates of non-radiating impurity gases esp. O_2 and
   hydrocarbons[27]. For a stable and reproducible operation, gas impurities should
   be kept on the level of $\leq$ 10 ppm. This needs some precautions, but rare gases
   are due to their inert character easy to purify by hot calcium or titanium and
   cerium getters. With adequate technology, large sealed Xe counters have been
   operated over periods of years with only a small cerium getter coupled to the
   gas volume[7].

g) Since the main emission is in the VUV, adequate coupling to light sensors is a
major problem. Usually wavelength shifters are applied (like p-terphenyl, p-TP)
to match gas luminescence to the spectral response of commercially available
phototubes. These shifters, however, reduce the originally high luminescence
yields and may influence fast timing response, and if applied inside the gas
volume may effect stability or poison gas purification systems. Shifting of the
spectra by gas additives and applying sensors with UV windows like Suprasil seem
to be a better joice, but here still some problems have to be solved.

4. Response of Gaseous Scintillators to Heavy Ions - Experimental Results

4.1 Experimental Details

In most of our studies we have used as scintillator a simple cylindrical gas chamber
(Fig. 4) of 80 mm diameter viewed by a 40 mm diameter quartz-window phototube. The
vessel is fabricated from nickel-plated brass, with standard Viton O-ring seals and
an entrance window from $\sim$ 70 $\mu g/cm^2$
polypropylene. The walls are covered
by reflector material[27]; in part of
the experiments, p-TP wavelength
shifter was evaporated onto walls
and PM window. As gases mainly Ar
and Xe, with purities of 99.9997 %
and 99.997 %, respectively, in the
range of pressure below 1 bar and Ar
with admixtures of N_2 below 15 %
were investigated. If the chamber
was carefully cleaned and evacuated
before filling, it was found sufficient to maintain a continuous gas

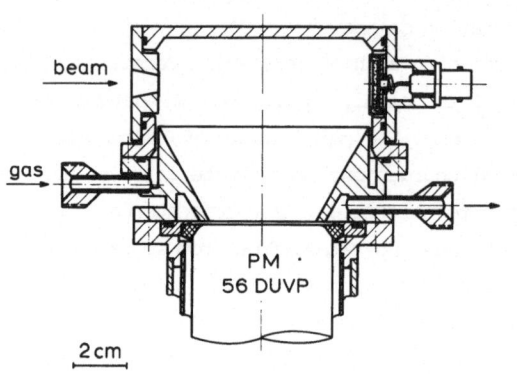

Fig. 4: Gas-scintillation counter with built-in surface-barrier detector

flow of $\sim$ 20 Nml/min for obtaining a stable operation. Xenon was generally pumped
in a closed circuit through a gas purifier[29]; the binary gas mixtures were obtained
from a gas supply regulated by thermal gas flow meters.
Test experiments were performed with [241]Am α-particles (5.48 MeV) and with heavy
ions at the accelerator UNILAC of GSI, Darmstadt, predominantly at the low-energy
beam facility with 1.4 MeV/amu ions.

4.2 Light Yields, Energy and Time Resolution

Relative light yields of scintillator gases measured both with and without p-TP
coatings are summarized in Table I. These data were obtained with α-particles, but
are approximately valid also for excitation by heavy ions.

Table I: Light yields of different gas filling at 400 torr, relative to the light yield of Ar with no wavelength shifter (p-TP) applied.

Gas:	no p-Tp	p-TP on reflector[28] and PM window
Ar	1	11.0
Kr	1.5	–
Xe	6.7	25.5
Ar + 0.5 % N_2	3.3	6.9
Ar + 2.0 % N_2	4.0	4.8
Ar + 10 % N_2	2.6	2.9

The large enhancement of the light amplitudes for the pure rare gases by p-TP demonstrate that the main luminescence is at wavelengths below the UV cut-off of the PM (~ 2000 Å). The luminescence observed without shifter is predominantly fast, with decay times in the ns range. For Ar-N_2, the relative yields with and without shifter show that with increasing N_2 concentration light emission is shifted towards the visible region.

With particle energies of ≥ 100 MeV, light signals from all scintillation gases studied were found to be large enough so that light statistics was no longer the limiting factor for the measured energy resolution. As an example, for ^{208}Pb ions of 1.4 MeV/amu resolutions of 1.6 % fwhm were measured both for Ar and Xe; and 2.3 % fwhm for Ar + 10 % N_2; here the main limitation is attributed to energy straggling in the window foil and the straggling of the electronic energy loss by the influence of nuclear collisions[30]. For ^{86}Kr ions of 8.1 MeV/amu, the scintillator with 1 bar of Xe has yielded a resolution of 0.7 % fwhm (Fig.5a) which compares to ionization chambers[31].

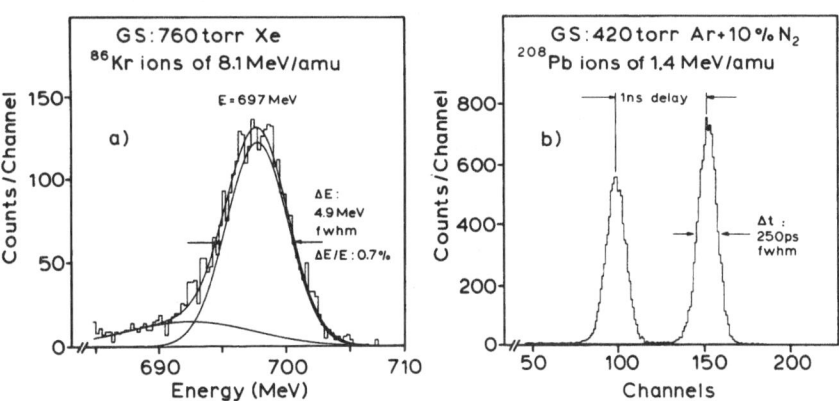

Fig. 5: Energy (a) and time (b) resolution capability of the gas scintillator shown in Fig. 4

With all gases, time resolutions of ~ 250 ps were obtained, using a secondary-electron time pick-off as reference[32] (Fig. 5b). The combined energy and timing capability of the gas scintillator is unmatched by any gas-ionization chamber.

With the detector of Fig. 4, good energy resolution is only obtained with a well localized (0.5 cm Ø) beam of particles. For practical larger-area designs, the problem of achieving uniform light collection efficiency must be considered. This is a general handicap of any scintillator which can be overcome by using several light sensors or specially designed light guides[33] or can be corrected for if trace positions are determined.

The high light-output of rare gases makes the use of non-amplifying light sensors feasible. We have tested a small scintillator where the PM was replaced by a 20 mm diameter saphire-window vacuum photodiode[34] (Fig. 6). Compared to a phototube, the diode has important advantages: It is a compact unit of excellent stability characteristics concerning temperature and high voltage, is highly independent of the counting rate and has a rise time of $\leq$ 0.25 ns.   The photodiode scintillator (Fig. 6), filled with 400 torr of Xe, has yielded an energy resolution of 3 % fwhm for ^{238}U ions of 1.4 MeV/amu; the limit in resolution by light statistics was determined to be 1.2 %, that of the electronic noise 0.3 %. This means that the diode can favourably replace the PM with respect to the energy resolution. This type of detector can surely be further developed by covering the scintillator walls with several diodes to improve light statistics and to determine trace positions by weighting the individual diode outputs. Nowadays also larger vacuum diodes are available, as well as UV sensitive silicon diodes[35]. It is also of great interest to combine heavy ions scintillators with photo-ionization detectors (PID) which came up in recent years as favourable light sensors[36] for large-area scintillation proportional counters.

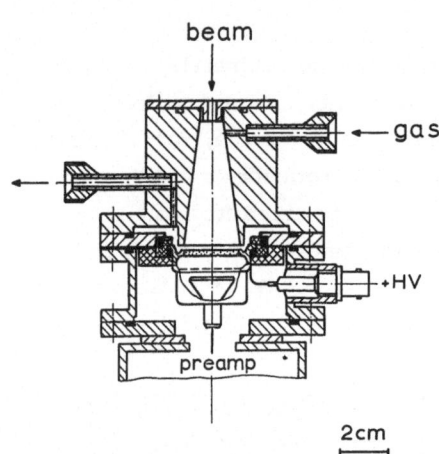

Fig. 6: Gas-scintillation counter with a saphire-window vacuum photo-diode (VALVO UVHC 20) as light sensor.

4.3 Scintillation Time Structure

Due to different processes contributing to radioluminescence and the large pressure dependence of the collision processes involved, the time response of gas scintillators is rather complicated and its investigation cumbersome and time consuming. On the other hand, a detailed study of decay times can single out the different mechanisms of luminescence, bringing gas scintillator development beyond the level

of mere empirical information. At present, only the main features of the time evolution of light emission in pure noble gases, i.e. the decays of the molecular components, are known and only weak informations about doped gases are found in the
literature.

In general, the time structure of the scintillation pulses exhibits at least three
different components which, dependent on pressure, are well separated or overlap. We
have first roughly measured the decay in Xe, Ar and Ar-N$_2$ by recording both prompt
and integrated PM current pulses with the aid of a 350 MHz oscilloscope and analyzing photographs of oscilloscope traces. In the case of the rare gases, the decay of
the most intense components coincide with the molecular process described in Chapt.
2. The results for Ar and Ar-N$_2$ are summarized in Fig. 7. Nitrogene admixture reduces the original long decay time of Ar (approaching the molecular decay time of
3.2 µs at 400 torr pressure) by two orders of magnitude. With an admixture of
10 % N$_2$, the known decay time of
~ 30 ns of the second-band emission
in N$_2$ is obtained, indicating a complete energy transfer to the dopant.
This is also apparent from the in-
effectiveness of wavelength shifter
in this case. Its fast decay makes
Ar-N$_2$ gas useful for high-rate
applications.

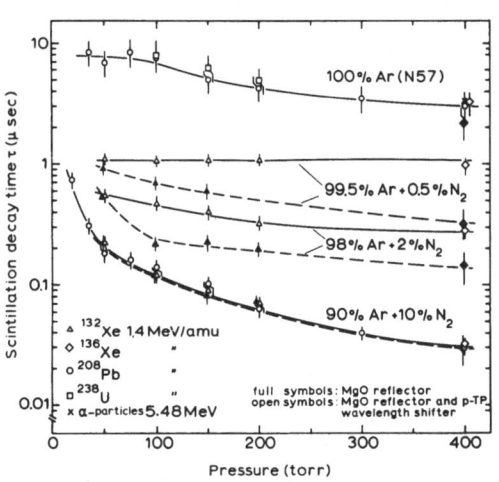

Fig. 7: Decay times of dominant scintillation
components in Ar and Ar-N$_2$ mixtures
versus gas pressure, measured with
α particles and various heavy ions

More accurate decay measurements were
performed by coupling via a light
guide a second small PM to the gas
volume and applying the single photon
method[37] (Fig. 8a) : a TAC ist started
by the fast pulses from the main PM
or by a built-in surface- barrier detector defining the interaction time
and stopped by single-photon pulses
recorded by the small PM. For these measurements, also a test chamber with internal
electrodes was used (Fig. 8b) to study simultaneously recombination luminescence.

A series of measurements was recently performed on fast decays in Ar and Ar-N$_2$ at
low pressure. In Ar (Fig. 9), a prompt component with a decay time of ≤ 1.8 ns was
found to be independent of pressure and wavelength shifter. It can probably be attributed to fast atomic transitions between higher states, emitting in the visible[20].
The slower component was fitted with the model of sequential decay by

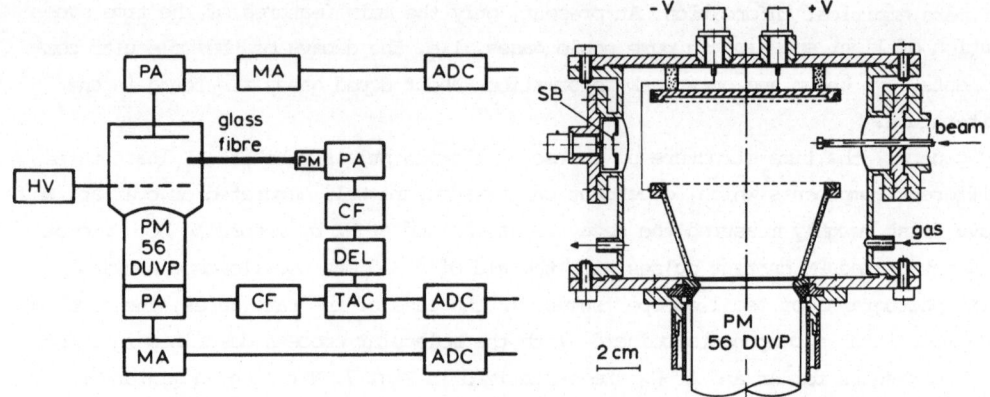

Fig. 8: (a) Decay time measurement by the single-photon method; (b) Scintillator with internal electrodes, to measure scintillations due to electron-ion recombinations

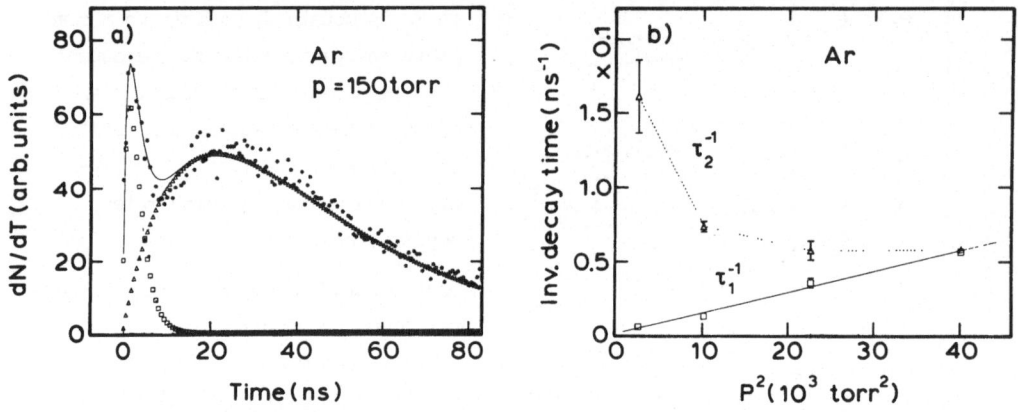

Fig. 9: Fast scintillation decay in Ar. (a) Time spectrum, and (b) decay time, according to Eq. (2).

$$f(t) = const \cdot \frac{1}{\tau_1 \tau_2} (e^{-t/\tau_1} - e^{-t/\tau_2}) \qquad (2)$$

yielding τ^{-1} to be proportional to the square of pressure:

$$\tau_1^{-1} = k_1 [Ar]^2 + C, \text{ with a reaction constant:}$$

$$k_1 = (1.2 \pm 0.1) \cdot 10^{-30} \text{ cm}^6 \text{ sec}^{-1}, \qquad (3)$$

characteristic for a three-body collision, and τ_2 varying from 6 ns at 25 torr to 17 ns at 200 torr (Fig. 9b). This is in line with a process similar to that of Eq. (1), but within the group of high excited states, i.e. third continuum decay:

$$Ar^{**} + 2\ Ar \xrightarrow{\tau_1} Ar_2^{**} + Ar \qquad\qquad (4)$$

$$Ar_2^{**} \xrightarrow{\tau_2} Ar_2^* + h\nu \qquad \text{with } \tau_2 = 6 \text{ ns from vibrational states}$$
$$\text{and } \tau_2 = 17 \text{ ns from relaxed states.}$$

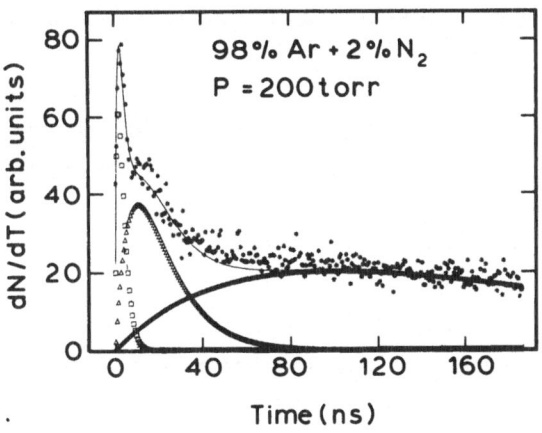

Fig. 10: Fast scintillation decays in Ar + 2 % N_2

The decay of Ar–N_2, at the same pressure (Fig. 10), exhibits two delayed components, and a prompt pulse with the same 1.8 ns decay as observed in Ar. Both delayed shapes can also be fitted by sequential decays; by varying both pressure and N_2 concentration allows to separate energy transfer processes that lead to Ar molecular states and to excited N_2 states, respectively:

$$\tau_1^{-1} = k_1\ [Ar]^2 + k_2\ [Ar]\ [N_2] + k_3\ [N_2] \qquad\qquad (5)$$

and with τ_2 being approximately constant.

With this model the faster component in Fig. 10 can be attributed to energy transfer processes on the level of higher excited states, and the slower component, which represents the main light intensity, to similar processes on the level of first excited Ar states. The reaction rates $k \cdot [\ \ ]$ represented in Fig. 11, confirm the expected dominance of a transfer to N_2 at higher N_2 concentrations, but also an enhancement of the feeding of the $Ar_2^{**}(Ar_2^*)$ states by collisons with N_2 molecules is noted. Details of these studies will be reported elsewhere.

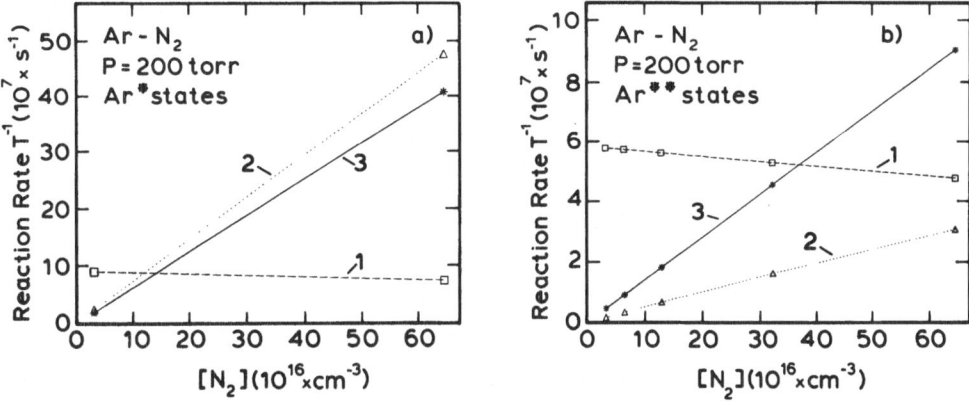

Fig. 11: Reaction rates of energy transfer processes in Ar–N_2, related with the depopulation of Ar^* (a) and Ar^{**} (b) states. Numbers indicate the sequence of terms in Eq. (5).

Recombination luminescence was studied in the range of pressure between 500 torr and 1 bar and excitation by [238]U ions to get the highest ionization density possible. Fig. 12 shows as example the result of Ar + 10 % N_2 at 1 bar. Recombinations contribute a delayed component which can be switched off with a moderate electric field applied perpenticular to the particles trace. With the field, the decay corresponds to the desintegration time measured by α particle excitation with no electric field, where the low ionization density does not induce recombination. It should be noted, that with the pure noble gases, esp. Ar, a clear identification of recombination terms in the time spectra is, due to the longer molecular decay times and the inset of electro-luminescence at fields around 0.3 $V \cdot cm^{-1}$ $torr^{-1}$, much more complicated, although here they are more important than in the Ar-N_2 case.

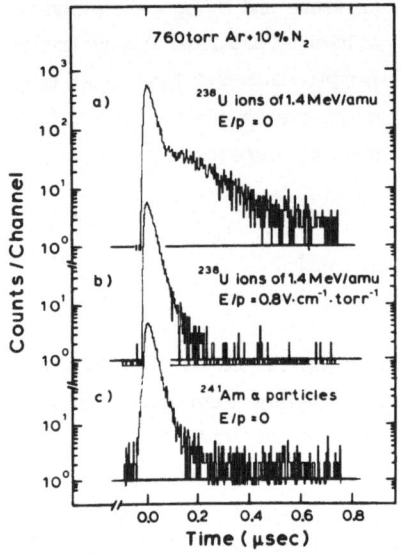

Fig. 12: Decay curves of scintillations in Ar + 10 % N_2 gas, including recombination luminescence

4.4 Linearity

The large dynamic range of heavy-ion experiments with respect to the particles mass and energy requires detectors for which the output signal is a linear function of the deposited energy, independent of ionization density. Gas scintillators were found to show no pulse-height defect of quenching, characteristical for most charge sensing devices, but conversely, a pulse-height excess for very heavy ions compared to α particles[13]. The pulse-height excess PHE defined by

$$PHE = (A_{HI}/A_\alpha)/(E_{HI}/E_\alpha) - 1 \qquad (6)$$

with A_{HI} and A_α being the measured pulse-heights for heavy ions and α particles, and E_{HI} (E_α) the corresponding energies deposited in the detector gas, was found to increase with mass number, pressure and electronic shaping constant. To relate this effect to the observed recombination luminescence, the device of Fig. 8b was used, and both light amplitude and charge collection were studied as function of electric field.

The variation of integrated charge and light signals with reduced electric field is shown in Fig. 13 for different argon-nitrogen mixtures. For Ar + 15 % N_2, the scintillation amplitude approaches a saturation value at the higher fields indicating a

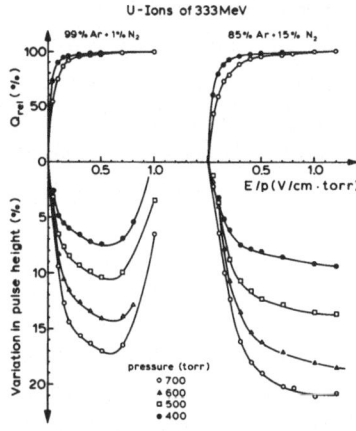

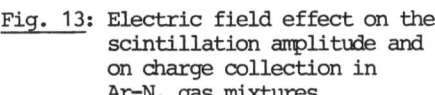

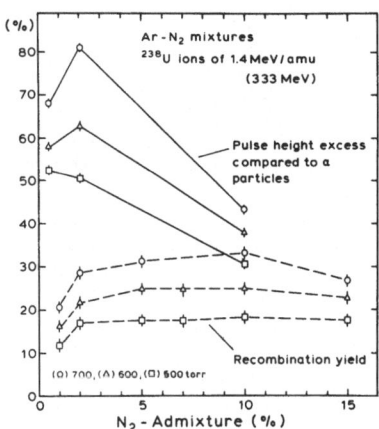

Fig. 13: Electric field effect on the
scintillation amplitude and
on charge collection in
Ar-N$_2$ gas mixtures

Fig. 14: Pulse-height excess and recombination yield in Ar-N$_2$ gas,
versus the N$_2$ concentration

complete separation of electrons and ions in the ionized column. For small N$_2$ admixtures and for the pure noble gases, the light amplitude does not saturate with E,
but increases again due to the beginning of secondary light emission by electro-
luminesence. The recombination yield defined by

$$Y_R = [A_{HI}(E = 0) - A_{HI}(E_{sat})]/A_{HI}(E_{sat}) , \qquad (7)$$

must in these cases be determined by extrapolating the ratios of charge over
light amplitudes to 100 % charge collection.

The recombination yields in argon-nitrogene mixtures are compared in Fig. 14 with
corresponding values ot the pulse-height excess. It turns out that for the higher
N$_2$ admixtures the pulse-height excess can be, to a high degree, explained by the recombination effect and consequently a scintillator with a weak electric field applied
to the interaction volume provides good linearity in the total range of masses. It is
to be noted that the average ionization densities involved in the absorption of ^{238}U
ions and α particles differ by about two orders of magnitude. For low N$_2$ admixtures
and for the pure noble gases, the increase of light output with ionization density
is only partly caused by recombinations. It is suggested that here an increase in
the population of higher excited states by the density of energetic electrons in the
column could occur. Further investigations, including spectroscopical methods, are
needed to explain this behaviour.

5. The Use of Both the Light and Electron Message in a Single Detector

For heavy ions it is an interesting aspect to use both the light and charge
message in a single detector, to increase the flux of information. To combine both
principles requires a counter gas with fast scintillation decay and with reasonable

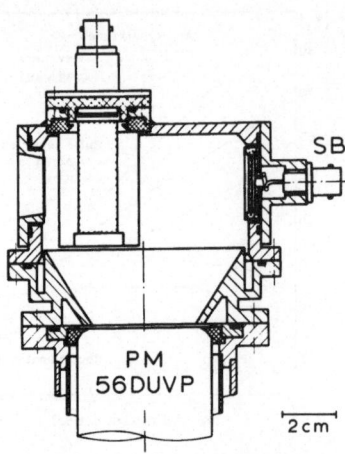

SB

PM
56DUVP

2 cm

Fig. 15: Prototype ΔE-Et detector
telescope. A gas-scintil-
lation drift chamber for
measuring ΔE and trace
positions is incorporated
into a scintillator which
measures E and t.

electron drift velocity and diffusion. Ar with
≳ 5 % N_2 is one candidate to fit these requirements. For pressures ≥ 400 torr, scintillation
decay is about 30 ns (Fig. 7) and drift velocities of ≥ 5 cm/µs are easily achieved[10].
Nitrogene effectively quenches the original
VUV radiation, so avoiding the problem of
afterpulses.

Fig. 15 shows as example a combined detector[13],
where a windowless scintillation drift chamber
[38,38] to measure ΔE and trace position is incorporated into a gas scintillator similar in
geometry to that in Fig. 4. Since all connec-
tions of the drift chambers wire cage are outside the scintillation volume, the performance
of the scintillator for the E and t measurement is not influenced. Trace position is determined in one dimension from the electron
drift time measured as the time delay between
the scintillation pulse and the light flash of secondary scintillations generated at
the end of the wire cage in a gap of high electric field between a grid and a plane
electrode. The second dimension is obtained by dividing the collected charge. Position accuracies of ~ 0.3 mm fwhm have been obtained for both dimensions. The ΔE
signal can be derived either by the total charge or by the amplitude of the secondary scintillations. With the scintillation drift chamber, high counting rates can
be achieved, since no charge amplification occurs. The ΔE signal is obtained completely independent of the E signal and has no influence on the E resolution.

6. Scintillation Response of Condensed Xenon to Heavy Ions

Condensed rare gases were recently found to be suitable scintillation media for
charged particle detectors[40,41]. With excitation by α and β particles, the light output is comparable to or slightly higher than that of NaI(Tl) and the scintillation
decay has intense fast components in the ns range for fast timing. The luminescence
of condensed rare gases is known to originate from the decay of self-trapped exciton
states ($^1\Sigma_u^+$ and $^3\Sigma_u^+$) emitting a continuous photon spectrum in the vacuum ultraviolet similar to the emission from the excimer states in the gaseous phase.
We have studied the scintillation response of solid Xe to α particles and heavy ions.
The detector cell located inside a high-vacuum chamber is a small stainless steel
cylinder (40 mm Ø) closed with an indium-sealed Suprasil window which faces a PM
tube covered with pTP wavelength shifter. Xenon was frozen onto the Suprasil disc

both from the gaseous and liquid phases by cooling the cell by a temperature controlled liquid nitrogene cryogenic system. Clear colourless several mm thick crystals could be grown by reducing temperature slowly in the region of the triple point (- 112°C). The crystals were then cooled down to - 175°C where the low residual gas pressure of 0.5 torr allows to use a thin particle entrance window. Measurements were again performed with 1.4 MeV/amu ions (^{50}Ti, ^{58}Ni and ^{238}U) at GSI and with ^{241}Am α particles. For comparison, liquid Xenon was also investigated, excited by an internal source of α particles from the decay of ^{212}Pb. The results are summarized as follows:

Energy resolution:

For α particles, a resolution between 6 and 9 % fwhm was obtained both for liquid and solid Xe. The resolution for heavy ions was between 8 and 10 %. The limitation in resolution is attributed partly to the low light geometry of the test device, the wavelength dependent transmission of Suprasil for VUV radiation and, for the heavy ion measurements, to a contamination of the open crystal surface.

Pulse-height defect:

The LET dependence of the luminosity of solid Xe determined by comparing measured pulse heights for heavy ions and α particles shows, contrary to low-pressure Xe gas, a pulse-height defect for the strongly ionizing heavy particles. Compared to α particle excitation, the luminosity is reduced to 0.50, 0.44 and 0.24 for ^{50}Ti, ^{58}Ni and ^{238}U ions, respectively. These values are in line with the pulse-height defect of liquid Ar[42] for fission fragments. Fig. 16 compares the LET dependence of con-

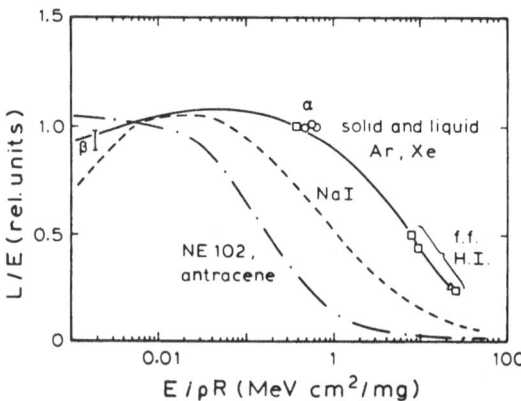

densed rare gas scintillators with inorganic[43] (NaI) and organic[44] (NE 102, antracene) scintillators, all curves being normalized in the region of weak ionizing particles. It is obvious, that the decrease in luminosity with increasing LET is much less for the noble gases than for all commonly used scintillators, an interesting aspect for the detection of ions in the subrelativistic energy region.

Fig. 16: Luminosity of condensed noble gas scintillators versus LET, compared to inorganic and organic scintillators. Fine structures in all curves are omitted. (□ this work, O , Δ Ref. 42)

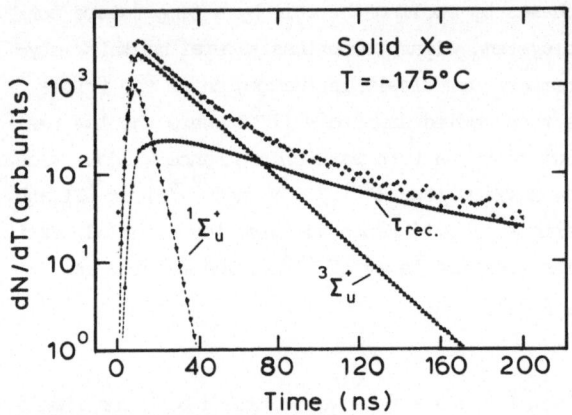

Fig. 17:

Scintillation decay of solid Xe, excited by ^{238}U ions of 1.4 MeV/amu

Scintillation time structure:

The scintillation decay in solid Xe (Fig. 17) measured with single-photon techniques can be well described by the fast exciton decays, and the somewhat longer tail by the thermalization of higher-energy delta electrons before recombination. The measured rise time of ~ 3 ns compares to the time response of the pTP wavelength shifter. An even faster rise time seems to be possible with a fast PM having a Suprasil window.

References

1) G. Charpak, IEEE NS-27 (1980) 118; Nuclear Instr. and Methods 196 (1982) 1

2) H. Stelzer, Proc. Conf. on Nuclear Physics, Aug. 24-30, 1980, Berkeley, Calif.

3) F. Sauli, Principles of operation of multi-wire proportional and drift chambers, CERN 77-09 (1977)

4) Excellent reviews of early results on noble-gas scintillators are given by J.A. Northrop, J.M. Gursky and A.E. Johnsrud, IRE NS-5, (1958) 81; J.-L. Teyssier and D. Blanc, L'onde electrique 40, (1964) 458

5) A. Policarpo, Space Sci. Instr. 3 (1977) 77

6) D.F. Anderson, Proc. Int. Symp. on Nuclear Radiation Detectors, March 23-26, 1981, ISN Tokyo, Japan

7) A. Peacock et.al., IEEE NS-26 (1979) 486; W.H.M. Ku,D.F.Anderson, T.T. Hamilton, R. Novick, IEEE NS-26 (1979) 490

8) H.E. Palmer and L.A. Braby, Nucl. Instr. and Methods 116 (1974) 587

9) M. Mutterer, J.P. Theobald and K.P. Schelhaas, Nucl. Instr. and Methods 144 (1977) 159

10) K.P. Schelhaas, M. Mutterer, J.P. Theobald, P.A. Schillack, G. Schrieder and P. Wastyn, Nucl. Instr. and Methods 154 (1978) 245

11) J.C. van Staden, J. Foh, M. Mutterer, J. Pannicke, K.P. Schelhaas and J.P. Theobald, Nucl. Instr. and Methods 157 (1978) 301

12) M. Mutterer, J. Pannicke, K.P. Schelhaas, J.P. Theobald and J.C. van Staden, IEEE NS-26 (1979) 382

13) M. Mutterer, J. Pannicke, K. Scheele, W. Spreng, J.P. Theobald and P. Wastyn, IEEE NS-27 (1980) 184

14) M. Mutterer, Nucl. Instr. and Methods 196 (1982) 73

15) M. Salete S.C.P. Leite, Portugal. Phys. 11 (1980) 53

16) A. Policarpo, Physics Scripta 23 (1981) 539

17) B. Pons-Germain, Thesis; Univ. Toulouse, France, 1978

18) G. Klein and M.J. Carvalho, Report CRN/CPR 80-17, Univ. Strasbourg

19) Y. Tanaka, A.J. Jursa and F.J. LeBlanc, J.Opt.Soc. Am. 98 (1975) 306

20) H.A. Koehler, L.J. Ferderber, D.L. Redhead and P.J. Ebert
Phys. Rev. A9 (1974) 768

21) P.E. Thiess and G.H. Miley, IEEE NS-21 (1974) 125;
S. Konno and T. Takahashi, Nucl. Instr. and Methods 150, (1978) 517

22) S. Kubota, M. Hishida, M. Suzuki and J. Ruan (Gen),
Phys. Rev. B20 (1979) 3486
T. Takahashi, J. Ruan (Gen), S. Kubota and F. Shiraishi,
Nucl. Instr. and Methods 196 (1982) 83

23) A. Gedanken, J. Jortner, B. Raz and A. Szöke, J. Chem. Phys. 57 (1972) 3456

24) T. Takahashi, S. Himi, M. Suzuki, J.Ruan (Gen) and S.Kubota,Rep.RUP 81-16 (1981)

25) T.D. Strickler and E.T. Arakawa, J. Chem. Phys. 41 (1964) 1783

26) G.S. Hurst and C.E. Klots, Advan. Rad. Chem. 5 (1976) 1;
J. Sequinot and T.Y. Ypsilantis, Nucl. Instr. and Methods 142 (1977) 377

27) T.J. Sumner, G.K. Rochester, P.D. Smith, J.P. Cooch and R.K. Sood,
IEEE NS-29 (1982) 1410

28) White reflectance coating, supplied by Eastman Kodak Co.,
Rochester, N.Y. 14650

29) BOC rare-gas purifier, supplied by British Oxygen Co. Ltd., London

30) U. Quade, Jahresbericht 1981, Beschleunigerlaboratorium U. und TU München,
p. 134

31) H. Stelzer, contribution to this conference

32) P. Wastyn, M. Mutterer and J.P. Theobald, GSI Annual Report 1978, p. 180

33) H. Klein and H. Schölermann, IEEE NS-26 (1979) 373

34) UVHC 20, supplied by VALVO, Hamburg

35) e.g. C 30842 by RCA

36) A. Policarpo, Nucl. Instr. and Methods 196 (1982) 53

37) S. Kubota, M. Hishida and A. Nohara, Nucl. Instr. and Methods 150 (1978) 561

38) A. Policarpo, M.A.F. Alves and M. Salete S.C.P. Leite,
Nucl. Instr. and Methods, 128 (1975) 49

39) G. Charpak, S. Majewski and F. Sauli, Nucl. Instr. and Methods 126, 381
(1975); IEEE NS-23, (1976) 202

40) T. Doke, Portugal. Phys. 12 (1981) 9; Nucl. Instr. and Methods 196 (1982) 87

41) S. Kubota, M. Hishida, M. Suzuki and J. Ruan (Gen), Nucl. Instr. and Methods
196 (1982) 101

42) A. Hitachi et. al., Nucl. Instr. and Methods 196 (1982) 97

43) R.B. Murray and A. Meyer; Phys. Rev. 122, (1961) 815

44) F.D. Becchetti, L.E. Thorn and M.J. Levine, Nucl. Instr. and Methods, 138
(1978) 93

THE GSI MAGNETIC SPECTROMETER

D. Schüll

GSI, Darmstadt

I. Introduction

Magnetic spectrographs were used for the spectroscopy of charged particles from the early days of nuclear physics. They have proven to be very powerful tools whenever the ultimate resolution or best background suppression is needed. For many years the development of best ion optical properties seemed to be the dominant design goal. Typical instruments of this type are the Q3D spectrographs (Eng79) which have extremely good ion optical resolutions, but in many cases the detector systems in the focal plane cannot make use of these features. Meanwhile several instruments were equipped with newly developed detector systems which are able to measure in addition to the position in the focal plane other geometrical quantities. Especially by measuring the angle of incidence in the reaction plane trajectory reconstruction is possible. This method allows to have much poorer focusing properties since the aberrations can be corrected later in the data analysis using the computer. Those detectors are complicated but the construction is much simplified when e.g. the incidence into the focal plane is nearly perpendicular or when there is enough space available between target and the first ion optical element. Therefore the design of the ion optics is strongly influenced in return by the requirements of the detectors.

If one works with lighter heavy ions the accurate position measurement, maybe combined wit a ΔE-E technique, reveals the whole information since the particles and their charge states are spatially well separated and the reaction channels well known. Using heavier beams (A>20) usually a large number of reaction products is emitted from the target and the identification of the particles becomes most important. In order to obtain a unique determination of nuclear charge Z and mass number A additional quantities like time of flight and a more accurate ΔE-E measurement have to be added to the magnetic analysis. Interesting reaction channels might have low production cross sections and the products have to be detected in the presence of strong elastic lines, so that high standards for the detection system are required.

These particles occur in many atomic charge states and the distributions can be 6 to 10 units wide for uranium ions. The position spectra in the focal plane are therefore represented by a superposition of different charge states, masses and excitation energies of the products. The need for a large momentum acceptance is therefore not only a question of counting efficiency but a substantial requirement from the physics application. A momentum acceptance of about 20 % seems to be necessary to avoid artificial cuts in the excitation spectra of the different isotopes.

The demands from the experimental program are very broad naturally when beams over the whole mass range can be provided and it is obvious that the experiments are not fixed to a special dedicated device. Some require the highest momentum resolution, for others a large momentum acceptance is more crucial. A variable dispersion seems to fit best into the conflicting demands. In addition the rapid detector development should be responded by a flexible geometry which enables future detectors to be integrated into the magnet system.

I will report here on a magnetic spectrograph built in the last years at GSI. It was strictly designed for use with the beams of very heavy ions (A>40) in the energy range 3-20 MeV/u. Before describing the details of the instrument, I will discuss short-ly the special problems and limitations when these very heavy ions have to be detected.

II. Limits to energy resolution

High energy resolution studies become more critical when going to higher projec-tile masses. Ford et al. (ForE77,Eng79) have made an analysis of system resolving pow-ers for a number of heavy-ion reactions. The different contributions are discussed in more detail by Walcher (Wal78). The limitations can be resumed by two different phe-nomena:

      a) target effects

      b) beam and magnetic spectrometer resolutions.

The energy loss straggling is proportional to the mean ionic charge and is nearly independent from energy. It can be estimated (Wil32) for a typical $100\mu gr/cm^2$ Sn-target and a 6 MeV/u Xe-beam to $\delta(\Delta E) = 250$ keV. For very thin targets the ener-gy loss straggling is relatively larger than estimated by a simple theory.

Target nonuniformities seem to be a very severe limitation for the heavy beams. With present target technologies (Fol82) thickness variations of less than 10 % for a typical $100\mu g/cm^2$ self supporting target of heavy material like lead and 30 % for thin layers of about $10\mu g/cm^2$ evaporated on carbon foils can be reached. It should be con-sidered that these thicknesses have to remain stable under bombardment over several hours and sputtering or clustering effects shall not deteriorate the original target structure. Since the total energy loss in the target amounts up to several MeV the tar-get inhomogeneity will cause an energy broadening of 200 - 300 keV.

Multiple small angle scattering in the target appears to influence the energy resol-ution in a more complicated way. Usually spectrographs are built such that all particles emitted from the same source point with equal momentum or equal momentum-angle cor-relation are focused on one spot in the focal plane. The variation of momentum with the reaction angle is expressed by the kinematical factor k

$$k = 1/p \cdot dp/d\theta \qquad\qquad 1)$$

For forward scattering and lighter projectiles k ranges between 0 and ~-0.5. Exper-iments of specific interest are reactions with nearly symmetric systems and bombarding energies around the Coulomb-barrier. In these cases the maximum transfer cross sec-tion is found around the grazing angle which is at at rather large backward angles. The k-factor for small excitation energies approximates then the tangent of the reaction angle in the laboratory system. For such an experiment or for bombardment of a lighter target with a heavier projectile k can assume large values. Typical for these reaction types is | k | ≥ 1.5. In principle this correlation can be respected in the data analysis as long as the scattering angle is measured accurately or it can be compensated by ion optical methods. The multiple scattering in the target disturbes this correlation and therefore no correction method is possible. The energy resolution is influenced through

$$(\delta E)_{MS}/E \;=\; 2 \cdot k \cdot \theta_{MS}. \qquad\qquad 2)$$

θ_{MS} is the mean multiple scattering angle. It can be calculated by theories of Meyer (Mey 71) and Sigmund and Winterbon (SigW74) and compared with data from Nickel et al. (NicM78) for our mass range. θ_{MS} can be evaluated for the above mentioned Xe beam and $100\mu/cm^2$ target to 1.5 mr. The contribution to the energy resolution amounts therefore to $2 \cdot 10^{-3}$.

The divergence of the beam contributes to the resolution in a similar way via the uncertainty in the reaction angle. In principal these effects can be corrected by a proper matching of beam handling system and spectrograph. The matching require-ments and the combined optimization procedure are described in the literature (Hen74, ReiM75, BohG78). They usually can be performed only with a complicated beam handl-ing system and the optimization and operation becomes rather difficult. Since the target effects limit the resolution anyhow to about 10^{-3} it seemed more economical for the available UNILAC -beam to cut down the emittance and the energy spread by about a factor of 3 using slits in the beam line than providing more optical elements. It should be concluded that the beam and the target effects give a severe limitation to the momen-tum resolution and the spectrograph must not be designed to a higher quality.

III. GSI-Spectrometer

III.1 Design Concept

The spectrometer was designed for use with very heavy ions with mass numbers above A=40. As pointed out already the *integration of ion optical properties* and the *characteristics of the detector system* was the design aim from the very beginning. The ability of the detectors beside the requirements for the particle identification to provide a measurement of the entrance angles into the spectrograph (θ and ϕ) released the need for complicated higher order corrections caused by geometrical or kinematical aber-rations. Those influences can be taken into account by an appropriate *software correction*. A solution could be found with simple magnets used in a *modular setup* which provides enough *flexibility* for the requirements of the experiments as well as adaption for further detector development.

III.2 Ion Optics

Fig. 1 shows the layout of the spectrometer. The elements are mounted on a turntable which can be rotated around the target within a range -45° to +135°
The ion optical elements are
   a) a simple 45° bending magnet with straight field boundaries perpendicular to the centre trajectory for obtaining momentum resolution. The radius of curvature is 2 m, the maximum field 1.7 T.
   b) the two quadrupoles Q1 and Q2 for focusing in both directions

c) the quadrupole Q3 with an aperture of 30 cm. With this element the dispersion can be varied within a range of 2 to 7 cm/%.

d) the sextupol Sex2 controls the focal plane tilt angle to be perpendicular to the centre trajectory.

e) Sextupol Sex1 is used to provide some second order correction.

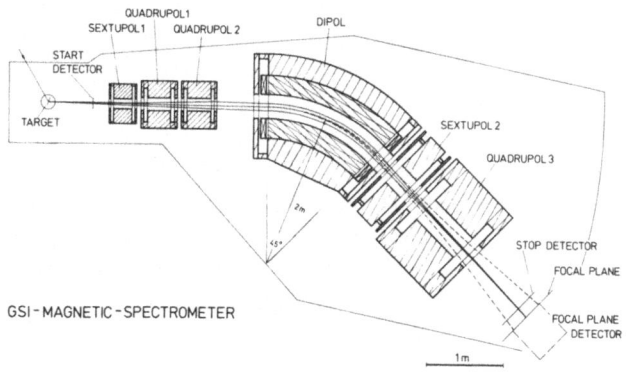

FIG 1: Schematic view of the spectrometer

We have chosen the relatively small deflection angle since the mass determination is done by a time of flight measurement. It is well known that the momentum resolving power and the path length differences are inversely proportional (Bro70) and one has

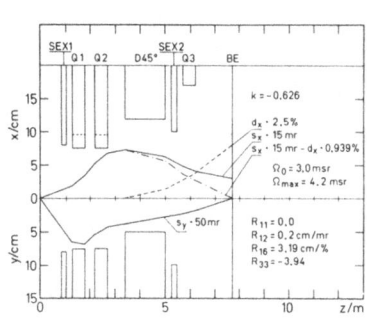

FIG 2: Characteristic trajectories of the spectrometer. TRANS-PORT- notations (Bro70) were used for the coefficients.

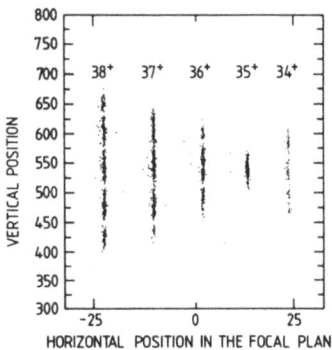

FIG 3: Two dimensional scatter plot of position in the focal plane vs. vertical position y for elastically scattered Xe- ions.

to compromise between mass and momentum resolving power. The latter will be limited to about 1/3000. Fig. 2 shows the characteristic trajectories for a typical case with medium dispersion (Wal74). But it should be mentioned that there are several other modes of operation: To minimize the contributions from the angle uncertainties caused by the divergence of the incoming beam , multiple scattering in the target, and position resolution of the start detector either to mass or energy resolution a kinematically compensated or uncompensated mode can be used. In cases when the transverse kinematics or the transverse optical aberrations become significant a defocused mode in the focal plane is superior to double focusing (fig. 2) since it allows a correction of these contibutions via a measurement of the entrance angle φ through the transverse coordinate y in the focal plane.

Fig. 3 demonstrates that the ion optics are not perfect in all cases. Shown is a scatter plot position in the focal plane vs. transverse coordinate. The lines correspond to the different charge states of elastically scattered Sn-ions. It can be seen that the envelop in the transverse coordinate is not constant which is due to a large second order coefficient for this rather low dispersion. Since these higher order corrections are for such a simple instrument principally poor we are restricted to a solid angle of about 3 msr which can be used only because of software correction in the data analysis. The gaps in y in fig. 3 are due to 2mm thick potential bars spaced 2cm apart behind the entrance window of the detector. The basic features of the instrument are compiled in table 1.

type	SQQDSQ
deflection angle	45°
Bρ	3.2 Tm
solid angle	3msr
dispersion variable	2....7 cm/%
focal plane	50 cm, ⊥
energy acceptance	50.....13 %
energy resolution	$5 \cdot 10^{-4}$ FWHM k = 0
	10^{-3} FWHM k = 1
flight path differences	$6 \cdot 10^{-3}$
kinematical correction	$\vert k \vert \leq 1.5$

Table 1: Spectrometer characteristics

III.3 Detector System

The detector system consists of a start detector at the entrance of the spectrometer, a scintillator stop detector and a hybrid ionisation chamber with a proportional counter as focal plane detector. These detectors measure the time of flight through the spectrograph, the entrance angle θ, the horizontal and vertical positions in the focal plane, the energy loss ΔE and the total energy E.

The start detector (BusP80) is a secondary emission channel plate detector with a carbon foil of about 25 µg/cm² mounted perpendicular to the particle tracks. The electrons from the surface of the carbon foil are accelerated and by means of an electrical mirror imaged to a channel plate with an active area of 20 x 40 mm². The catcher anode consists of 2 mm wide strips. The position signal is determined by the charge division method. Fig. 4 shows a schematic drawing of the detector. The timing resolution is about 200 ps, the position resolution 1.2 mm which corresponds to about 2 mr angle resolution.

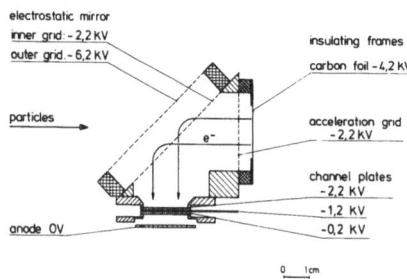

FIG 4: Cross section of start detector

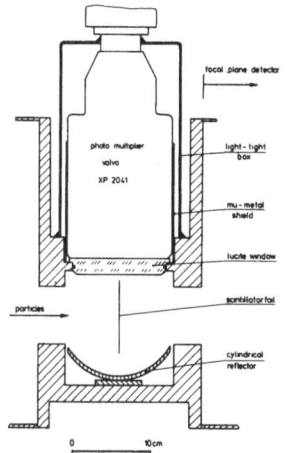

FIG 5: Cross section of scintillation stop detector

Fig. 5 shows a cut through the stop detector. A thin self supporting scintillator foil (~150 μg/cm²) with dimensions 50 x 7 cm² is placed close before the entrance window of the focal plane detector. The light emitted from the surface after passage of a heavy ion is detected by four 3" phototubes mounted in series on top of the vacuum chamber. A cylindrical reflector on the bottom improves the detection efficiency and levels out the flight path differences of the light from the different vertical and horizontal positions. The output signals of the four phototubes are summed by a cable and put on the input of a constant fraction discriminator. The timing resolution of about 400 ps is limited already by the path differences of the light from the various foil positions.

The focal plane detector is a hybrid ionisation chamber with two proportional counters integrated into the anode group. It is similar to the one used in Rochester and Argonne (ShaV75, ErsB76). The length and depth are 50 cm each. The entrance window is supported by a grid with a mesh width of 3 mm and a transmission of 75 %. The anode is subdivided into 9 sections for the ΔE-E measurement. Usually the length of the ΔE-section and the operating pressure are chosen that the ΔE signal corresponds to about 40 % of the full energy. The ΔE-resolution for 800 MeV - Xe ions is 1.2 %. The ΔE-signals and the rest energy are summed by the computer to obtain the total energy E. For E a resolution of 0.5 % was reached. The resolution of the summed signal was comparable to the cathode signal.

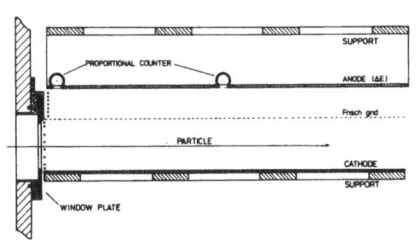

FIG 6: Schematic view of focal plane detector

The detector is operated with pure methane gas with pressures of 100 - 500 mbar. For higher bombarding energies we used pure tetrafluor-methane (CF_4) which has a four times larger stopping power. We could achieve similar E-resolutions with the advantage of much lower operating voltages compared to methane or iso-butane. The proportional counter consists of a high resistance carbon coated Si-wire. The position signal is derived from the right and left side by the charge division technique. A resolution of 1.2 mm is obtained.

We have applied two proportional counters to do track reconstruction in the focal plane. This is only of importance when no start detector is used. The timing signal is then derived from the microstructure of the beam pulse, the θ - information from the position differences of the two proportinal counters. The detector signals and the corresponding resolutions are summarized in table 2.

focal plane detector		
position	X	1.2 mm
total energy	E	0.8 %
energy loss	ΔE	1.2 %
vert. position	y	~1 mm
start detector		
time signal	T	150ps
entrance angle	θ	2mr
scintillation stop detector		
time signal	T	400ps

Table 2:   Detector signals and accuracies for 800 MeV Xe-ions

III.3 Data analysis and performance

The data processing of the different signals from the detectors is straight forward. They are fed into Analog to Digital Converters connected to the PDP 11/45 online computer. Online analysis is performed via a fast link to the IBM 3081 computer. Thus a direct check of the experiment is possible since e.g. atomic charge state distribution, mass, Z and energy are no direct output signals from the detectors. They are derived from combinations of the measured signals. This online analysis is especially of advantage for the tuning procedure which can be performed in a short time by watching two dimensional scatter plots e.g. position vs. θ (see fig. 7) or ΔE vs. E. The adjustment of the right field settings of the quadrupols or sextupols is very easy and quick since it is not necessary to check the resolutions by determining line widths of projected parameters but just looking to slopes or curvatures in two dimensional plots and try to correct them by hardware or - when the intrinsic resolution is sufficient - decide to do it by software in the final analysis. Fig. 7 shows as an example the scatter plot horizontal position x in the focal plane vs. entrance angle θ. The inclined lines correspond to different charge states of elastically scattered Xe-ions. The slope directly measures

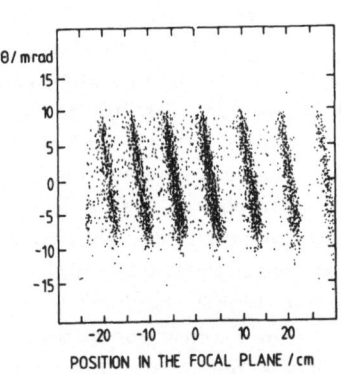

FIG 7: Two dimensional scatter plot of position in the focal plane vs. entrance angle θ. The lines correspond to different charge states of elastically scattered ions.

the kinematical factor K. A change of the quadrupols Q1 and Q2 rotates these lines to be vertical in fig. 7. This is equivalent to a shift of the horizontal focus to the detector position. Different slopes on the left or right edges mean that the focal plane tilt angle is not 90°. It can be adjusted by sextupol Sex2. What mode will be chosen will be better understandable from the chapter below.

For the particle identification the four basic relations between the measured quantities energy loss ΔE, energy E, flight time T and position in the focal plane x (s = path length through spectrograph) and the physical quantities nuclear charge Z, mass number A, velocity v and atomic charge state q are used:

a) $\Delta E \sim Z^2 \cdot A/E$; b) $T = S(\theta)/v$;

c) $B\rho \sim x \sim A \cdot v/q$; d) $E = A \cdot v^2/2$;

From these follows

$$q = \frac{2}{S(x,\theta)} \cdot \frac{E \cdot T}{B\rho(x,\theta)} ; \qquad 3)$$

In spite of the fact that three signals have to be combined the resolution is mainly determined by the energy resolution of the focal plane detector. It accounts to $\delta q/q = \delta E/E \sim 1/120$ for 800 MeV Xe ions and 1/200 for 1500 MeV Kr ions. These resolutions seem to be much better than required for any element, but one should keep in mind that very often transfer products have to be determined in the presence of a very strong elastic line and then contributions on the 10^{-4} level from a wrong charge state can deteriorate the mass spectra. When q is analyzed, a window can be set for the next step of the mass determination

$$\frac{A}{q} = \frac{B\rho(x,\theta) \cdot T}{s(x,\theta)} \qquad 4)$$

Here the uncertainty in the time of flight measurement limits the mass resolution $\delta A/A = \delta T/T = 0.45ns/180ns = 1/400$. To obtain these resolutions corrections for variation of dispersion over the focal plane and flight path differences have to be applied. They can be calculated by TRANSPORT-code (Bro70) and are found to be small for our instrument. They account to 10^{-3} for an uncertainty of 2 mr in the θ determination. Once q and A are fixed the particle energy is calculated from the relation

$$E_p = (B\rho)^2 \cdot q/(A/q) ; \qquad 5)$$

Since for A and q discrete values are assumed the energy resolution is mainly due to the intrinsic Bρ -resolution (ion optics, detector position resolution) and the target effects. For the standard mode resolutions of 1000 are typical.

I should mention that for some applications when the mass resolution seems to be the most critical parameter in the experiment a spectrometer setting ('kinematically uncompensated') can be used. In this case the contributions from θ cancel out in equ. 5 optimizing the mass resolution on the account of a somewhat lower energy resolution since the corrections have then to be applied in equ. 6.

Finally the ΔE - E functions

$$Z \sim f\,(\Delta E, E/A) \qquad\qquad\qquad 6)$$

are linearized in order to obtain the nuclear charge Z in the usual technique. The resolution $\delta Z/Z \sim 1/90$ for 800 MeV Xe-like ions is demonstrated in fig. 8. It shows the Z-distribution of quasielastic reaction products.

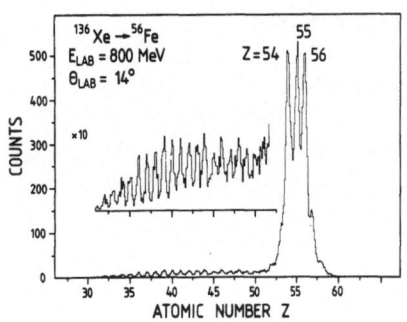

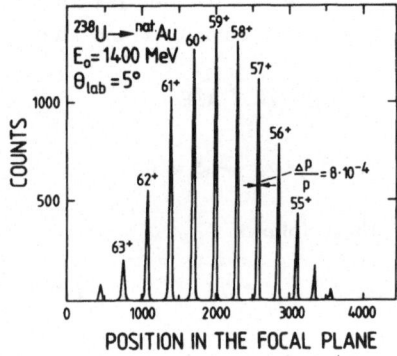

FIG 8: Z-spectrum of Xe-like reaction products

FIG 9: Charge state distribution of U-ions seen as position spectrum in the focal plane.

In the following I will show some examples which shall illustrate further more the data analysis procedure and give some insight into the performance of the apparatus as well as the field of experiments done until now. Fig. 9 shows a raw position spectrum of elastically scattered uranium ions. The peaks correspond to different charge states. It is obvious that a large momentum acceptance is required to collect at least the dominant charge states. The distribution is centered for this energy at 59 and is about 5.5 units wide (FWHM). In order to get absolute cross sections or even relative yields between reaction products of different Z the charge state distribution has to be measured for all these products. For high resolution work and consequently a lower momentum acceptance the charge state distribution can be measured by a separate field setting using the variable dispersion feature. The momentum resolution $\delta p/p = 0.08\%$ for this very heavy mass is quite reasonable bearing in mind that the beam line, the target and the detector foils contribute considerably. Beside the more complex data analysis procedure, the many charge states have one advantage: there are always four or more calibration lines in the spectra which can be used for the energy and mass calibration and for a check of the ion optical imaging coefficients.

Another example is the scatter plot charge state vs. A/q (fig. 10). The picture shows the products from Xe induced quasielastic reactions on a Fe-target. The individual masses for each charge state can be seen. The projections on the two axes reveal the charge state distribution and - selecting out each charge state - the mass distribution. For the final mass distribution these mass spectra have to be stretched, shifted and then summed. Such a spectrum is shown in fig. 11.

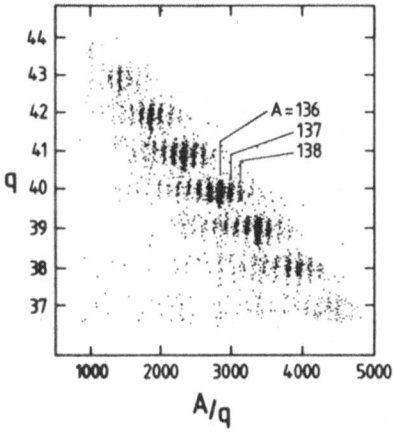

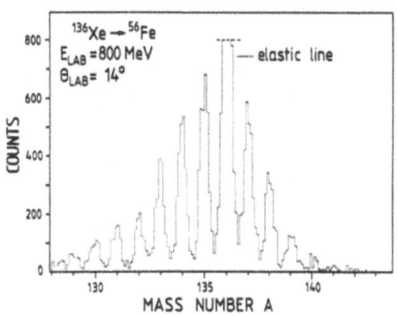

FIG 10: Two dimensional scatter plot of charge state vs. A/q. The vertical lines correspond to individual masses.

FIG 11: Mass spectrum of quasielastic products from Xe-induced reactions.

The resolution appears to be $\delta A/A$ = 400. This extremely good resolution is achieved routinely in an uncompensated mode for all k-values. When the energy resolution has to be optimized to ≤ 0.3 % for k-values ≥ 0.5 a slightly worse resolution is achieved.

The importance of a good mass charge state and energy resolution becomes obvious in an experiment ^{120}Sn on ^{112}Sn-target. The goal of this experiment is the observation of neutron pair transfer compared to a multistep mechanism for bombarding energies below the Coulomb threshold (OerG83). Fig. 12 shows a scatter plot in the A vs. Bρ-plane for a forward angle θ_{lab} = 28° in the laboratory system. The strong line corresponds to the forward scattered ^{120}Sn. Clearly separated individual mass lines can be seen from mass 110 up to 116. A contamination from another ionic charge state of mass 120 events close to the line of mass 116 is

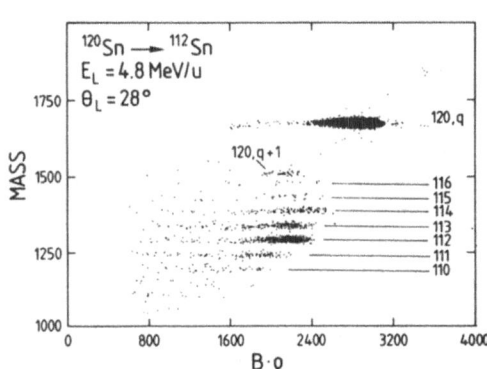

FIG 12: Scatter plot of mass vs. Bρ for transfer products and elastically scattered Sn-isotopes.

visible because of insufficient separation in the charge spectrum. In spite of the fact that the spectrum is broadened due to target inhomogeneities it can be seen that the cross section is concentrated around the ground state transition. It shows that the transfer probability for the two neutron transfer is about as high as for the one neutron transfer.

IV. Summary

I have presented a view on an instrument which takes advantage of the recent detector developments for very heavy ions. The features of these detectors were considered already in the design phase of the ion optical and geometrical layout of the spectrograph. The result is a modular setup with separated function magnets, with variable dispersion and a flexibility to adapt to the various requirements of the experiments. The instrument was designed to be used exclusively for very heavy ions (A > 40). The combination of a time of flight measurement and magnetic analysis, together with a determination of ΔE, E and some geometrical quantities gives good particle identification properties. I have shown some examples of excellent mass- and Z-resolutions for the mass range 100 and above. The energy resolution of about 0.1 % is moderate, in fact, in most cases of appliance the limitation is not introduced by the performance of the spectrometer and detector system but by effects in the target and by the divergence of the incoming beam.

The instrument is used for the investigation of elastic and inelastic scattering in the heavy mass range (JiaC81) where the separation of the transfer channels seems to be absolutely necessary for the detailed understanding of the nuclear potentials. But I should mention, that the domain of application is the study of quasielastic transfer reactions (SchS81, BraS82, BreC82) to investigate the reaction mechanism for short interaction times. The excellent particle identification together with a good energy resolution allows to see structures in the nuclide distributions due to shell effects in the nuclear potential as well as structures in the spectra of the different transfer channels. These experiments are still under way.

The project of the magnetic spectrograph was suggested by F. Pühlhofer and set up by a joint effort of groups from the University of Marburg, Max-Planck-Institut für Kernphysik Heidelberg and GSI. The ion optical design is due to Th. Walcher. I should like to thank F. Pühlhofer, W. Pfeffer, F. Busch, Th. Walcher, B. Langenbeck, K. Blasche, W.C. Shen and R. Bock for many fruitful discussions during the setup time of the project and while preparing the manuscript. The project was supported by the Bundesminister für Forschung und Technologie.

References:

BohG78 D.H. Bohlen, G. Gebauer, and W. von Oertzen, Hahn Meitner Institut, Berlin report HMI-B 171

BraH82 P. Braun-Munzinger, K.D. Hildenbrand, W.F.W. Schneider, D. Schüll, H. Stelzer, H. Freiesleben, M. Marinescu, and F. Pühlhofer, GSI annual report 1981, p 26

BreC82 C. Brendel, J. Carter, G. Delic, A. Richter, G. Schrieder, and D. Schüll, GSI annual report 1981, p 10,11

Bro70 K.L. Brown, Stanford Linear Acc. Centre, Stanford, SLAC report No. 75

BusP80   F. Busch, W. Pfeffer, B. Kohlmeyer, D. Schüll, and F. Pühlhofer, Nucl. Instr. and Meth. 171 (1980) 71

Eng79   H.A. Enge, Nucl. Instr. Meth. 162 (1979) 161

ErsB76   J.R. Erskine, T.H. Braid, and J.C. Stolzfus, Nucl. Instr. and Meth. 135 (1976) 67

Fol82   H. Folger private communication

ForE77   J.L.C. Ford, Jr., H.A. Enge, J.R. Erskine, D.L. Hendrie, and M.J. LeVine ORNL/TM-5687

JiaC81   Jiang Cheng-Lie, P.R. Christensen, O. Hansen, S. Pontoppidan, F. Videbaeck, D. Schüll, Shen Wen-Qing, A.J. Baltz, P.D. Bond, H. Freiesleben, F. Busch, and E.R. Flynn, Phys. Rev. Lett. 47 (1981) 1039

Hen74   D.L. Hendrie in Nucl. Spectroscopy and Reactions, Part A, p 365 ff, ed. by J. Cerny, Academic Pr., New York and London 1974

Mey71   L. Meyer, Phys. Stat. Sol. 44 (1971) 253

OerG83   W. von Oertzen, B. Gebauer, A. Gamp, H.G. Bohlen, F. Busch, and D. Schüll to be published

ReiM75   J. Reich, S. Martin, D. Protic, G. Riepe, Proc. 7. Int. Conf. Cyclotrons and their Applications, Zürich 1975, p 245, Birkhäuser Basel 1975

SchS81   D. Schüll, W.C. Shen, H. Freiesleben, R. Bock, F. Busch, D. Bangert, W. Pfeffer, and F. Pühlhofer Phys. Lett. 102B (1981) 116

ShaV75   D. Shapira, R.M. DeVries, M.R. Clover, R.N. Boyd, and R.N. Sherry, Jr., Phys. Lett. 71B (1977) 293

SigW74   P. Sigmund, and K. Winterbon, Nucl. Instr. Meth. 119 (1974) 541

Wal74   Th. Walcher, MPI-Sonderdruck MPI-4-1974-V25, Heidelberg 1974

Wal78   Th. Walcher, in Experimental Methods in Heavy Ion Physics, p 236, ed. by K. Bethge (Ed.), Springer Heidelberg 1978

Wil32   E.J. Williams, Phys. Rev. Sol. A 135 (1932) 105

STUDY OF 12C INTERACTIONS AT HISS

H.J. Crawford

HISS Group

Lawrence Berkeley Laboratory,

Space Sciences Laboratory,

University of California

Berkeley, California 94720

Single particle inclusive measurements in high energy nuclear physics have provided the foundation for a number of models of interacting nuclear fluids. Such measurements yield information on the endpoints of the evolution of highly excited nuclear systems. However, they suffer from the fact that observed particles can be formed in a large number of very different evolutionary paths. To learn more about how interactions proceed we have performed a series of experiments in which all fast nuclear fragments are analysed for each individual interaction. These experiments were performed at the LBL Bevalac HISS (Heavy Ion Spectrometer System) facility where we studied the interaction of 1 GeV/nuc 12C nuclei with targets of C, CH2, Cu, and U. In this paper we describe HISS and present some preliminary results of the experiment.

We first review some of the results of single particle inclusive measurements to illustrate the motivation for a facility such as HISS. These are summarized in fig. 1. Such properties suggest a schematic representation for relativistic heavy ion reactions such as shown in

SINGLE PARTICLE INCLUSIVE

HAS SHOWN US:

PERSISTENCE OF VELOCITY	MAGNETIC SPECTROMETERS BECOME Z SPECTROMETERS M
FACTORIZATION (TARGET INDEPENDENCE)	TARGET AND PROJECTILE FRAME PHYSICS IS EQUIVALENT

PROBLEMS ARE:

ALL SPECTRA LOOK LIKE PHASE SPACE (THERMAL MODEL)

WE NEED BETTER DEFINITION OF THE FINAL STATE IN ORDER TO AVOID INTEGRATION OVER THE UNSEEN PRODUCTS.

Figure 1

fig. 2. A question relevant to all three diagrams concerns the relation
of the energy and momentum transferred at the vertex. It is possible
that the vertex is complex, such as in a nuclear cascade, or single,
such as in electromagnetic excitation of a giant dipole resonance. To
measure this energy-momentum transfer we utilize the kinematic focussing
property to investigate final states of the incident 12C nucleons. Theoretical predictions of the energy transfer spectrum from three different
models are shown in fig. 3.

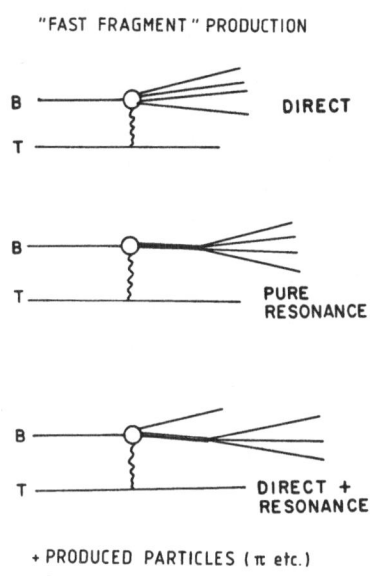

Figure 2: Reaction schematics

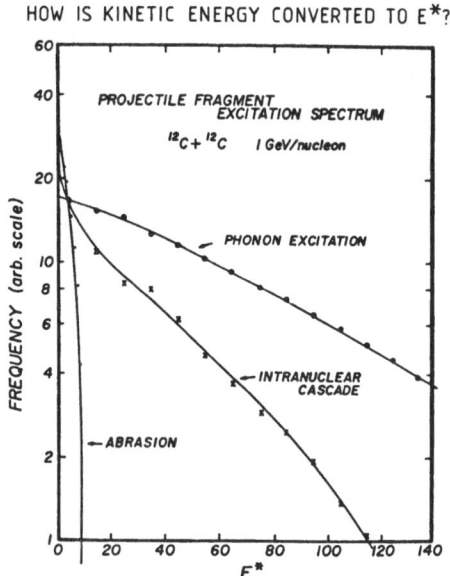

Figure 3: Predictions of excitation
spectrum

The HISS facility was constructed to allow particle identification
(charge Z, mass M, and vector momentum P) for many particles from the
same interaction. The facility consists of:

1. Large superconducting dipole;
2. Multi CPU computer system;
3. Trajectory measuring devices-Drift chambers;
4. Velocity measuring devices-Time-of-flight (TOF) array;
5. Charge measuring devices-Scintillators and MUSIC.

The floor layout of HISS at the Bevalac showing the dipole, cave area,
electronics house, and VAX house is given in fig. 4. I would like to
thank the many people whose names are shown in fig. 5 for making the
HISS facility work.

The first five experiments accepted for running at HISS are summarized in fig. 6. These range in experimental complexity from single

HEAVY ION SPECTROMETER SYSTEM GROUP

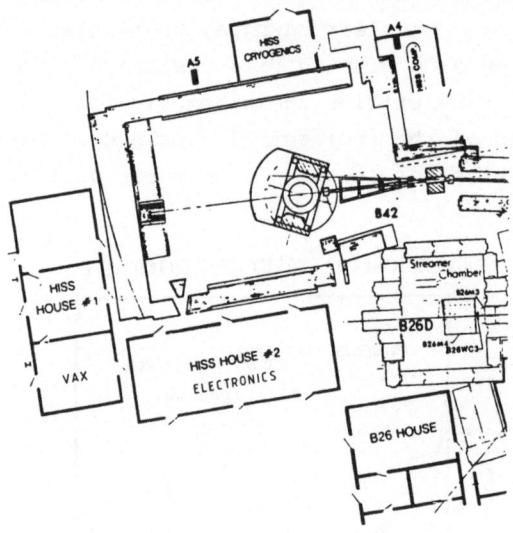

E. BELEAL	LBL
F. BIESER	LBL
M. BRONSON	LBL
H. CRAWFORD	UC-SSI
J. ENGELAGE	LSU
I. FLORES	UC-SSL
D. GREINER	LBL
M. JOHNSON	UC DAVIS
P. LINDSTROM	LBL
C. McPARLAND	UC-SSL
J. PURTER	LBL
D. OLSON	LLNL
H. SANN	GSI
R. WADA	LBL

Figure 5

Figure 4: HISS area at BEVALAC

HISS PHASE I EXPERIMENTS

COULOMB DISSOCIATION OF $^{16,18}O$--LLL, SSL, LBL

   GOALS: EXCITATION OF GIANT RESONANCES IN $^{16,18}O$
      DECAY

INVARIANT MASS SPECTRA FROM ^{12}C--LSU, UCD, NRL, LBL

   GOALS: EXCITATION SPECTRA FROM EXCLUSIVE MODES
      SEARCH FOR STRUCTURE IN INVARIANT MASS
      SPECTRA

FRAGMENTATION OF ^{56}FE--SSL, NRL, LBL

   GOALS: SEARCH FOR COLLECTIVE EFFECTS EVIDENCED BY
      STRUCTURE IN INVARIANT MASS SPECTRA

MEASUREMENTS OF LARGE P_T FRAGMENTS--INS, LBL

   GOALS: STUDY OF COLLECTIVE EFFECTS NEAR KINEMATIC
      LIMIT AND THE ASSOCIATED MULTIPLICITIES

TWO PARTICLE CORRELATIONS AT SMALL ΔP--UCLA, UCD, LBL

   GOALS: USE SECOND ORDER INTERFERENCE BETWEEN
      IDENTICAL PARTICLES TO DETERMINE INTERACTION
      VOLUME AND TIME

Figure 6

particle inclusive spectra to full event reconstruction. The HISS facility is equipped with a set of facility detectors described below and
shown in fig. 7. It is often the case that a measurement will require
additional detector systems. In fig. 8 we show the set up for the high
Pperp experiment which has, in addition to the facility detectors, a set
of MWPC and scintillators that operate in the field region of the dipole

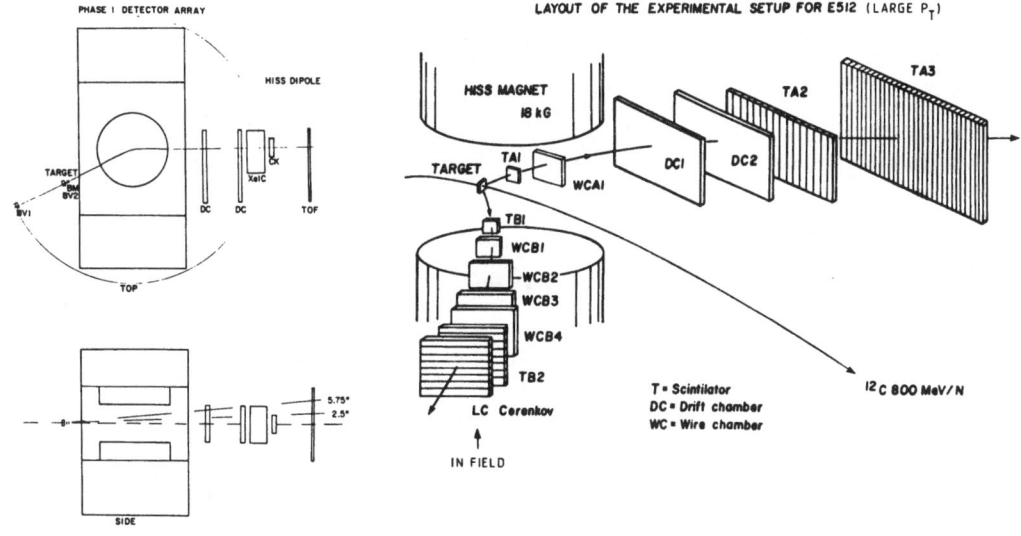

PHASE I DETECTOR ARRAY

LAYOUT OF THE EXPERIMENTAL SETUP FOR E512 (LARGE P_T)

T = Scintilator
DC = Drift chamber
WC = Wire chamber

^{12}C 800 MeV/N

Figure 8

Figure 7

that were constructed by the INS group.

The heart of this HISS facility is a large (1 m gap, 2 m poletips), superconducting dipole (30kG field), capable of rotating through 360 degrees. We measure rigidity by determining particle trajectories through this field. The field is surface mapped (1cm grid) to an accuracy of 1 gauss. The surface map is converted to a volume map using LaPlace's equation in a VAX computer. A large number of sample trajectories are then sent through the volume and numerically integrated to return a set of Chebychev coefficients which give the vector momentum, rigidity, and pathlength from a set of input x, y, and z coordinates.

A multi CPU computer system is used at HISS to control the dipole, monitor the beam line during experiments, and gather and analyse data. The HISS computer system is shown in fig. 9. The basic data flow is from detectors through CAMAC into a micro-programmable branch driver (MBD) and then into a front-end acquisition CPU (11/45). There the data is spooled onto tape and onto an intermediate disk which has a port connection to the VAX so that complex on-line analyis can be performed (e.g. event reconstruction).

Particle identification at HISS is accomplished through measurements of rigidity, R (= momentum/charge), charge, Z, and velocity, β. We describe below the performance of the prototype Phase I detectors shown in fig. 6 and discuss some of the parameters of our upgraded Phase II array.

HISS COMPUTER SYSTEM

online event reconstruction

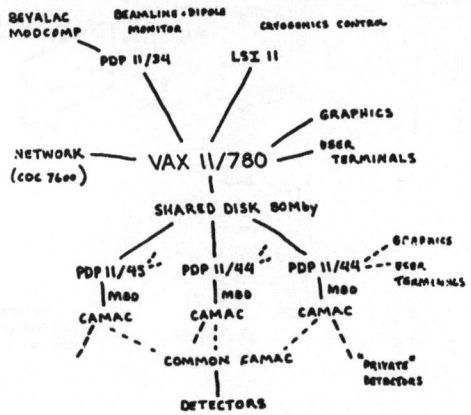

Figure 9

We use drift chambers upstream of the target to determine the incident beam trajectory and downstream to determine the trajectories of all fast charged fragments produced in the forward 5 degree cone. A schematic diagram of such a drift chamber is shown in fig. 10. Note that there are two vertical (or "S" plane) wires, two "T" plane (+60 degrees), and two "U" plane (-60 degrees) wires to remove the left-right ambiguity of a single wire measurement.

HISS DRIFT CHAMBER

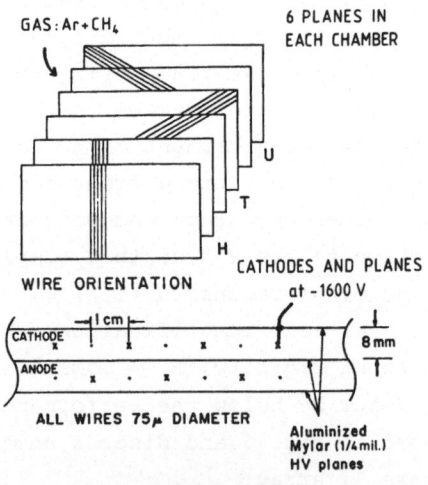

Figure 10

These chambers are designed to give locations for all charged fragments. Since the chambers must locate protons as well as heavy ions
they must be sensitive to high energy delta rays as well. The track of
a heavy ion must be determined in the presence of a large halo of high
energy delta rays. To determine the location of the core ionization in
a heavy ion track we use a dynamically set threshold on the timing discriminator for each wire as shown in fig. 11. This timing pulse is sent
as a stop signal to a channel of LeCroy 4290 system TDC. The analog
fraction of the signal is sent to a LeCroy 2880 system ADC for all of
the "S" plane wires.

HEAVY ION DRIFT CHAMBER FRONT END

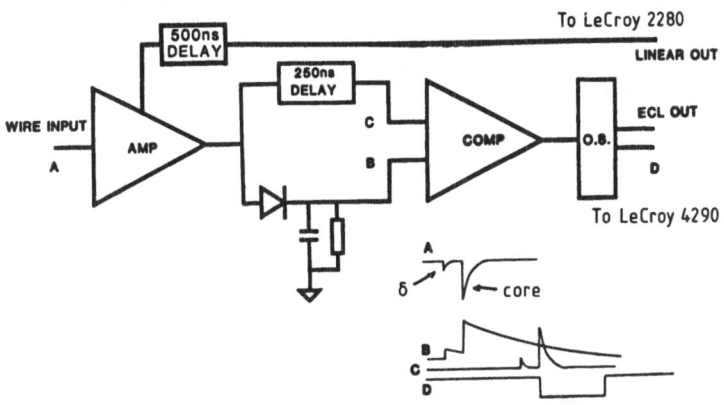

Figure 11

Each of the large drift chambers has six planes so that a good
fragment track has 12 TDC values to be used to determine two x, y locations. To calibrate the system the time to distance function must be
determined for each TDC-wire combination. The nonlinear time to distance function for a sample cell is seen in the scatter plot of TDC1
and TDC2 from the "S" plane of DC1. The width of the sum of the linearized TDC signals from such a pair gives us our single cell spatial
resolution, shown in fig. 12 to be a FWHM of 200μ. This leads to a resolution for beam particles of dP/P = 0.001.

The time-of-flight (TOF) scintillator array is designed to measure
the charge and flight time for fragments of charge 1 to 6. Each scintillator slat is 2.5 cm thick, 10 cm wide, and either 200 or 300 cm high.
Each slat is viewed at both ends by an Amperex XP2230 phototube whose
base was designed to optimize charge and time resolution over this
charge range. The electronic schematic for the TOF array is shown in
fig. 13.

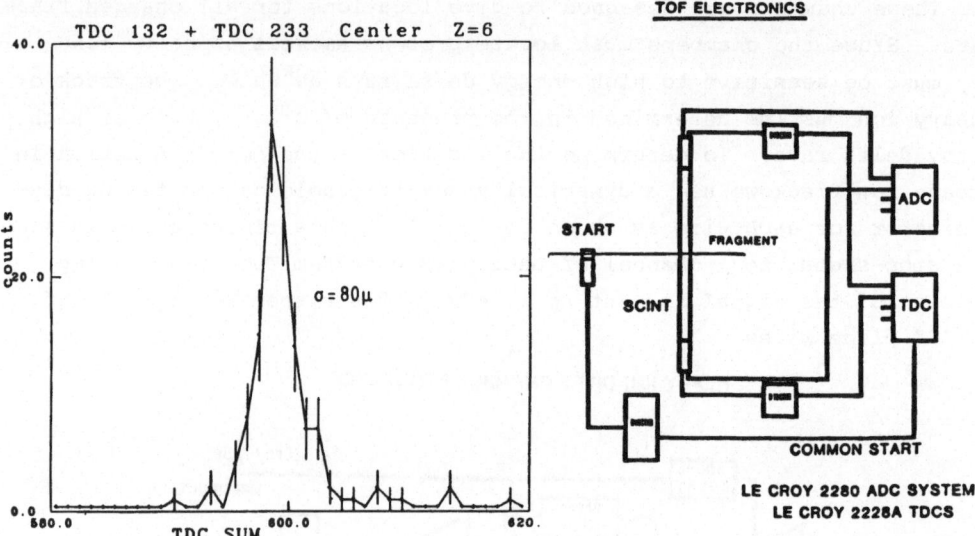

Figure 12: Drift chamber TDC sum   Figure 13

The charge calibration of each scintillator is accomplished by selecting a set of fragments whose trajectories have been determined and establishing the relation between charge and the product of the ADC signals from the two tubes on each slat using the function:

$$Prod(Z) = s*Z**2/(a*Z**2+b*Z+c) * Prod(Z=6).$$

We use the ADC product to cancel out any simple exponential attenuation effects in the scintillator. We use this form to take account of the saturation properties of the scintillator material. The resulting charge distribution for a single slat is shown in fig. 14.

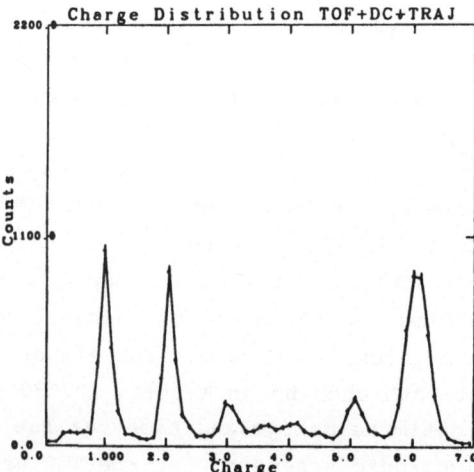

Figure 14

To determine the mass of a fragment we must calibrate the TOF TDCs. We determine the flight time from the TDC value, start time offset (t0), flight pathlength (L), and time-channel function. To determine t0 we sprayed a constant rigidity beam across all slats. Since we are using leading edge rather than constant fraction discriminators, we must correct the observed TDC channel for pulse height effects. The ADC corrected TDC spectrum for constant R beam particles is shown in fig. 15, showing a resolution of 200ps(FWHM).

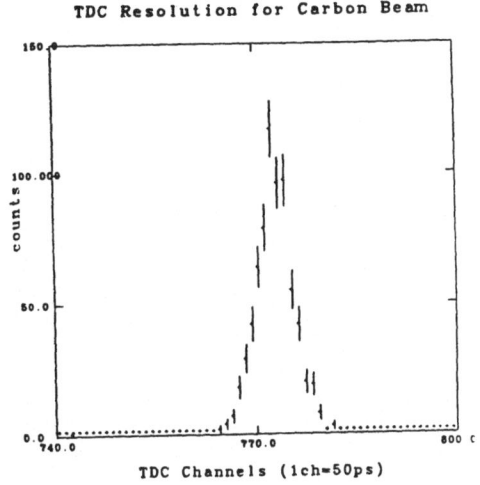

Figure 15

To complete the particle identification we obtain mass, m, from the following relation:

$$m = RZ\frac{c}{L} \left((t+t0)^2 - (L/c)^2 \right)^{\frac{1}{2}}.$$

The raw rigidity versus slat distribution for charge 1 particles is shown in fig. 16. The mass distribution for charge 1 fragments is shown in fig. 17.

Before going into the results of our experiment we present here some of the plans for Phase II detector systems. In confining our measurements to charged particles we are unable to reconstruct channels having a free neutron in them and we miss all of the energy transferred to the fragments that is dissipated as photons. We would like to augment our facility detector system with a large acceptance neutrals detector and we have borrowed a 28cm x 28cm x 30cm segmented NaI array for testing. A sample output showing hits in 3 adjacent 4cm x 4cm cells is shown in fig. 18.

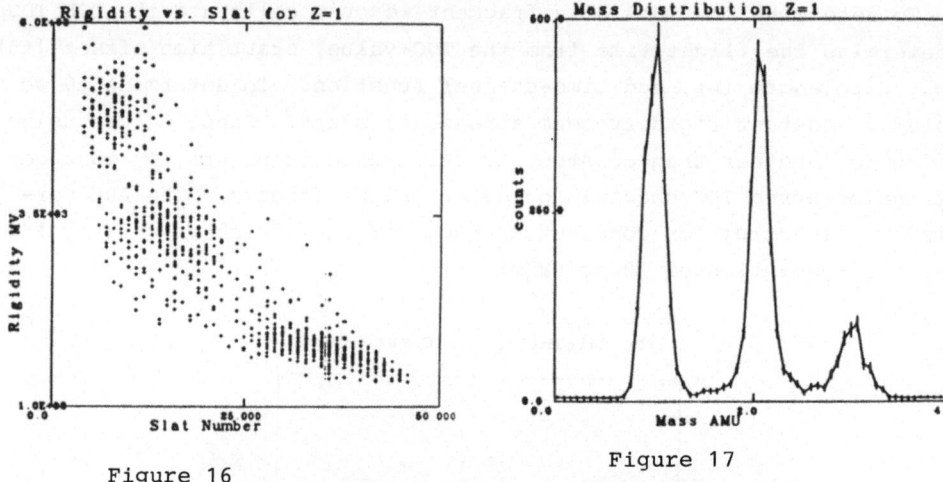

Figure 16

Figure 17

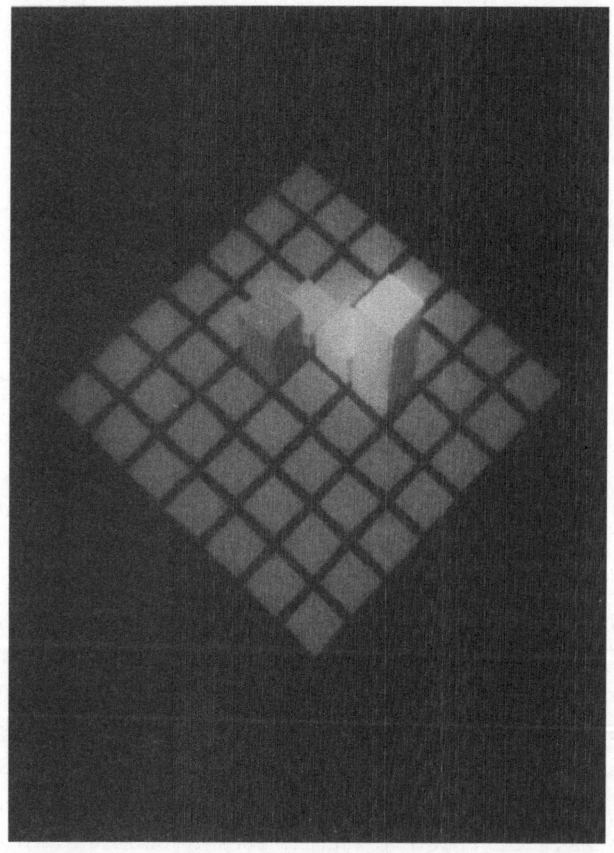

Figure 18:   Neutron response of NaI detector.

We expect to use the HISS facility to investigate very high charge and mass fragments from U interactions. In velocity regions of small delta ray production we expect the charge resolution of the TOF scintillators to be greater than 1. To provide the large acceptance needed at HISS we chose to develop a gas based identification system. Consequently we are developing a very large area multiple sampling ionization chamber (MUSIC) detector to give single charge resolution at U. At the Bevalac we work in a region of charge-velocity where the Landau distribution for thin detectors suggests that the charge resolution at constant velocity for a 50mg/cm**2 detector is worse than for 50 measurements from 1mg/cm**2 detectors. An ionization detector responds to energy lost in traversing it, the difference arising from high energy delta rays which exit the detector.

We have tested the chamber shown schematically in fig. 19. The electron cloud around a track drifts downward in the 300V/cm field, through a grounded 95% transmitting Frisch grid and onto 64 anode wires spaced 1 cm apart. The signal from each anode is strobed onto a CCD array being shifted at 10MHz. On receipt of a trigger signal the CCD collection is halted and the array is shifted out to an ADC array at 20kHz, where the computer takes it and places it in the data stream. The electronics, based on the TPC system, are shown in fig. 20.

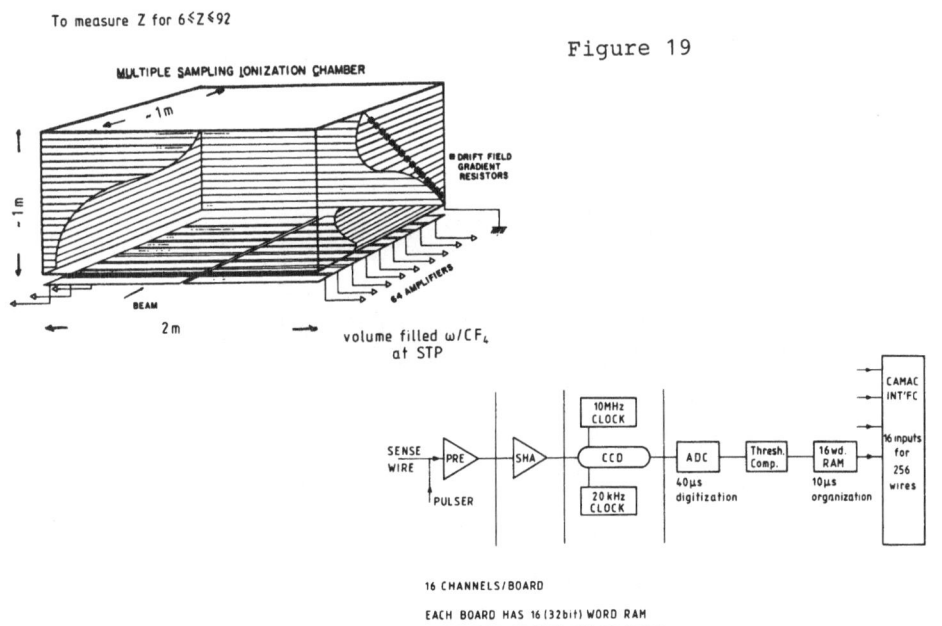

Figure 19

Figure 20: Music electronics

To complete the HISS Phase II detector arrays we must also increase
our acceptance for light fragments. We were limited in the 12C experiment to +/-140Mev/c perpendicular momentum, causing an acceptance for
protons of only 50%. Therefore we are adding 40 more TOF slats and
building a single gas volume drift chamber 2m x 5m x 1.2m. The Phase II
detector array is shown in fig. 21. The cave area showing the Phase I
drift chambers, TOF well, and MUSIC detector is shown in fig. 22.

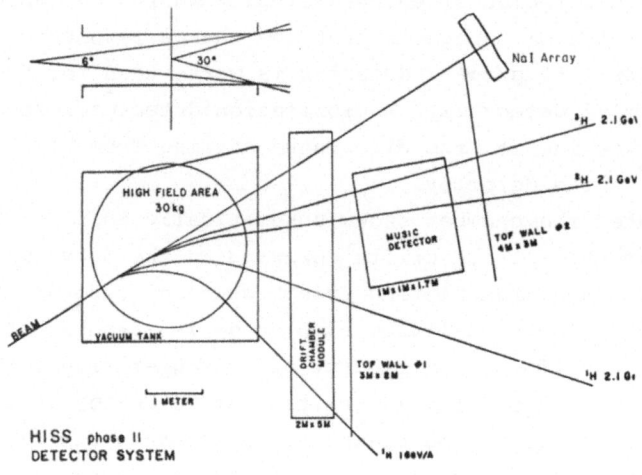

Figure 21

CBB 828-7538

Figure 22:   HISS cave with TOF walls and MUSIC detector.

We now present some preliminary results from the 1GeV/nuc 12C experiment. The reaction we have studied can be written schematically as:

12C + T → (channel) + T´

where channel means any final state for the 12 nucleons originally in the C nucleus (e.g. 11B+p or 3 alphas) and T´ is the final state of all target nucleons. The 11B fragments can be formed only in the 11B+p channel but the 4He fragments can be formed in a large number of different channels. In our experiment we attempt to measure complete (i.e. 12 nucleon) channels. After correcting for the acceptance of our system we are thus able to measure the relative probability of forming each separate channel in the interaction. We then form the invarient mass for each interaction channel and subtract from the mass of the 12C parent to determine the amount of energy transferred to the fragment channel in each interaction. By analysing many fragments from a single interaction we can also take subgroups of fragments and calculate invarient masses for these subgroups to look for preferred intermediate states (e.g. 2 alphas from an 8Be).

The experimental layout is shown in fig. 23 including beam definition devices, large gap superconducting dipole, fragment trajectory drift chambers, and time-of-flight (TOF) scintillator array. The event trigger required a single 12C hit on the target and no 12C signature at DS, indicating that the projectile had been scattered or destroyed.

MULTI-PARTICLE FINAL STATES FROM ^{12}C at I GeV/n

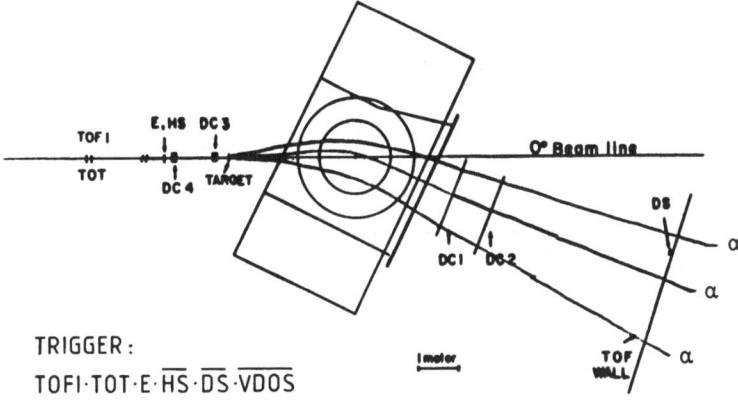

TRIGGER:
TOFI·TOT·E·H̄S·D̄S·V̄DO̅S

Figure 23

The first results from our experiment are shown in fig. 24 where we have used all channels having $\Sigma Z f = 6$ and $\Sigma A f = 12$ giving a global excitation energy spectrum. The prominant peak near 30 MeV excitation leads to fragments having 50-200 MeV/c momentum in the projectile rest frame, fragments primarily responsible for observed single particle inclusive momentum distributions. Data out to 400 MeV excitation are

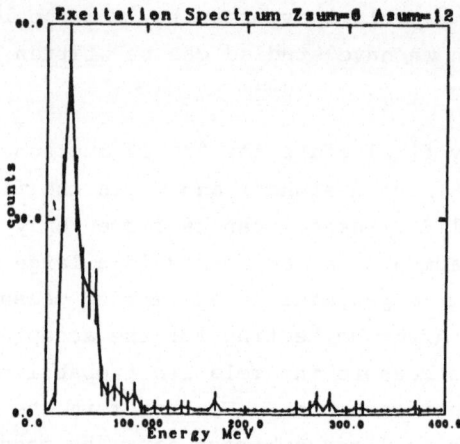

Figure 24

from interactions in which large energies are transferred to the 12C
system, the energy then being shared among a few multi-nucleon fragments.   Energies of 300 MeV shared among 3 alpha particles have been
seen suggesting a collective mode transfer of energy such as would be
expected from a phonon like exchange mechanism.

In conclusion I would like to say that we have observed the excitation spectrum for fully reconstructed events and seen energies up to
400 MeV transferred to bound fragment systems.   We have data for all
charged particle channels and for invarient mass calculations for ar-
bitraty particle groupings.   Finally, the Phase II detector system
should be available at HISS by spring 1983.

This work was supported by the Director, Office of Energy Research, Division of Nuclear Physics of the Office of High Energy and Nuclear Physics
of the U.S. Department of Energy under Contract DE-AC03-76SF00098 and
National Aeronautics and Space Administration grant number NGR 05-003-
513.

The Magnetic Spectrometer at Vicksi

H. G. Bohlen

Hahn-Meitner-Institut für Kernforschung, Berlin

Dr. Horst Lettau died on 30. September 1982 at the age of 41 years. He was a devoted experimental physicist, who worked with imagination and in a consequent way. As a member of the spectrometer group he did the main work for the construction of the very versatile detector system, which is now in use. Due to his untimely death he is not able to share the fruits of his work. We all miss him as a colleague, friend and physicist.

The physics department of the Hahn-Meitner-Institut

Abstract

The performance of the Q3D-spectrometer at Vicksi is described with
the emphasis on the design of the detector system. It consists of a
focal plane detector, a parallel plate avalanche counter at the
entrance of the spectrometer and a coincidence detector at the scattering chamber. The method of particle identification from the
measured parameters is discussed in detail. A short review of the
experimental program is given at the end.

I. Introduction

Q3D-magnetic spectrometers have been widely used for light and heavy ion spectroscopy. The high momentum resolution and large solid angle are very attractive for many
experiments. For the application of such an instrument to heavy ion reactions it is
of equal importance to install a good particle detection system as to have a good
ion-optical design of the magnets. Detectors of rather large length have to be used,
which should allow the identification of heavy ion reaction products at any position
in the focal plane. The spectrometer installed at VICKSI is a Q3D-instrument of type
QMG/2 (1). The complete system is shown in fig. 1. The different tasks of the two
subsystem (magnets and detectors) are characterized in the following way:
1) The magnetic system converts the energy spectrum of a particle group into a position spectrum with high momentum resolution
2) The detector system has to
   a) measure the position of incidence in the focal plane
   b) identify mass A, charge Z and charge state q of particles
   c) measure the scattering angle Θ for large aperture openings
Since the measured parameters for a) and b) may be dependent on the angular coordinates Θ and ϕ of the solid angle, it is additionally necessary to measure these dependences for good resolution.

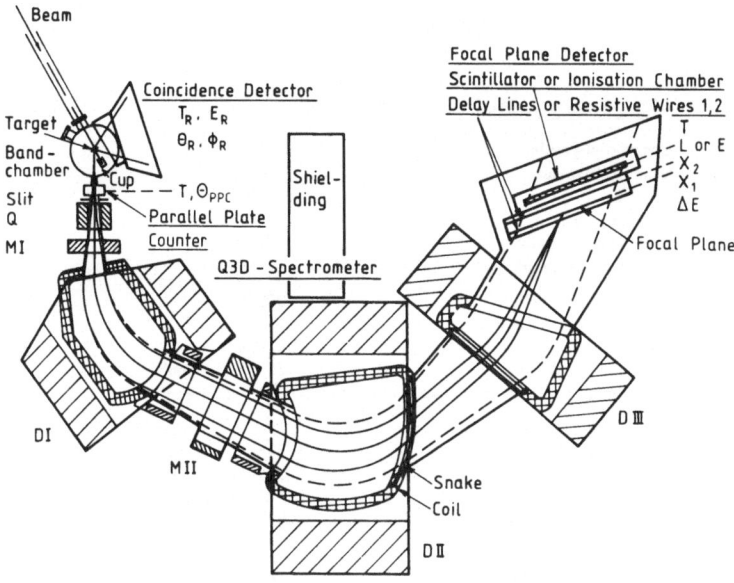

Fig. 1   The Q3D-magnetic spectrometer and the detector system

The different detectors at the spectrometer can be viewed as the sense organs of the whole instrument. There are 1) the focal plane detector, 2) the parallel plate counter at the entrance of the spectrometer and 3) the coincidence detector for special purposes. They are described in detail in section III, after a short out-line of the magnetic system in section II. The way, how a unique particle identification is derived from the measured parameters, is shown in section IV. Finally a short review of the experimental program is given in section V.

II. Magnets and Ion Optics

The characteristics of the magnetic system are summarized in table 1. The following view points have been the basis of the ion optical design (1):
1) Solid angle 10 msr
2) 20% energy range on a straight focal plane of about 1 m length
3) Vertical cross over, to keep the pole gap small (6 cm)
4) Corrections of the kinematical momentum dependence $K=\frac{1}{p}\frac{dp}{d\theta}$ up to K=-0.3 ‰ /mrad for $\theta=\pm$ 50 mrad (this is achieved with multipoles I and II)

Table 1

Q3D Magnetic Spectrometer, Type QMG/2

Magnetic elements	Q M D M D D
Mean orbit radius	100 cm
Deflection angle	155°
Maximum $B\rho$	1.7 T·m
Mass—energy product $A \cdot E/q^2$	140 MeV
Maximum energy per nucleon E/A for $q=A/2$	35 MeV/N
Energy range	20%
Energy resolution (design)	5000
Dispersion along focal plane	9.5 m (constant)
Length of focal plane	95 cm (straight)
Angle of incidence in focal plane	45° (constant)
Angular acceptance	
horizontal	±50 mr ≙ 5.7°
vertical	±52 mr ≙ 6.0°
solid angle	10.4 msr
Kinematical correction	$k = 0.0 - -0.3$ ‰/mr
Path length of rays	
for full solid angle and $\rho = 100$ cm	$746 \begin{smallmatrix} + 43 \\ - 30 \end{smallmatrix}$ cm
for full momentum range, $\rho = 100±5$ cm	$746 ± 30$ cm
Magnification, horizontal	−1.0
vertical	4.5 − 7.1

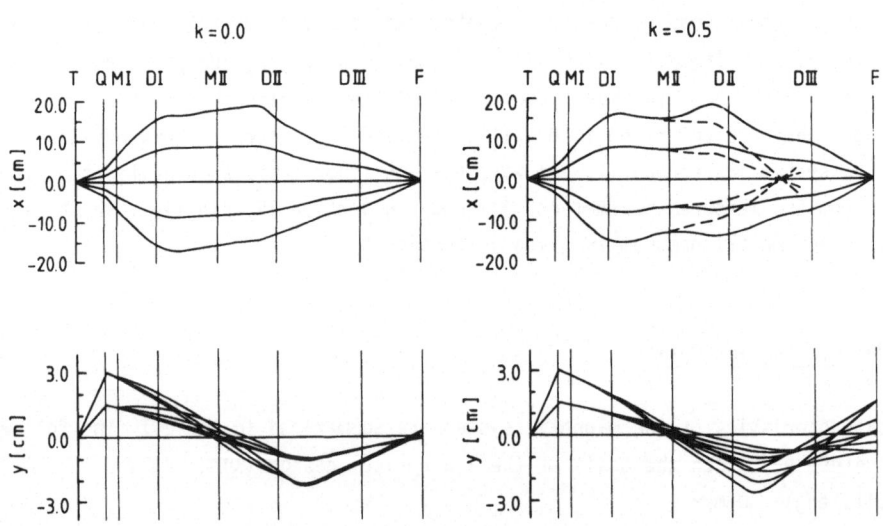

Fig. 2 Ray trace calculations for the kinematical factors K = 0.0 and K = -0.5 ‰/mrad. The dashed lines at the upper right show the shift of the focus if the quadrupole component of MII is switched off.

The imaging is shown in fig. 2 with ray tracing calculations. The quadrupole focusses vertically such that a vertical waiste is obtained at MII. The pole gap can be kept small for this reason. The corresponding defocussing in horizontal direction is compensated by the edge focussing of the DI field boundary. Further focussing in both directions is obtained by the edge angles of DII and DIII for K=-0.1. For other K-values it is necessary to excite MII. The quadrupole component has to defocus the outer rays for $|K| > 0.1$ (fig. 2). Multipole II is constructed in such a way that the different components have their multipole field distribution only in the x-direction, therefore, the y-focussing is only little affected.

Up to now the ion optic of the instrument is calibrated in first order for different K-values and in dependence on the main dipole field. This was achieved by experimental ray tracing with two position measurements close to the focal plane for each particle trajectory. The design properties of the focal plane, namely the straight shape and constant dispersion, have been confirmed. We measured a resolution of dp/p=$4 \cdot 10^{-4}$ with a ^{12}C-beam of 80 MeV. It was mainly determined by the energy straggling in the target and angular straggling in the detector entrance foil.

The optimization of the second order imaging is just in progress, because detectors are now available to measure these dependences accurately. Fig. 3 shows the scatter-

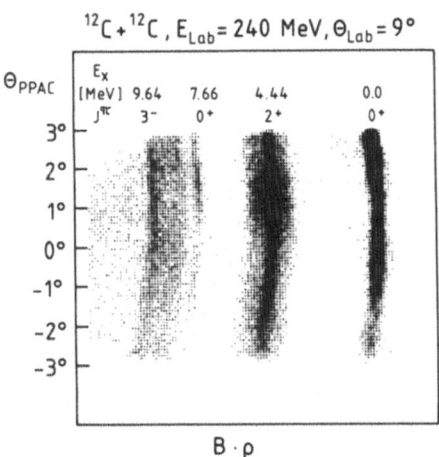

^{12}C + ^{12}C , E_{Lab}= 240 MeV, Θ_{Lab} = 9°

Fig. 3 Scatter plot of events measured in the Bρ-Θ-plane. The curvature of the image lines is due to the second order imaging error. With proper setting of MI and MII, straight lines can be obtained.

ing of ^{12}C on ^{12}C at 240 MeV and 9° measured with full solid angle. The Θ-dependence of Bρ of the different lines, observed as a bended shape, indicates an imaging error of second order for the coefficient $d^2x/d\theta^2$. The same effect may occur for $d^2x/d\phi^2$. These dependences can be corrected now off-line. But also an on-line optimization of the 2nd order magnetic components (MI,MII) can be achieved. When the first and second order fields are set correctly, a single excited state appears as an upright straight line in the focal plane for the Θ- and φ-dependences.

III. Detectors

1. The first HMI focal plane detector

After the installation of the spectrometer a relatively simple heavy ion detector was used in the focal plane with a sensitive lenth $L_0=72$ cm, which was easy to handle and very reliable. It consists of two parts (fig. 4a):

The first part contains two position-sensitive resistive wires at a distance of 4.5 cm, from which we obtained the energy loss signal ΔE and the position x from signals at the right and left ends of each wire; $S_r=x \cdot \Delta E$ and $S_\ell=(L_0-x) \cdot \Delta E$ gives $\Delta E \approx S_r+S_\ell$ and $x=S_r/(S_r+S_\ell)$. The angle of incidence is calculated from the positions x_1, x_2 of the first and second wire. This set up was well suited for the experimental ray tracing (fig. 4b) and for first experiments, and reasonable resolutions have been obtained: $\delta x=1.5$ mm, $\delta\theta=1.3^0$, $\delta\Delta E/\Delta E=4.4\%$. The second part contains a scintillator bar, which stops the incident particles. The light output L was measured with photomultipliers on both ends of the bar. Typical resolutions of $\delta L/L=3.7-4.5\%$ have been achieved, e.g. for ^{20}Ne, over the whole length after correction for the light attenuation along the scintillator. According to Becchetti et al. (2) the light output L is related to the energy of the particles by $L \approx E^{1.63}/(Z^{0.08} \cdot A)$ and it is used for particle identification. The time of flight T of the particles to the detector is

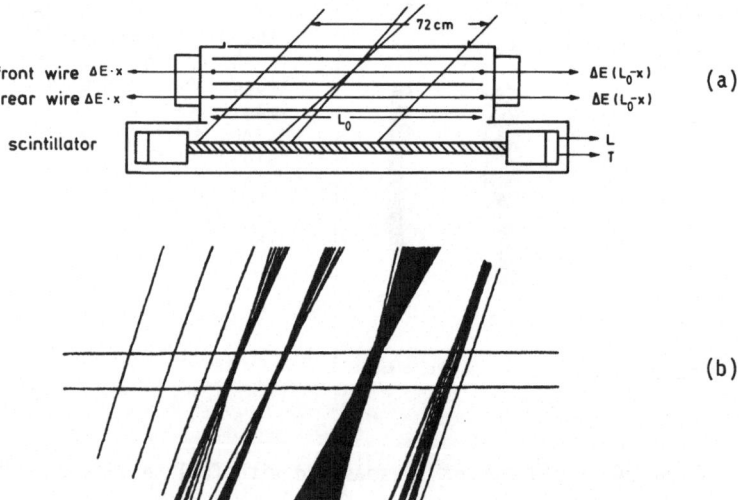

Fig. 4a) Focal plane detector with resistive wires for position and energy loss measurement and with a scintillator bar for light output and time-of-flight measurement.
b) Experimental ray tracing of the elastic and inelastic scattering of ^{12}C on ^{88}Sr.

measured with the start signal from the scintillator and the stop signal from the
cyclotron frequency. Since the beam bunches have a width of 1 nsec the resolution
of T is about 1.2 nsec. Typical flight times are 150-200 nsec. The T-signal was used
for particle identification with a small horizontal aperture opening, because for
large openings the flight path differences are rather large and can not be corrected sufficiently with the Θ-measurement due to the low Θ-resolution. If the particle
identification was already unique with ΔE and L, the correlation between T and Θ
(due to the path length differences) was used for a Θ-measurement with the T-signal;
a resolution of about 0.4^0 was standard.

2. The modular focal plane detector system

The new focal plane detector is being designed for the measurement of the position
in the focal plane for a length of 1 m and the identification of A, Z and q in the
mass region up to mass 50 (^{40}Ar is the heaviest beam particle at VICKSI at the mo-
ment, except ^{86}Kr). The measured quantities for particle identification are ΔE, E,
L, T, but it is also necessary for corrections and background suppression to measure
the angle of incidence in the focal plane θ_F, and the Y-coordinate (perpendicular
to the reaction plane). The calculation of A, Z and q from these quantities is discussed in the next section (particle identification).

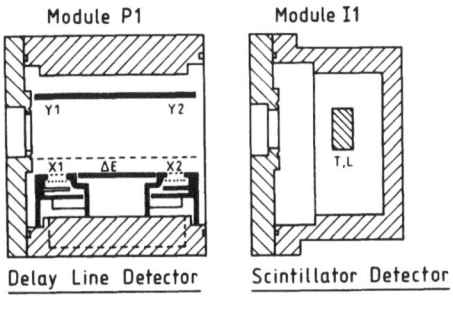

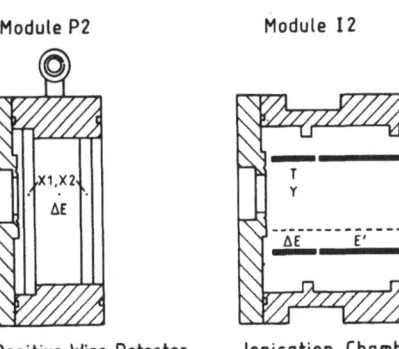

Fig. 5 Cross section view of the modular focal plane detector system (sensitive length 92.5 cm). The measured parameters are indicated.

The detector is constructed in a modular way (fig. 5) and consists of a position part and a particle identification part. There are two different modules for each part: The position is either measured with the delay line detector (P1) or with the resistive wire detector (P2), and the second part is either a scintillator detector (I1) or an ionisation chamber (I2).

The delay line detector and the scintillator are in use now since half a year, while the other two modules are not yet ready. The assembled combination P1-I1 is used for masses up to A=25. The combination of modules P1-I2 is designed for masses up to A=50 and P2-I1 is suited for all purpose without high resolution.

Each module can be operated with its own gas pressure, because the window flanges fit also between the detectors. The large window of 3.5x100 cm^2 is closed with foils of 2.5 or 10 μ thickness, depending on the pressure (typically 50 to 150 Torr of isobutane).

The Delay Line Detector

The designed momentum resolution of 1·10^{-4} in the focal plane corresponds to a line width of 0.95 mm. This requires a position resolution of 0.5 mm corresponding to 5·10^{-4} with respect to 1 m detector length (for 10% momentum range), if the line width should not be increased appreciably by the position resolution. This high precision is achieved only with timing techniques. Fig. 6 shows the cross section view of the delay line detector. Two position measurements are incorporated in an ionisa-

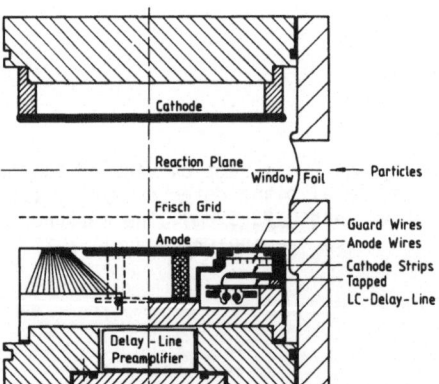

Fig. 6 Cross section view of the delay line detector. The tapped delay line is located below the cathode strip structure. The anode and guard wires are positioned in slits at both ends of the chamber and fixed on contact strips, as shown at the left part of the figure. The delay line preamplifiers are placed close to the ends of the delay line.

tion chamber (for ΔE-measurement). The primary electrons are drifting through guard wires to the 6 anode wires, which are operated with a high multiplication factor. The guard wires have to shield the ΔE anode plate from this high space charge. The signal is influenced on the cathode strip structure and fed into a tapped delay line. The strips are etched on a printed circuit board and are parallel to the angle of incidence ($\approx 45°$), they have a width of 1 mm and a distance of 0.5 mm. The LC-delay line is constructed at the HMI Electronics Department with selected capacitors of 56 pF and inductivities of 0.57 µH (22 windings on Ferrit rings) with tolerances less than 0.5 % ($Z = 85\Omega$). Each LC-element has a delay time of 4.8 nsec, which corresponds to a delay of 3.2 nsec per mm for a strip structure distance of 1.5 mm. Two delay line pieces of 54 cm length have been connected. The 722 delay elements have a total delay time of $t_D = 3.52$ µsec and a rise time of $t_R = 80$ nsec, a quality factor $t_D/t_R = 44$ is achieved. The response of a 54 cm piece to a fast signal is shown in fig. 7. Special preamplifiers have been designed (HMI, IV49) with a rise time of 13 nsec, which allow the impedance matching at the delay line. The time measurement is started with the signal from the anode wires and stopped with one end of the delay line output.

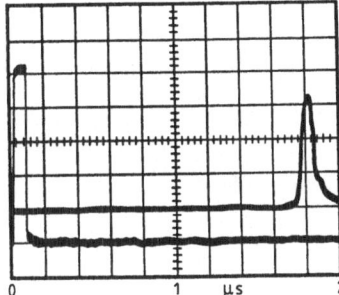

Fig. 7 Delay line response (right) to a fast input signal (left)

The discrete strip structure imposes a regular modulation on the measured position due to the finite width and number of strips. The measured position x deviates from the true position x_t, because the average value is obtained usually only from 4-5 strips: $x = \frac{1}{Q} \Sigma q_n x_n$ (Q total charge, q_n charge influenced on the strip at position x_n), while the true position is given by $x_t = \frac{1}{Q} \int q(x)\, dx$. The difference $\Delta x = x - x_t$ oscillates along the position with a period given by the strip structure distance s with an amplitude A, which depends on s and the wire-to-strip distance h. We have calculated for our geometry (s=1.5 mm, h=4.0 mm) A=±0.02 mm. Such a small variation can be neglected as compared to the position resolution. But this effect may be important for discrete structure read-out with high position resolution.

The delay line detector was tested with the elastic scattering of ^{22}Ne, 120 MeV, on ^{197}Au at various gas pressures. A vertical slit of 1 mm width was placed in front of the detector. The following resolutions have been obtained in dependence on the gas pressure of isobutane and with a 2.5 µ Hostafan foil:

P	δx(DL1)	δx(DL2)	δθ$_F$	ΔE	δY
30 Torr	0.55 mm	1.25 mm	0.7^0		
50 "	0.70 "	1.57 "	0.9^0	3.9 %	1 mm
70 "	1.32 "	1.85 "	1.3^0		

The position resolution of DL2 is much worse than for DL1 due to the angular straggling in the detector gas. The same effect is responsible for the decrease of resolution at DL1 with increasing pressure. The Y-coordinate is measured by the drift
time of the electrons to the anode wires, the start signal is taken from the scintillator.

The scintillator detector, used together with the delay line detector, is operated
in the same way as for the preliminary focal plane detector and has the same resolutions for L and T.

The ionisation chamber is not yet assembled, but we expect in connection with the
delay line module a resolution of δE/E = 2%. The prompt timing signal is than taken
from one segment of the cathode plate, or from a thin scintillator foil in front of
the detector similar to the system at the GSI-spectrometer (3).

3. The Parallel Plate Avalanche Counter

The measurement of the entrance angle to the Q3D-spectrometer is very important for
the large size of the aperture opening. A position sensitive parallel plate avalanche counter (PPAC) was therefore installed between the scattering chamber and quadrupole magnet, which allows to measure the angle with high precision. The construction and performance of the counter is described in detail by H. Lettau et al.(4).
It is possible with the PPAC to measure angular distributions in pieces of 5.7^0
with a resolution of better than 0.1^0. In addition the time-of-flight measurement
can be improved even in two ways: 1) the stop signal is taken from the PPAC-anode
(instead of the cyclotron-frequency), which has a time resolution of better than
500 psec, 2) the time-of-flight variation due to the flight path differences of the
trajectories through the spectrometer can be corrected, which is very important for
a good mass resolution. Finally all other parameters, which depend on θ, can be
corrected (e.g. second order aberration). The energy and angular straggling in the
detector is rather small and can be accepted in many experiments, because the detector thickness is only about 300 μg/cm^2 (C-equivalent).

A schematic view of the detector design is shown in fig. 8. It consists of two pressure foils with a window size of 60 x 60 mm^2 and an anode and cathode foil at a distance of 2 mm. The sensitive area of 36 x 40 mm^2 is defined by an evaporated gold layer on the anode and cathode foils. The cathode layer is evaporated as a strip structure, which is connected to a tapped delay line with the same delay elements as used for the focal plane detector.

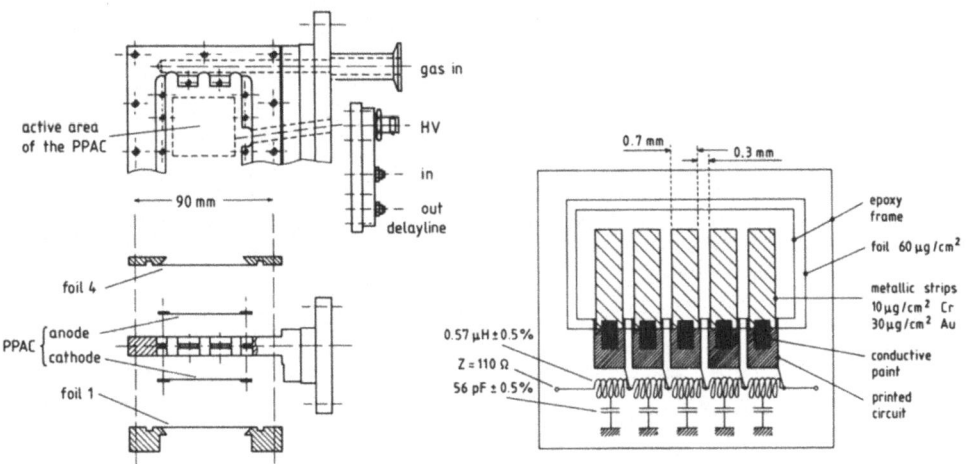

Fig. 8 Side and top view of the position sensitive parallel plate avalanche counter (left) and schematic view of the delay line read out of the cathode strip structure (right).

The foils of 60 µg/cm^2 thickness are produced on a special foil stretching machine (5), which has been build especially for PPAC detectors.

The evaporation of 10 µg/cm^2 of Au is also a very delicate operation. After assembly the detector was tested with an α-source. The start signal for the position measurement is taken from the anode, the stop from the output of the delay line. The efficiency is shown in fig. 9 as a function of voltage for 17 mbar of isobutane. These curves can be used to fix the operation conditions for 100 % efficiency for heavy ions according to their energy loss in the counter. A position resolution of 0.16 mm is achieved; this corresponds to an angular resolution of 0.033^0 at the location of the PPAC.

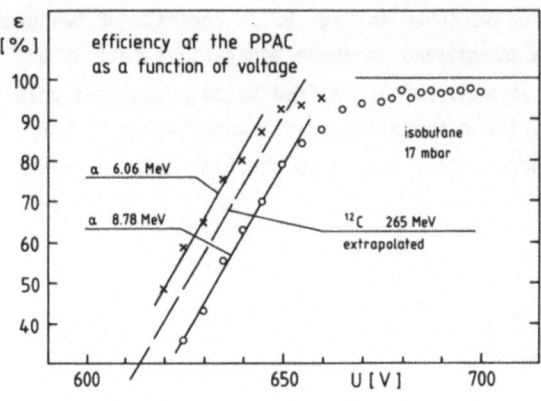

Fig. 9 Efficiency of the parallel plate avalanche counter independence on the applied voltage at 17 mbar of isobutane.

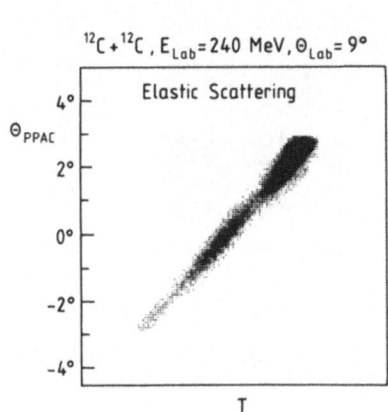

Fig. 10 Correlation between the entrance angle to the Q3D-spectrometer Θ_{PPAC} and the time-of-flight T through the spectrometer for the elastic scattering of ^{12}C on ^{12}C at 240 MeV and 9^0 with an aperture opening of 6^0.

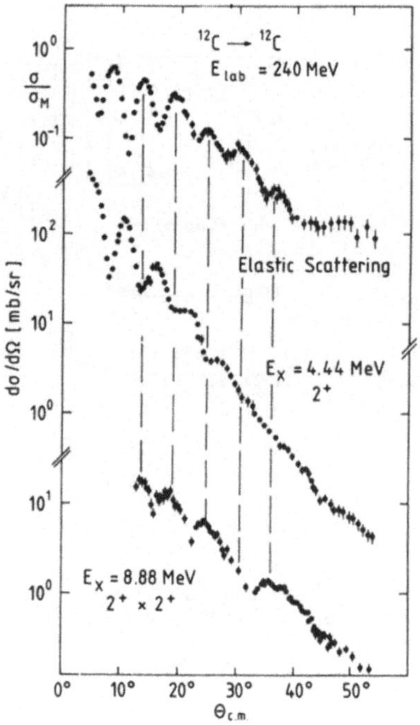

Fig. 11 Angular distributions of the elastic and inelstic scsattering of ^{12}C on ^{12}C at 240 MeV, measured with the PPAC in 4 angular settings with a width of 6^0.

This high resolution allows to correct fully the angular dependence of the time-of-flight due to the path length differences. The remaining width is the intrinsic T-resolution (1/150). Fig. 10 shows the relation between T and Θ_{PPAC} for the elastic scattering of $^{12}C + ^{12}C$ at 240 MeV and 9^0. Another application has been shown already in fig. 3 for second order corrections of ρ. The angular distributions shown in fig. 11 are measured only with 4 angular settings of the spectrometer, each subtending over 11^0 in the c.m. system (some additional points have been measured at forward angles).

4. Coincidence detectors

Coincidence detectors will be used to select a special reaction channel due to the reaction mechanism or kinematics. They are located at the scattering chamber and designed either 1) to measure kinematical coincidences between the recoil nucleus and the ejectile detected in the focal plane (this detector is called Co1 in the following) or 2) to measure coincidences between light particles and the ejectile (called Co2).

Detector Co1 consists of a large area PPAC for time measurement with a two-dimensional position read-out with wire grids. The measured quantities are T_R, Θ_R, ϕ_R. The sensitive area is $250 \cdot 130$ mm^2, which corresponds to an angular range of $\Theta = \pm 14^0$ and $\phi = \pm 7.4^0$ at a distance of 50 cm.

Detector Co2 has 8 scintillator pieces, each having an angular width of $\Delta\Theta = 12.2^0$ and $\Delta\phi = 44^0$, this covers a solid angle of 1.15 sr. The time-of-flight T_R and the light output L_R are measured at the scintillator. A resistive wire is mounted in a gas cell in front of each scintillator to measure the position dependences of T_R and L_R for corrections. A schematic view of the detector is given in fig. 12. Both detectors are now under construction and will be soon assembled.

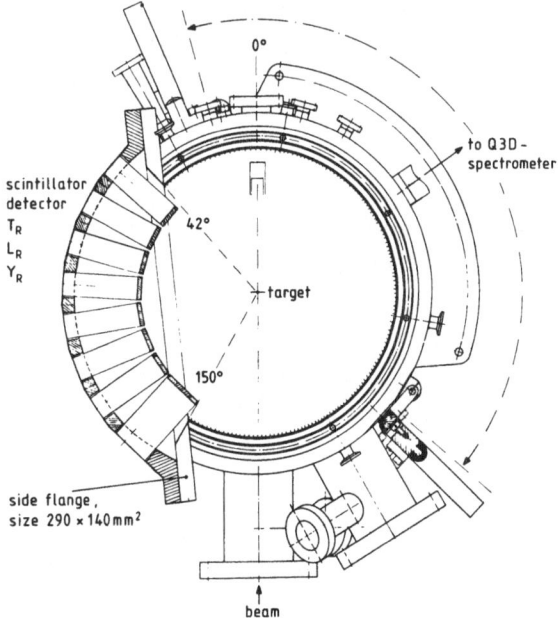

Fig. 12 Scintillator coincidence detector for light charged particles.

IV. Particle identification

In heavy-ion reactions many nuclide are produced at medium incident energy. The focal plane detector has the very important task to allow the identification of particles according to the mass and charge number A and Z, and also the charge state q.
The measured parameters x (or bending radius ρ), T and E are related to A, q and
velocity v by the 3 exact equations, given in first line of table 2. There exist
semi-empirical formulae for the dependence of L and ΔE on the velocity v, which are
given in the lines below.

The measured parameters are dependent on the position x and the angle of incidence
Θ_F. Since the particle identification requires position independent parameters, it
is necessary to divide out the positional variation of T, E, L, ΔE. This can be done
simultaneously for all particles with the correct functional dependence. The resolution can be further improved, if the variation on Θ_F ($\pm 3^0$ with respect to the
central ray) is also corrected. This correction is stringent at least for T, because
the flight-path-differences within the full solid angle amount to $\Delta S/S = 73/746 =$
9.8%. This is equivalent to setting $B\rho$ = constant in the first equation, and a
variation of v occurs only proportional to the identification parameter q/A. After
replacing v by q/A in the general equations, we obtain the particle identification
parameters A/q ~ T, q^2/A ~ E etc. The A-identification can either be done with T or
with E or L. T can be measured with the best resolution of these three identification parameters, however, A/q is often ambiguous. The E- and L-resolution is not as
good as for T, but here the different charge states q of a sequence of isotopes are
better separated due to the large exponent of q, e.g. q^2 for E (see table 2, identification: A). As we will see, the better resolution of T is more important,
because the ambiguity of A/q can be removed with a corrected ΔE_c, which is recal-
culated from the relation between ΔE and E or L (see below).

The charge state q is identified in a plot of E versus T because E~q/T is independent of A and Z. The relation between L and T is not as well suited due to the worse
L-resolution and the lower exponent on q: $L \sim q^{0.63}/T^{2.63}$.

The Bethe-Bloch formula gives a Z^2/v^2-dependence of the energy loss ΔE for fully
stripped ions. For heavy ions of moderate energy the dependence on Z is replaced by
an effective charge Z_{eff}. The calculation of Z_{eff} is parametrized in the literature
by semi-empirical formulas (6). We found from our data, that at medium energies the
energy loss shows a simple Z^2/v-dependence. This is shown in fig. 13 for heavy ions
up to Na in the energy range from 5-20 MeV/N, where ΔE is plotted versus T. The linear dependence $\Delta E \sim Z^2(T-k)$ reflects the relation $\Delta E \sim Z^2(1/v-k)$. This plot allows a

Table 2

General relations:

$$Mv = q \cdot B\rho \qquad\qquad T = \frac{s}{v} \qquad\qquad E = \frac{1}{2} Mv^2$$

$$L \sim \frac{1}{Z^{0.08}_A} E^{1.63} \approx \frac{A^{0.63}}{Z^{0.08}} v^{3.26} \qquad (\text{Becchetti et al.})$$

$$\Delta E \sim \frac{Z^2_{eff}}{v^2} \ln\left(\frac{2m}{I} v\right) \sim \begin{cases} Z^2 \dfrac{1}{v^2} & (1)\ \text{high energy} \\[2ex] Z^2 \left(\dfrac{1}{v} - k\right) & (2)\ \text{medium "} \\[2ex] Z^2 \left(\dfrac{1}{v} - k\right) \ln\left(\dfrac{2m}{I} v\right) & (3)\ \text{low "} \end{cases}$$

Bρ constant: $v \sim \frac{q}{A}$	Identification:
$T \sim \dfrac{A}{q}$	
$E \sim \dfrac{q^2}{A} \qquad \sim \dfrac{q}{T}$	$\left.\begin{array}{c} \\ \\ \end{array}\right\}$ A q
$L \sim \dfrac{q^{3.26}}{Z^{0.08} A^{2.63}} \sim \dfrac{q^{0.63}}{Z^{0.08} T^{2.63}}$	
$\Delta E \sim \begin{cases} Z^2 T^2 \\ Z^2 (T-k) \\ Z^2 (T-k) \ln\left(\frac{2m}{I} \frac{1}{T}\right) \end{cases}$	$\left.\begin{array}{c} \\ \\ \\ \end{array}\right\}$ Z
$\Delta E \sim \begin{cases} \dfrac{A Z^2}{E} \\[1ex] Z^2 \left(\frac{q}{E} - k\right) \\[1ex] Z^2 \left(\frac{q}{E} - k\right) \ln\left(\frac{2m}{I} \frac{E}{q}\right) \end{cases}$	$\left.\begin{array}{c} \\ \\ \\ \end{array}\right\}$ q, Z

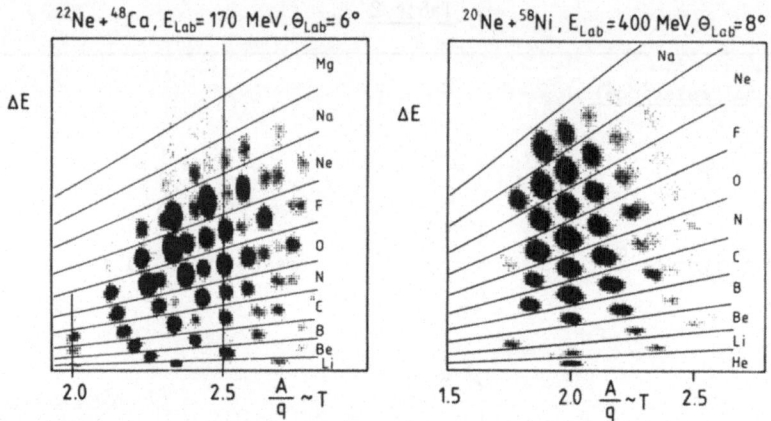

Fig. 13 Correlation between ΔE and T in the energy range of 5-10 MeV/N (left side) and 10-20 MeV/N (right side) for reaction products in the scattering of ^{22}Ne on ^{48}Ca and ^{20}Ne on ^{58}Ni, respectively.

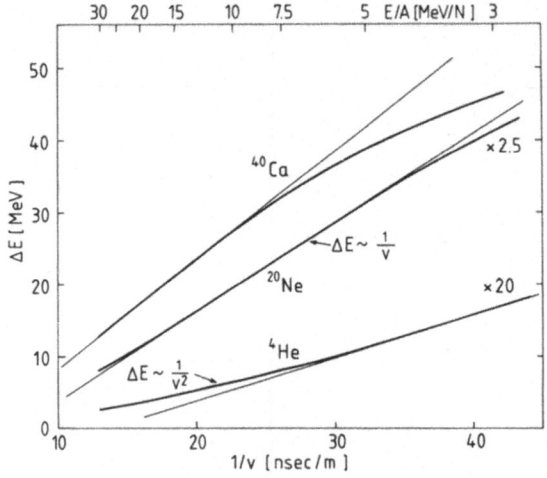

Fig. 14 Energy loss in carbon, 5 mg/cm^2, in dependence on the velocity v for ^{4}He-, ^{20}Ne- and ^{40}Ca-ions calculated with a semi-empirical formula (6). The energy loss of ^{20}Ne is a linear function of 1/v in the range between 4 and 25 MeV/N.

very safe Z-identification for any particle group observed, e.g. all observed Ne-isotopes from A = 20 to 25, each with charge states 8$^+$, 9$^+$, 10$^+$ and corresponding different energies, are arranged on a straight line. If the ion velocity is high enough, that all electrons are stripped, than this 1/v-dependence changes to the well known 1/v^2-dependence. The semi-empirical parametrization of Z_{eff} describes this transition, as is shown in fig. 14 for ^{20}Ne.

The $1/v$-dependence of ΔE can be derived, if Z^2_{eff} is parametrized as $Z^2_{eff} = Z^2[1-\exp(-\frac{v}{u_0})]$ and the exponential function is expanded as $1 - \frac{v}{u_0} + \frac{v^2}{2u_0^2} - \ldots$ We obtain $\Delta E \sim Z^2_{eff}/v^2 = Z^2[\frac{1}{vu_0} - \frac{1}{2u_0^2}]$. The velocity parameter u_0 is dependent on Z. The bend of the curves in fig. 14 for ^{20}Ne and ^{40}Ca at the low energy side is due to the combined action of Z_{eff} and the logarithmic term in the Bethe-Bloch formula.

The light output L of the scintillator is a function of all 3 identification parameters: $L \sim \frac{q^{3.26}}{Z^{0.08} A^{2.63}}$. In a ΔE-L-plot the different isotopes with the same q and Z fall on a hyperbola with $\Delta E \sim Z^2(\frac{q^{0.24}}{L^{0.38}} - k)$, fig. 15. But due to the non-

^{22}Ne + ^{48}Ca, E_{Lab}=170 MeV, Θ_{Lab}= 6°

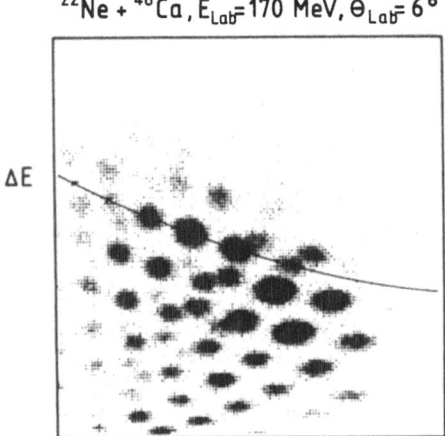

ΔE

Scint. Signal

Fig. 15 Particle groups in a ΔE-L-plot for reaction products of the system ^{22}Ne + ^{48}Ca, E_{Lab} = 170 MeV. The solid line represents the mass hyperbola for Ne-isotopes with q = 9+, the mass number of the observed groups ranges from A = 20 (right) to A = 24 (left) on this line.

linear dependence on Z and q it is difficult in this complicated matrix to predict the exact position of an isotope, which is produced with low cross section or which may partly overlap with other isotopes. A more systematic ordering is achieved, if the $1/L^{0.38}$-dependence of ΔE on L is divided out. The corrected ΔE_c-parameter is constant for an isotope chain with the same Z and q. The same behaviour is observed also in a plot of ΔE_c versus T. But ΔE_c still contains the $Z^2q^{0.24}$-dependence, which now removes the A/q-ambiguity. If we take e.g. ^{25}Ne^{10+} and ^{20}Ne^{8+}, we have on the T-axis A/q=2.5 in both cases and on the ΔE-axis $\Delta E \sim Z^2(T-k)$, the same value for both Ne-isotopes. The particle groups fall on top of each other in the ΔE-T-plot. The corrected ΔE_c now shifts isotopes with different charge states proportional to $q^{0.24}$, and ^{25}Ne^{10+} is well separated from ^{20}Ne^{8+} (fig. 16). The position of any isotope is given by the crossing of straight lines in an isotope chain horizontally and the A/q-line vertically. If there is still an overlap of isotope groups, one can put a gate on these isotopes and plot them in a ΔE_c-L-matrix, where they will appear well separated.

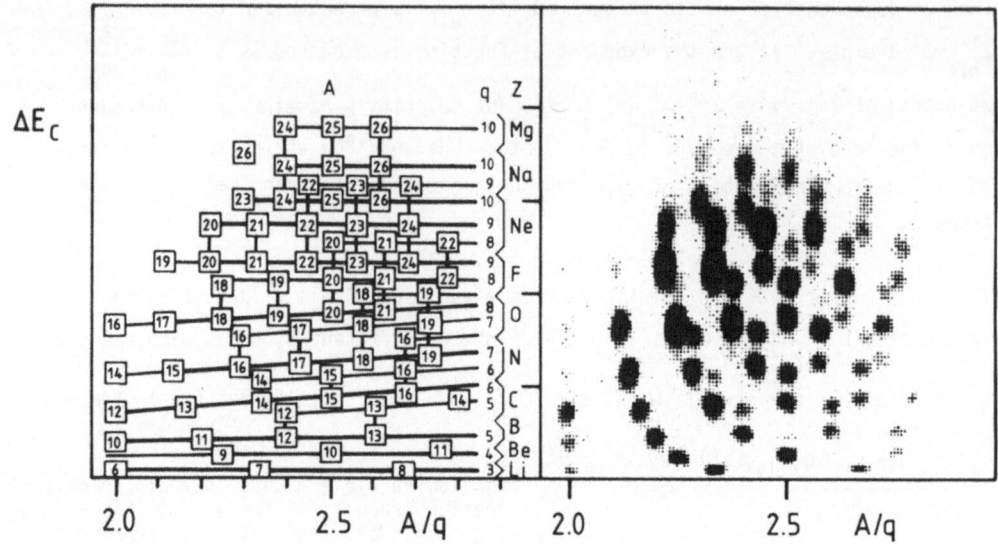

Fig. 16 Particle groups (right) and identification of A, Z and q (left) in a ΔE_c-T-plot for reaction products of the system $^{22}Ne + ^{48}Ca$, E_{Lab} = 170 MeV. Isotope chains with the same Z and q can be connected with straight lines (horizontal), this is also the case for element chains with the same A and q (vertical lines).

The same correction procedure can also be applied to a ΔE-E-plot, where the 1/E-dependence of ΔE can be divided out, but the q-dependence of ΔE is kept in ΔE_c.

V. The experimental program

In the following some examples are shown of the experimental work done at the spectrometer within the three years of operation.

1. M-substate-population of excited states below the particle threshold

Excited states of the ejectile, which decay by γ-emission, appear as broad lines in the focal plane due to the energy shift of the γ-recoil: $\Delta E = \beta \cdot E_\gamma \cdot \cos \Theta_\gamma$ (β=v/c velocity in units of speed of light, E_γ γ-transition energy, Θ_γ emission angle with respect to the particle path). The γ-angular distribution is transformed into an energy distribution by this equation. The unfolding of the measured line shape with characteristic line shapes of the different m-substates allows the extraction of the spin alignment. The line shape has to be measured with high resolution and the spectrometer is well suited for this purpose. Fig. 17 shows as an example the line shape of the first excited 2^+-state of ^{12}C in the inelastic excitation $^{88}Sr(^{12}C, ^{12}C')$ at E_{lab} = 80 MeV. The measured angular distribution of the m-substates are well described by DWBA-calculations (7).

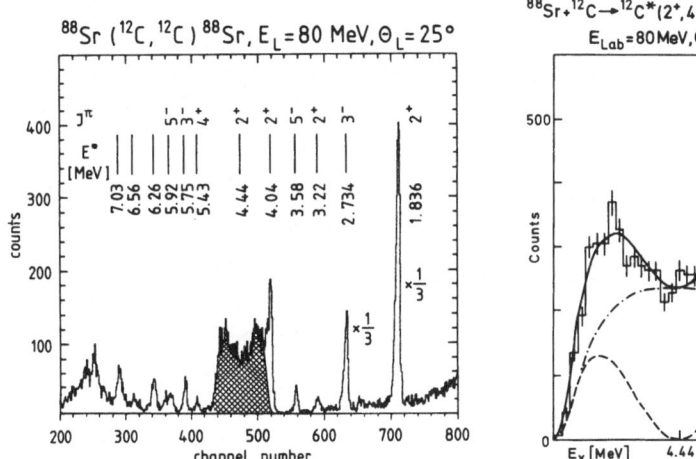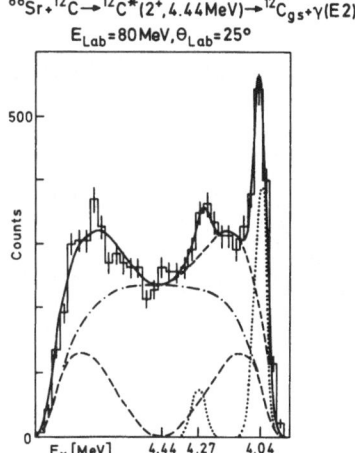

Fig. 17 Spectrum of the inelastic scattering of ^{12}C on ^{88}Sr at E_{Lab} = 80 MeV (left part) and decomposition of the γ-recoil broadened line of the ^{12}C-2^+-state at E_x = 4.44 MeV with line shapes characteristic for the m-substates (right part).

2. Determination of the L·S-potential in heavy ion reactions

Final states in one-nucleon transfer reactions are polarized for a large ℓ-value mismatch between entrance and exit channel, which occurs e.g. for large Q-values (the nuclei in the entrance channel are assumed to have spin zero). The spin polarisation "up" or "down" depends on the fact, whether the spin j_f of the excited state is larger or lower than the orbital angular momentum ℓ of this state. The L·S-potential is generated with the angular momentum L_0 of the grazing collision and the spin S of the ejectile and it changes the shape of the angular distribution. Sign and strength can be extracted by calculation of the reaction asymmetry A = $(\sigma_1-\sigma_2)/(\sigma_1+\sigma_2)$ for states 1 and 2 with different polarisation. This method has been applied to the reactions $^{88}Sr(^{16}O, ^{15}N)^{89}Y$ (8) and $^{208}Pb(^{12}C, ^{13}C)^{207}Pb$.

3. Charge exchange reactions

The reaction $(^{13}C, ^{13}N)$ is investigated with the targets ^{26}Mg and ^{58}Ni, which are well suited for spin-isospin-flip transitions in the d- and f-shell, respectively. Preliminary experimental results show a selective population of 1^+ and 3^+ states.

The transition strength and reaction mechanism will be studied at different energies.

4. Nuclear rainbow scattering for $^{12}C + ^{12}C$

The elastic and inelastic scattering of $^{12}C + ^{12}C$ was investigated at $E_{lab}=240$ MeV and 300 MeV in a broad angular range up to $\Theta_{cm}=70^\circ$ (fig. 18). The analysis showed that the potential ambiguity is almost removed and the depth of the nuclear potential for a Woods-Saxon shape is of the order of 180 - 250 MeV at $E_L = 300$ MeV (9). Best fits are obtained with a folded potential in a coupled channel calculation. The semi-classical deflection function, calculated from the S-matrix, shows a pronounced nuclear rainbow scattering minimum at $\Theta_\ell = 55^\circ$ for $\ell=20-28$ (grazing ℓ-value is $\ell = 38$). The total reaction cross section is calculated from the S-matrix, but no reduced value is observed as compared to the geometrical value.

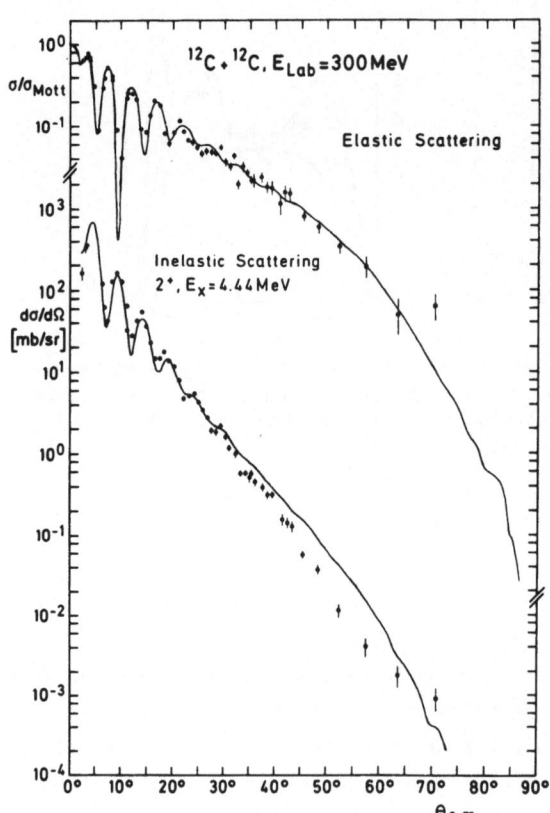

Fig. 18 Angular distributions of the scattering of ^{12}C on ^{12}C at 300 MeV. The solid lines represent coupled channel calculations.

5. Excitation of high lying inelastic modes

This investigation has been started with the excitation of the E2-resonance at $E_x=16$ MeV in ^{58}Ni with ^{20}Ne at 290 MeV and 392 MeV. An E2-strength of $\approx 55\%$ EWSR is observed, which is the same as compared to results of e^-- or α-scattering. Higher lying collective modes were obscured by strong background processes (fig. 19). Contributions of these processes can be removed in the heavy ion scattering by coincidence measurements due to the different kinematics of the background reaction mechanism as compared to the target excitation. After installation of the coincidence detector the excitation of E3- and isovector modes will be studied.

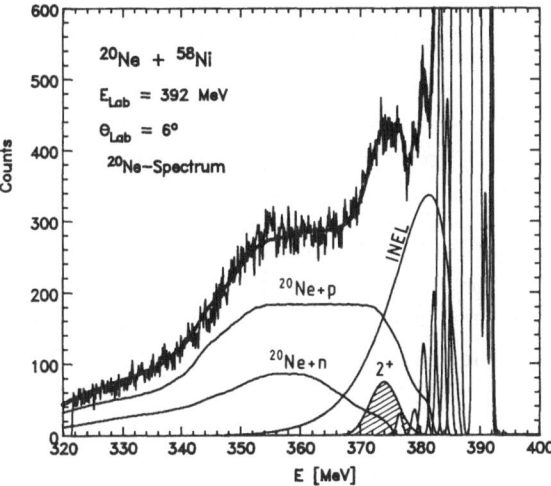

Fig. 19 Spectrum of the inelastic scattering of ^{20}Ne on ^{58}Ni at 392 MeV. The shaded area represents the quadrupole resonance. The unfolding of the spectrum shows that the high energy part is dominated by background processes, e.g., the 3-body channel ^{20}Ne+p+^{57}Co.

The experimental program includes further investigations on mutual excitations of projectile and target, mass measurements of neutron-rich nuclei and many outside user activities. All these experiments show that a Q3D-spectrometer together with a good detector equipment is a very versatile instrument. The advantage of a large solid angle with a high resolution angular detector should be especially emphasized.

References

1) A.G. Drentje, H.A. Enge and S.B. Kowalski, Nucl. Instr. Meth. 122 (1974) 485

2) F. D. Becchetti, C.E. Thorn and M.J. Levine, Nucl. Instr. Meth. 138 (1976) 93

3) F. Busch, W. Pfeffer, F. Pühlhofer, D. Schüll, Annual Report, GSI-80-3, p. 185

4) H. Lettau, H.G. Bohlen, H. Rossner, W. von Oertzen and M. Martin, submitted to Nucl. Instr. Meth.

5) D.M. Barrus and R. L. Blake, Rev. Sci. Instr. 48 (1977)

6) J.S. Forster, D. Ward, H.R. Andrews, G.C. Ball, G.J. Costa, W.G. Davies, I.V. Mitchell, Nucl. Instr. Meth. 136 (1976) 349

7) G. Ingold, H.G. Bohlen, M. Clover, H. Lettau, H. Ossenbrink and W. von Oertzen, Z. Phys. A305 (1982) 135

8) P. Wust, W. von Oertzen, H. Ossenbrink, H. Lettau and H.G. Bohlen, Z. Phys. A291 (1979) 151

9) H.G. Bohlen, M.R. Clover, G. Ingold, H. Lettau and W. von Oertzen, Z. Physik A308 (1982) 121

SPEG, A SPECTROMETER FOR GANIL

J. Gastebois

DPh-N/BE, CEN Saclay, 91191 Gif-sur-Yvette Cedex, France

INTRODUCTION

SPEG is an energy loss spectrometer under construction for GANIL. "Energy loss" means that one intends to study two-body reactions with a resolution better than the incident beam energy width (thought to be $\sim 10^{-3}$), and consequently, that one must couple a spectrometer to a dispersive beam line satisfying the adequate requirements.

On Fig. 1 are shown the main magnetic elements. On the beam line, starting from an object point down to the target position, one will find three quadrupoles, one dipole, again two quadrupoles, and two sextupoles. Beyond the target, the spectrometer will consist of one quadrupole, one sextupole, two dipoles and one quadrupole. The whole spectrometer is described in detail in ref.[1].

I. Specific requirements.

They are of two kinds : 1) to achieve a "good" resolution ; 2) to be able to do measurements at zero degree and very small angles.

I.1. A "good" resolution.
First of all, what is the meaning of "good" at GANIL energies? As already mentioned, the incident beam is expected to have an energy spread of about 10^{-3}, which corresponds, at the top energies, to ΔE values larger than 1 MeV, such as shown on the few following examples :

$$^{12}C \text{ at } 100 \text{ MeV/amu} \qquad \Delta E = 1.2 \text{ MeV}$$
$$^{20}Ne \text{ at } 100 \text{ MeV/amu} \qquad \Delta E = 2.0 \text{ MeV}$$
$$^{40}Ar \text{ at } 50 \text{ MeV/amu} \qquad \Delta E = 2.0 \text{ MeV.}$$

One sees immediately that this is already too much, in almost all cases, to distinguish between elastic and inelastic events. A clean separation can be expected provided these figures are improved by a factor between 5 and 10.

So the requirement is to achieve a momentum resolution of 10^{-4} (i.e. 2×10^{-4} in energy) of course with a solid angle of a few millisteradians. That implies a spectrometer working in the energy loss mode.

I.1. Zero degree and small angle measurements.
Another important requirement is to be able to do measurements at very small laboratory angles, including zero degree, in view of the very small grazing angles at high energies. One would like to catch the incident beam as far as possible from the detectors, those which will be around the focal position as well as those which will be in the vicinity of the target, to avoid a too high neutron level. Each experiment is, of course, a special case, and no general solution can be given. However, two major possibilities will be offered, provided

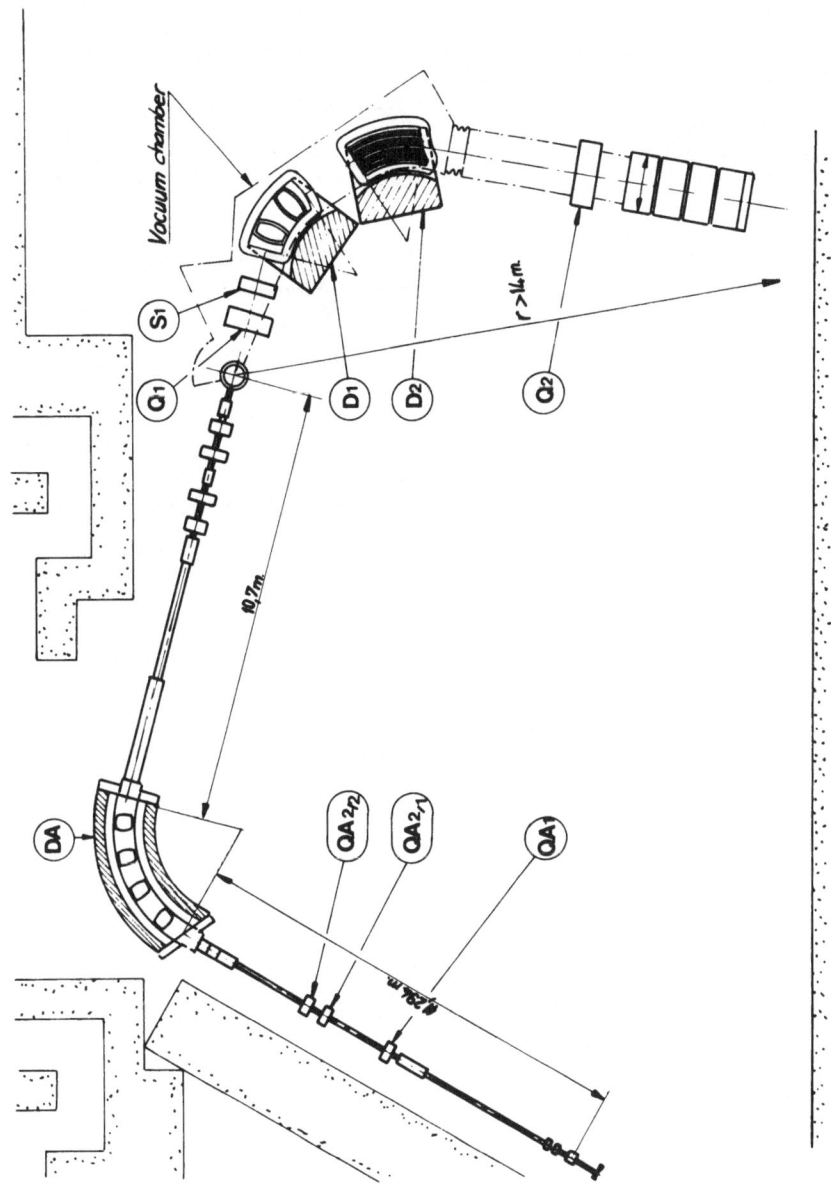

Fig. 1 - General view of SPEG.

that the magnetic rigidity of the incident ions (beyond the target) is either large (a) or small enough (b) compared to the one of the selected ions to be detected.

Case a : The two dipoles of the spectrometer will have a C shape. In other words there will be no return yoke on one side (the left-hand side, with respect to the optical axis oriented along the ion velocity), so that the vacuum chamber can be extended outside, providing some room to use a beam catcher. That facility can be used for zero degree experiments, when adequate conditions are fulfilled. At small angles (different from zero degree), the beam can also escape on the left hand side, thanks to the same C

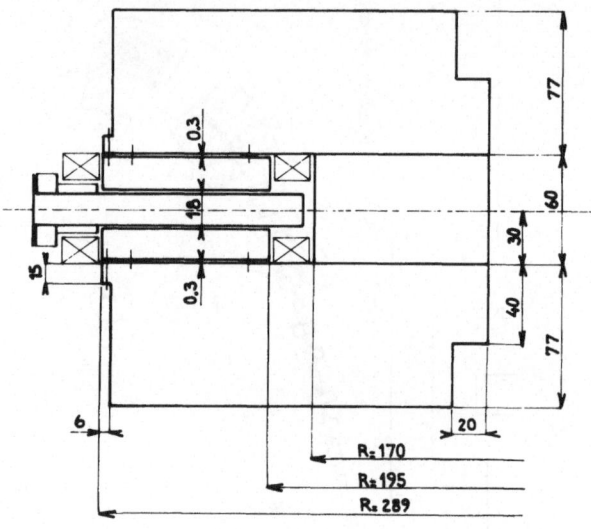

Fig. 2 - Cross section of one dipole.

shape adopted for the entrance lenses (quadrupole and sextupole). See Figs. 2 and 3. [2], [3].

Case b : Room is provided between the two dipoles to set a beam catcher. That will allow to stop the incident ions having the most probable charge state beyond the target (again, when adequate conditions are fulfilled concerning magnetic rigidities).

II. Role of the various magnetic components.

II.1. The beam line. The beam line must satisfy the following requirements : i) to have a momentum resolution lower or equal to 10^{-4} ; ii) to keep a vertical beam spot on the target $\leqslant 4$ mm ; iii) to allow an horizontal focus located at the target position for K=0 (K is the usual first order kinematic coefficient) and an adjustable position beyond the target according to K values ; iv) to match, at the target position, given values of the angular dispersion and of the spatial dispersion. These two values are those given by the spectrometer considered in the inverse way, that is to say starting from the focal position up to the target.

II.1.a. First order tuning. The values of these two dispersions remain constant whatever are the kinematic conditions of the reaction under study. So, in practice, the values are achieved in tuning the two quadrupole lenses in front of the target, in accordance with the magnetic rigidity of the incident ions.

When kinematic conditions are varied, one has to adjust the horizontal focus of the analyzing line. That can be done in tuning the quadrupole lenses in front of the dipole. The calculated tuning of the line leads to an increase of the momentum resolution, from 3.4×10^{-5} (FWHM) for K=0 up to 10^{-4} for K=0.5. To keep a constant value of the resolution one needs to cut down in size the initial horizontal object, and to loose intensity.

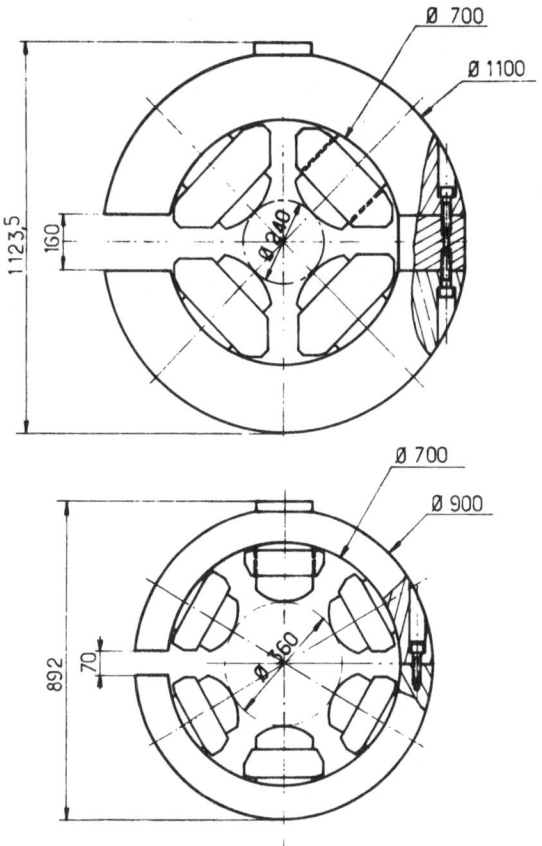

Fig. 3 - Entrance lenses.

II.1.b. _Second order corrections_.

Second order corrections (including the tilt angle of the horizontal focal plane of the line) will be adjusted in using two sextupole lenses (in front of the target) together with mechanical and electrical shims installed inside the dipole itself (the electrical shims are needed to simulate variations of the thickness of the mechanical shims, according to kinematic conditions).

II.2. The_spectrometer.

II.2.1. _First order tuning_. Vertical focusing is obtained mainly with the entrance quadrupole and the inclined exit face of the second dipole (D2). The last quadrupole (Q2) is only very slightly defocusing in the vertical direction. The main role of this lens is to insure a zero value for the angular dispersion ($\theta/\delta = 0$), so minimizing the focal recoil with kinematics ($\sim$ 70 cm for $\Delta K = 0.1$). At the same time, the dispersion along the focal plane is kept to a constant value (8 meters), whatever its position is.

This quadrupole is of Collins type, and will provide a pure quadrupole field, once integrated over a trajectory.

II.2.2. _Second (and third) order corrections_. Three second order corrections will be obtained with : i) the entrance sextupole (S1) ; ii) the curvature of the exit face of the first dipole (D1) ; iii) the curvature of the exit face of the second dipole

(D2). The third one will govern the tilt angle of the focal plane. In principle, we must get a tilt angle around 10°.

The three effects are well decoupled one from the other, because located at rather different places. The second one acts mainly on the $(X|\theta^2)$ term. In order to simulate a third order correction, the exit face of the first dipole will be machined with two slightly different radii on each side of the optical axis. As on the beam line, second order corrections will depend upon kinematic conditions. The only adjustable parameter is the field in the entrance sextupole, but it is not enough to take care of the main second order aberrations. So, in order to simulate variations of the curvature of the exit face of the first dipole, we will use two pairs of coils, installed inside the first dipole.

II.2.3. *Additive correction*. Due to the C shape of the dipoles, there will be a transverse field gradient effect inside each one, depending upon the induction. The tuning of the entrance quadrupole lens may not be enough to take care of it, so we may need to use, inside the second dipole, a set of conductors, parallel to the optical axis, to generate an additive gradient.

III. Main characteristics.

The main characteristics of the beam line are given in Table I, and those of the spectrometer in Table II.

Table I

Basic optical characteristics of the analyzing line (nominal values, i.e. for K= 0 tunings)

Dispersion on target	10 m
Full width of the target beam spot, for a monochromatic beam*	0.7 mm
Full height of the target beam spot*	4 mm
Dipole :	
Mean radius	3 m
Nominal induction	1 Tesla
Mean deviation	75°
Entrance face tilt angle	23°5
Exit face tilt angle	0°

*For a nominal object at the object slits of 4 mm × 4 mm.

Table II

Basic optical characteristics of the spectrometer

Dispersion at the focal plane (can be varied from 6 to 10 using different tunings for the last quadrupole lens)	8.1 m
Horizontal magnification	0.8
Vertical magnification	4.7
Solid angle (± 35 mrd in each plane)	4.9 sr.
Dipoles :	
Mean radius	2.4 m
Nominal induction	1.2 Tesla
Mean deviation (for each dipole)	42°5
Exit face tilt angle in the second dipole	24°
Exit face radius of curvature : . of the first dipole	R_1=1.46 m R_2=1.57 m
. of the second dipole	R=1.6 m
Energy range	14 %
Focal length	60 cm
Focal plane tilt angle	8°

IV. Additive possibilities.

IV.1. Advantage of the angular achromaticity.

As already mentioned, the coupling of the beam line to the spectrometer is so that one achieves not only the spatial achromaticity (needed for the energy resolution), but also the angular achromaticity. This property insures a univoque relation between the angle of the exit trajectories and the reaction angle, which extends within the angular aperture of the spectrometer ($|\Delta\theta| \leqslant 35$ mrd, i.e. 2°). So, the precise value of the reaction angle can be known in measuring angles at the exit of the spectrometer. Of course, the accuracy of the reaction angle is limited by the angular spread of the incident beam direction at the target position.

IV.2. Ion_identification_experiments.

There are experiments where the main purpose will only be the identification of ion species, without any good resolution requirement, for instance when more than two bodies are present in the exit channel. The spectrometer will then be used only to measure magnetic rigidities, with an accuracy of $\sim 10^{-3}$. In that case, the beam spot on the target is considered as an emitting source of nuclei. The requirement is then to limit the size of the beam spot. So, the beam transport system must be about achromatic (no beam dispersion on the target). That will be obtained in tuning differently the analyzing line and the deviation in front of it (deviation between the central beam line and the entrance section of the analyzing line in the experimental room).

V. Detector studies

I will not review in detail all kinds of detectors which are thought to be used : channel plates, parallel plates, proportional counters, drift chambers, ionization chambers etc. I will limit myself to the last improvements obtained in the study of position measurement detectors and low pressure ionization chambers already built and tested, and to be used around the focal region for good resolution experiments with *light energetic* ions (masses $\leqslant 40$, (E/A) between 50 and 100 MeV/amu).

In that case, we are in the following conditions : i) 0.8 mm along focal plane corresponds to 10^{-4} momentum spread ; ii) we want to measure exit angles so we need two position measurements, horizontally as well as vertically ; iii) we need one (or several) ΔE signals, with an accuracy lower or equal to 3 % ; iv) we need a time of flight measurement. In most cases, a time resolution ~ 2 nsec must be enough.

Several detectors are to be used in the focal region. A typical arrangement is schematically shown on Fig. 4. It includes : i) 2 position measurement (in both planes) counters ; ii) 1 parallel plate (t signal) ; iii) 1 ionization chamber (including 2 horizontal position measurements, and at least two ΔE signals).

V.1. Position_measurement_counters.

Two of them (P_1 and P_2, see Fig. 4) are to be used, on each side of the focal plane, each one giving two position (X and Y) signals.

Each counter consists of two parts (see Fig. 5) : i) a drift chamber, in which the drift time of electrons gives a measurement of Y ; ii) a proportional counter. The

132

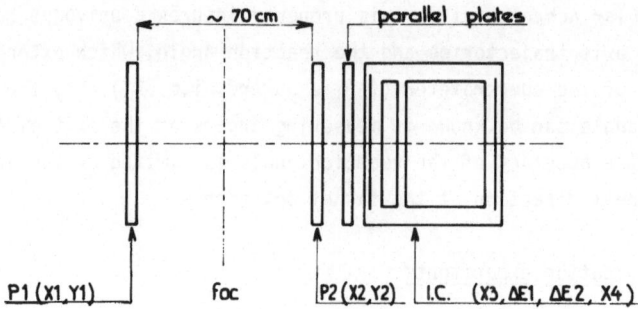

Fig. 4 - A typical arrangement of detectors.

cathode is made of thin strips (1 mm width, every 2.54 mm), connected to one another via a delay line. A time measurement of the signals induced on the cathode, and collected to each end of the delay line, provides a measurement of X.

One such counter has been built and checked with alpha sources [4]. The characteristics are as follows : active length : 70 cm ; active height : 12 cm ; window thickness : 2.5 μm (to that the angular straggling is kept around 0.1 or 0.2 mrd for light energetic ions) ; counter thickness : 2 cm ; running pressure : ≤ 30 torrs (isobutane) that is enough for position measurements).

The measured performances of this counter, after taking into account the alpha-source contribution, is 0.25 mm in each direction (including contributions due to electronics). Under those conditions, ta-

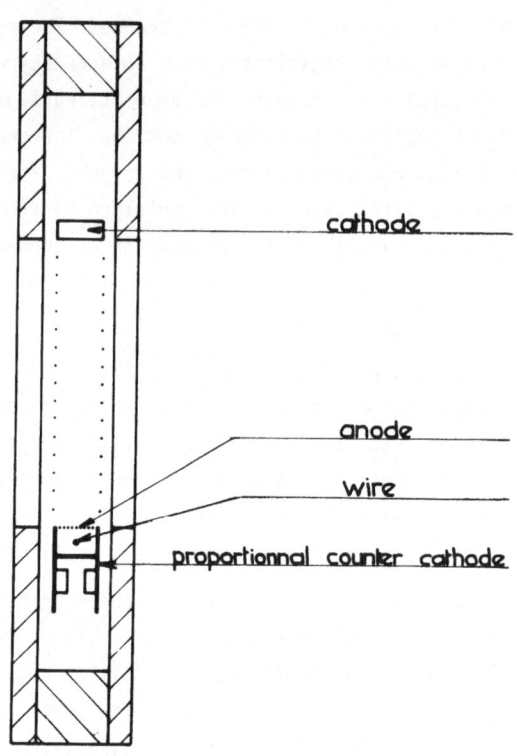

Fig. 5 - A position measurement counter.

king into account the estimated angular straggling effects in the various foils, one must deduce exit angles with accuracies around ± 1 mrd, and positions along the focal plane with an accuracy around 0.2 mm.

V.2. ΔE measurements

They will be obtained using an ionization chamber. One prototype has already been built and checked [4]. The characteristics are the following :

- useful length : 60 cm ;
- useful height : 12 cm ;
- depth : 50 cm ;
- anode : one plate, in which are included two proportional counters used for position signals, one close to the entrance window, one close to the exit window. One
uses resistive wires (7 kΩ/m) and charge division method for position measurements
(resolution between 1 and 1.3 mm). These position measurements are used to correct ΔE
signals for variations of trajectory lengths across the gas, and to eliminate spurious
signals when compared to the positions measured with the two other counters (see V.1),
in order to remove background ;
- gas : isobutane.

This arrangement has been used for several experiments at Saturne [5], around the focal plane of SPES I spectrometer. To limit myself to a single example, it has been
used in the study of elastic and inelastic scattering (giant resonance studies) of
480 MeV alpha particles on ^{208}Pb. In that case, the energy loss was too small (even
at a pressure of one atmosphere) to provide any significant signal and only position
signals were used.

The energy resolution was checked using a thorium source (energies around 6 and 9
MeV). The pressure was 150 torr, high enough to stop all particles. The energy resolution was then measured to be 2.5 %.

Efforts are made to build such ionization chambers for much higher running pressures
(at least 1 atmosphere). People from C.R.N. Strasbourg are implied in such developments, but no new result has been obtained so far.

VI. First proposals for experiments.

A spectrometer provides the possibility of doing experiments which could not practically be done using classical arrangements of detectors inside a scattering chamber.
Apart from the fact that the measurement of a magnetic rigidity always helps for ion
identification and that the use of a magnetic field contributes to background reduction, there are specific advantages offered by an energy loss spectrometer such as
SPEG, when associated with a performant detection system. It must be a very interesting tool, considering all together the main following characteristics :

. Large solid angle, allowing measurements of very small cross sections.
. No limitation in energy resolution due to the incident beam energy spread.
. Compensation of kinematic effects.
. Measurements of very small angles, including zero degree, provided adequate condi-
  tions are fulfilled concerning magnetic rigidities.
. Measurement of a bit of angular distribution at once, within the full angular aper-
  ture (4°), each minimum angular slice being governed only by the incident beam angular
  aperture.

Several kinds of experiments are already intended to be done. They are mentioned below, with very short comments.

VI.1. Elastic scattering

. Incident ions from ^{16}O to ^{84}Kr.
. Incident energies between 20 and 100 MeV/amu.
. Study of the absorption radius (so called surface transparency problem).
. Behaviour of Im f(0) at high incident energies.

VI.2. Inelastic scattering

Study of giant resonance excitation, with various projectiles (^{12}C, ^{13}C, ^{16}O, ^{20}Ne), on several targets (Ni, Sn, Pb) to be compared with results obtained at Saturne using incident alpha particles on the same targets.

VI.3. High spin states

i) Study of the possibility to populate high spin states via one- or two-nucleon transfer, using projectiles such as ^{50}Ti (2-neutron transfer from a closed shell, and having their spin and angular momenta almost parallel).

ii) To look whether such states can be populated via inelastic scattering followed (or preceded) by particle transfer. Because of the large solid angle, one could even try to do coincidence experiments with gamma rays. A first test is proposed, which consists to use SPEG in bombarding a ^{208}Pb target with incident ^{152}Sm ions, detecting the recoil nuclei close to zero degree, in order to try to reach high spin states in the ejectile. A Ge counter will probably be used (without any coincidence) to look at the same time at gamma-ray spectra.

VI.4. Production of neutron rich isotopes

For instance, via the fragmentation of incident ^{48}Ca and ^{50}Ti ions, between 60 and 80 MeV/amu incident energies, taking advantage of the large solid angle of SPEG and the expected possibility of measurements around zero degree without too much background. Rather thick targets can be used together with rather high beam intensities (10^{11} to 10^{12} ions/sec). Produced light isotopes ($Z \lesssim 20$) could be identified without too many difficulties using already built counters, such as described in paragraph V.

VI.5. Reaction cross section measurements

This kind of measurement, using the attenuation method at zero degree, could be done with SPEG, provided that two main problems are solved : i) reduction of beam intensities down to $\sim 10^4$ ions/sec ; ii) a set of detectors allowing a fast counting rate. If the second one is thought to be solved without too much difficulty, the first one has still no obvious answer.

VII. Conclusion

SPEG is now under construction. It is expected to be completed before the end of 1984, and every body hopes that the first beam tests could take place during fall 1984.

References

[1] P. Birien et S. Valéro, Projet de spectrométrie magnétique à haute résolution pour ions lourds, Note CEA-N-2215, mai 1981.
[2] A. Daël, Internal report DPh-N/BE 80-312, AD/MN du 4.11.1980.
[3] A. Daël, Internal report DPh-N/BE 80-254, AD/MN du 12.08.1980.
[4] Compte rendu d'activité du Département de Physique Nucléaire 1980-1981, Note CEA-N-2276, p. 237.
[5] Compte rendu d'activité du Département de Physique Nucléaire 1979-1980, Note CEA-N-2207, p. 254.

4π PHYSICS WITH THE PLASTIC BALL

H.H. GUTBROD, H. LÖHNER, A.M. POSKANZER, T. RENNER,
H. RIEDESEL, H.G. RITTER, A. WARWICK, F. WEIK, H. WIEMAN

Gesellschaft für Schwerionenforschung, Planckstrasse 1
D–6300 Darmstadt 11, West Germany
and
Nuclear Science Division, Lawrence Berkeley Laboratory
University of California, Berkeley, CA 94720

Abstract:

   4π data taken with the Plastic Ball show that cluster production in
relativistic nuclear collisions depends on both the size of the participant volume
and the finite size of the cluster. The measurement of the degree of thermalization
and the search for collective flow will permit the study of the applicability of
macroscopic concepts such as temperature and density.

1. Introduction

   After several years of studying relativistic nuclear collisions one of the main
goals still is to learn about the equation of state or even to find extreme states
of nuclear matter. Most of the time past was used to get oriented and to learn
about the environment of problem areas and about the phase space of these
reactions. At energies of few hundred MeV/nucleon to 2 GeV/nucleon one soon
realized that no pool but a tiny one of proton–nucleus data exists where one could
dwell on for comparison with nucleus–nucleus collisions.

   With their huge number of exit channels the nuclear collisions viewed by single
particle detection schemes looked "thermal" in all channels. Early multi–particle
detection schemes paved the way to nowadays 4π detectors that reflect our respect
for the complexity of relativistic nuclear collisions.

   The design criteria for the Plastic Ball were to measure so much of an event
that dynamics in the event could be separated from the background of phase space
population. Large dynamical effects are predicted on the one side by hydrodynamic
models whereas on the other the nuclear fireball model totally ignores them. So the
measurement of particles over a wide energy range under all angles is needed for
coping with the observed large multiplicities of more than 100 particles.

   Since we are looking for matter properties and since large cluster production
has been observed in earlier experiments a good particle identification was
mandatory of a 4π detector for nuclear collisions. Positive pion data had focused
the attention onto their production mechanism. Therefore a 4π detector had to allow
to observe the pion channels as well. At the time of conception of the Plastic Ball
design one could summarize those goals by only one:

"Measure as much of the event as possible."
During the construction period and provoked by streamer chamber results, the global
analysis concept for 4π data was introduced into this field from particle physics,
and clear differences between intranuclear cascade calculations and hydrodynamical
model calculations did show up due to a collective flow predicted by the latter
model.

Before the 4π physics is discussed on the first experiments an extensive report
on the Plastic Ball is given to familiarize the reader with the details of this
first electronic 4π detector with particle identification capability.

2. The Plastic Ball

A concept of a 4π detector consisting of many individual ΔE-E and
time-of-flight telescopes was chosen that promises fast data analysis. The number
of counters necessary to cover 4π is related strongly to the multiplicities of the
events to be studied, since too few large counters would result in a high
probability of multiple firing. Too many small counters for a given size of the 4π
detector volume, however, would cause a large percentage of the particles to scatter
out of the detectors into neighboring ones.

For the coverage of most of 4π the Plastic Ball was built, completely
surrounding the target except for the very backward angles where the beam enters the
system and the extreme forward angles. These forward angles of 0°-9° are covered
with a multielement time-of-flight system (called the Plastic Wall) taking into
account the large fragment velocities at forward angles. The following gives a
detailed description of the whole system, as it is schematically shown in Fig. 1.

2.1 GEOMETRY

The geometry of the Plastic Ball was selected with specific consideration of
earlier multiplicity distribution measurements[1]). Since the 4π detector is to be
used at various incident energies, its spatial resolution, and thus the number and
dimensions of individual counters, must be suitable to resolve the strongly
forward-peaked multiplicity distribution of the reaction products.

The SLAC Crystal Ball[2]) design was adopted, which is based on the geometry of
an icosahedron: a 20-faced solid figure in which each face is an equilateral
triangle of the same dimensions. Each face is divided into 36 triangles, resulting
in the division of the surface into 720 triangles, with only 11 different
two-dimensional shapes. For the Plastic Ball modifications were made in details of
the entrance and exit ports. Modules in the backward cone between 160° and 180° and

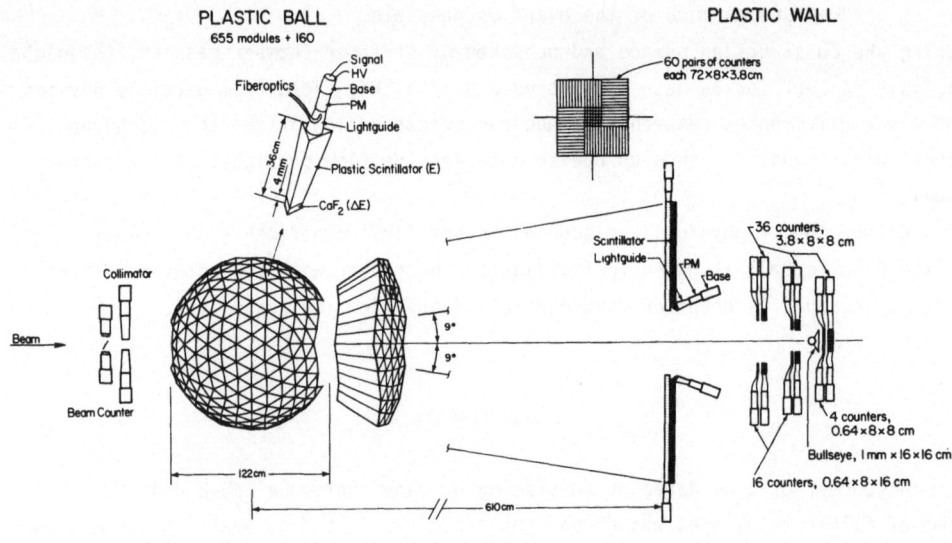

Fig. 1.   Schematic view of the Plastic Ball and the Plastic Wall.

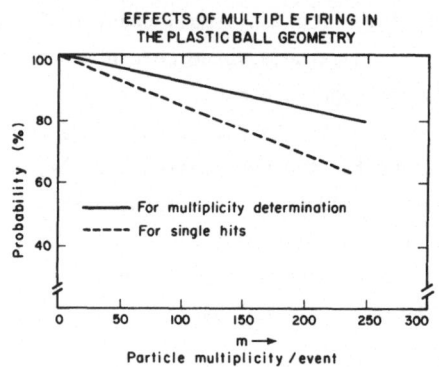

Fig. 2.   Probability for single hit of a Plastic Ball module as a function of multiplicity.

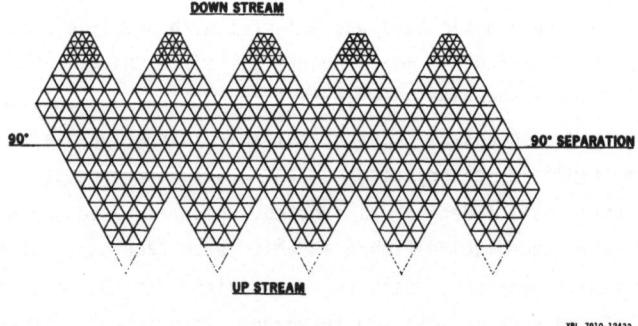

Fig. 3.   Mercator projection of the Plastic Ball.

in the forward cone between 0° and 10° had to be omitted in order to allow room for the beam pipe and prevent the beam halo from hitting the detector modules.

A further modification was necessary for the region between 10° and 30° in order to guarantee good particle identification despite the high multiplicities expected at forward angles. This region (known as the Mall) is subdivided into 160 modules instead of the 40 in the original design and positioned at a larger radius from the target position. This final geometry was chosen based on the results of Monte Carlo calculations using earlier measurements of multiplicity distributions[1]). In fig. 2 the probability for single hits in the Plastic Ball geometry is plotted (dotted line) versus the event multiplicity. As an example, for a multiplicity of 100 particles, the Plastic Ball will have typically 92 detectors fire, out of which 84 will have seen a single particle only.

The geometrical relationship between the triangular detector faces of the Plastic Ball, unfolded in a plane, is shown in fig. 3. The central spherical cavity of the ball has a radius of 25.4 cm, the outer radius is 61.4 cm. The aforementioned surface division scheme is applied to both the outer and inner surfaces. For assembly and access to the target, the ball is divided into two half-spheres along the $\theta = 90°$ plane. A thin sheet of 1.2 mm steel with a hole 50 cm in diameter forms the baseplate of each hemisphere.

2.2 INDIVIDUAL DETECTOR MODULE

2.2.1 General Characteristics

Each of the 815 detector modules represents a particle identifying telescope with a ΔE and E detector using a slow and a fast scintillator read out via one photomultiplier[3]) (fig. 1). The ΔE counter is a $CaF_2(Eu)$ crystal[4]) with a characteristic decay time of 1 μs for the emission of the scintillation light. Unlike NaI the $CaF_2(Eu)$ is not hygroscopic and therefore needs no bulky canning, which would have jeopardized the system goal of a total coverage of 4π of solid angle. Each module is only wrapped with a double layer of aluminized mylar foil as an optical separator between adjacent modules. The light output of CaF_2 compared to that of anthracene is 100 to 120% and the crystal is extremely free of afterglow compared to NaI. The ΔE light is read out through the E counter (i.e., the E counter serves as a lightguide to the ΔE counter), which consists of a plastic scintillator[5]) with about 45% of the light output of anthracene. The light emission of the plastic scintillator is approximately 100 times faster than that of the $CaF_2(Eu)$ so that 90% of the E signal is collected within 10 ns. Such a big difference between the time characteristics of the two scintillators is desirable because the plastic scintillator has, besides the dominant short decay time, some components with long decay times, which result, after 120 ns, in a _pulse height_ ratio of only 1000:1 between fast and slow plastic decay components, but in ratios of about 10:1 for integration of the fast and slow components. This determines our choice for the onset of light collection for the ΔE signal, which is affected by these long decay times from the plastic scintillator. An optimum has been found for the ΔE-E resolution when the ΔE light integration lasts for 1.5 μs and starts about 240 ns after the start of the E signal.

The thickness of the ΔE CaF_2(Eu) crystal was chosen to be 4 mm with 35.6 cm as the length of the E plastic scintillator. This allowed good ΔE signals even for minimum ionizing particles and also a minimum low energy cutoff for particles stopping in the ΔE counter. In order to obtain clean proton spectra up to 200 MeV, the detector length was chosen to stop 240 MeV protons. This additional length assures that in the ΔE-E diagram the punch through deuterons do not disturb the proton spectra below 200 MeV. It also allows some toleration of dead zones in the outer corners of the modules, where mounting studs for the assembly were placed. The total length chosen represents a physical upper limit for particle identification for stopped particles due to the onset of reaction losses.

The readout of the ΔE-E module is done via a conically shaped lightguide (lucite), which couples to a 2-inch, 10-stage photomultiplier[6]. The phototube was selected for good gain and high linearity up to the high currents that are necessary for the subsequent pulse shape analysis. The gain of the phototube is monitored by a light pulse fed into the light guide via an optical fiber[7] from a pulsed Xe-light flasher.

2.2.2 Neutron response

Since plastic scintillators are efficient neutron detectors, the effects of neutrons emitted in coincidence with a charged particle into the same detector module need to be considered to avoid misleading results in particle identification and energy determination.

Calculations were performed[8] for a module shape similar to a Plastic Ball module taking into account recent measurements of proton kinetic energy and angular distributions at Bevalac energies, neutron multiplicities, and neutron efficiencies of plastic scintillators[9]. The average deposited neutron energy turned out to be ~10 MeV for neutron kinetic energies between 100 and 800 MeV and was slightly lower for lower energy neutrons. Thus the angle averaged probability for single hits (see fig. 2) will be reduced from 84% (considering only charged particles) to 81% when neutrons are included in the example of a multiplicity 100 event. There is no effect of neutrons on altering the charge identifcation for Z = 1,2. However, the particle mass may be assigned incorrectly if the deposited neutron energy is comparable to the separation of the particle identification curves in the ΔE-E plot. For charged particle multiplicities of 100 about 2% of protons will therefore be misidentified as deuterons, 2% of deuterons will be shifted to tritons, and 1% of protons will be misidentified as tritons. The ^{3}He and ^{4}He identification suffers approximately 0.5 to 1% misidentification.

2.3 PION IDENTIFICATION

In designing the Plastic Ball, special care was taken to detect the positive pions. As the yield of the π^+ is only about 10% of the proton yield at beam energies well above the pion production threshold, in a pure ΔE–E identification scheme the pions would be overshadowed by the background produced by heavier particles. Therefore, the π^+ are additionally identified by their delayed decay. Stopped π^+ decay into a μ^+ and a neutrino with a mean life of 26 ns and a Q–value of 4.12 MeV. The μ^+ subsequently decays into a positron and two neutrinos with a mean life of 2.2 µs. In this decay the e^+ is emitted with an energy of up to 53 MeV and produces therefore a signal that is easily detectable in the plastic scintillators, as schematically shown in fig. 4. A discriminator in the electronics (Ball Box) detects the occurrence of this second spike and a TDC measures the delay time. The TDC covers the range between 250 ns and 10 µs so that about 90% of all $\pi^+ \to \mu^+ \to e^+$ decays can be recorded, provided the e^+ deposits an energy greater than 1 to 2 MeV in the scintillator. This requirement accounts for an additional loss of a few percent of the pions. If the decay occurs during the integration time of the ΔE–ADC (250 ns–1750 ns), the ΔE signal can be spoiled, as

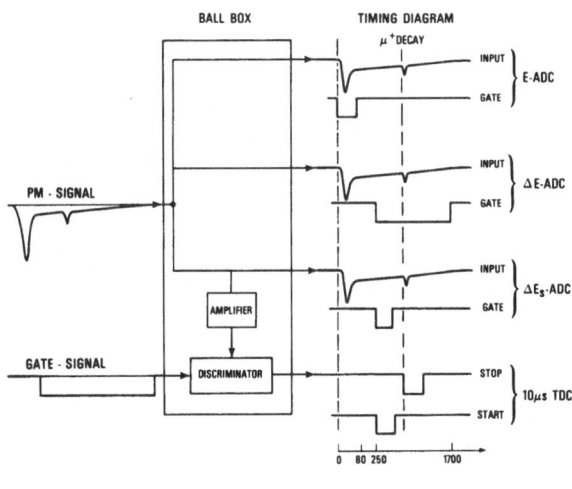

XBL 8210 1224

Fig. 4.   Scheme of the pulse shape analysis of the signal from the Plastic Ball detector for π^+ particle identification.

the positron energy can be much higher than the energy loss of the pion in the CaF_2 (typically 2-3 MeV). Therefore, a separate signal is given to a third ADC, called ΔE_s, with a shorter gate (700 ns) so that, depending on the time of the decay, one or the other ΔE signal or a combination of both can be used to extract the ΔE information.

This method can not be used to detect negative pions, as the π^- are promptly absorbed by a nucleus in the detector and the 140 MeV rest mass of the pion is released as kinetic energy. The efficiency for π^+ detection was studied in detail using a pion beam at LAMPF. At very high incident energies this scheme can be used for K^+ detection.

2.4 GAIN MONITOR XENON STROBE SYSTEM

A xenon strobe flash tube system was adapted from a design used on the Stanford Crystal Ball[2]) to provide a monitor of the gains of the phototubes in the Plastic Ball. Eight flashers, each coupled to a bundle of about 100 light fiber cables[7]), deliver light pulses to the 815 phototubes. A silicon PIN photo diode[10]) in each flasher module records the pulse by pulse amplitude. During the beam spill the eight flasher modules, together with the wall pulsers are triggered cyclically with a 16-output-pulse-sequencer driven at a rate proportional to the counting rate in the beam start detector.

3. The Plastic Wall

The Plastic Wall (Pilot F) covers an area of 192 cm x 192 cm and provides fine position resolution coverage of the angular region between $\theta = 0°$ and the forwardmost sections of the Plastic Ball ($\theta \sim 10°$). It comprises two sections of scintillators, the outer Wall and the inner Wall (fig. 1).

The outer Wall serves to extend the acceptance of the Ball-Wall system from $\theta = 10°$ to $\theta = 2.5°$ detecting particles in the same range of mass and charge as the Ball itself, but measuring velocity and Z instead of energy, Z and A as in the Ball. The inner Wall extends the angular measurement to 0°. Besides measuring light fragments like pions, hydrogen, and helium the inner Wall is used to observe heavier fragments (Li, B...), which have small velocities in the projectile frame. In addition, the centre region of the inner Wall makes up the event trigger in the first round of experiments.

3.1 THE OUTER WALL

The outer Wall region consists of 60 pairs of counters, each 72 x 8 x 3.8 cm, providing coincidence units sensitive to charged particles passing through both bars of a pair. Neutron and gamma background is rejected by this coincidence requirement.

A time resolution of 350 psec (FWHM) was achieved for minimum ionizing protons. The mean time of the pair gives the flight time from the target and hence the particle velocity.

Each scintillator signal is fed to an ADC, and Z identification is achieved (as suggested by the Bethe Block equation) by generating a function $v^2 \Delta E$, where v is the measured particle velocity and ΔE is one of the two ΔE signals.

Corrections to the scintillator pulse height were necessary because of

a)   attenuation of light along the length of the scintillator and

b)   non-linearity at the end of the rear scintillator, due to the bent light guide.

Since the yield of fragments with Z > 3 was negligible the phototube gain was set so that heavy fragments caused an ADC overflow, hence they were counted but not identified. On-line gain monitoring was carried out with light from a N_2-laser.

Calibration of the position spectra was done by means of eight narrow (3.8 cm) bar scintillators mounted as a third layer, perpendicular to the length of the pairs at a fixed position from one end. By means of a threefold coincidence caused by particles passing through the mid-point of the pair and through the calibration bar, a position calibration was achieved to an accuracy of 2 mm.

3.2 THE INNER WALL

The inner Wall covered the region within 2° of the beam and was finely divided because of the high multiplicity of fragments near the beam. Its main purpose was to form the fast trigger with the upstream beam counter and collimator to decide which events were to be recorded. Of course, it also gave nuclear charge and velocity information for the particles it detected and could be used for centering the beam on the Wall.

The inner Wall consisted of a 6 x 6 array of 36 thick scintillators 8 x 8 cm^2 each, 3.8 cm thick. In front of each was a 0.64 cm thin scintillator. For the center four counters the thin and thick scintillators were of the same area and in one to one correspondence, but for the remaining 32 thick scintillators, only 16 thin scintillators were used, each 8 x 16 cm^2 in area and covering a pair of thick detectors. The 56 scintillators had light guides and phototubes coupled on one edge. In addition, in front of the center four pairs of scintillators was placed one 16 x 16 cm^2 "Bull's Eye" scintillator, only 1 mm thick. The purpose of the bull's eye was to reject beam particles, and it was made thin to minimize nuclear interactions in the detector itself. The Bull's Eye had light guides and phototubes coupled onto two opposite edges.

4. Upstream Beam Counter

The upstream counter is a thin plastic film scintillator with a projected diameter of 2.5 cm and together with an active collimator 3 m upstream defines the beam acceptance of the ball detector system. This counter establishes the timing reference for the fast trigger and also the start signal for the TOF measurement in the Plastic Wall. It was designed to have very low mass and also for optimal light collection in order to achieve good time resolution while minimizing the background from material in the beam. A sandwich of three scintillator films with a combined thickness of roughly 3.6 mg/cm^2 were supported in the beam at 45° on a stretched film of 0.275 mg/cm^2 (2.5 μm) hostaphan. The frame for this is a cylinder of aluminized mylar (125 μm), which doubles as a reflector. Two Phototubes (RCA 8575) view the scintillator from each end of the cylinder. A coincidence requirement between the two tubes eliminates a small background caused by beam halo striking a photocathode. The timing reference is extracted from the analogue sum of the two phototube signals with a constant fraction discriminator. The time resolution achieved between this detector and the inner Plastic Wall is 350 ps. A large area (30 cm x 30 cm) active collimator, which is sensitive to minimum ionizing particles, largely rejects nuclear interactions in the upstream thin film beam counter, which would otherwise be confused with target interactions. Although the rejection of beam in the active collimator is done on-line with hardware, the rejection of minimum ionizing particles is done using its TDC off-line so as not to reject particles coming backwards from the target.

5. Mechanical systems

The geometry chosen for the Plastic Ball made for convenient construction in two hemispheres, permitting easy access to the interior of the Ball for the target installation. The use of thin targets, required to avoid losses at large angles, made the construction of a full vacuum vessel mandatory to ensure that beam particles interact only in the target. Figure 5 shows the inside of the Plastic Ball with the 360 μm thick Am-vacuum chamber and the target wheel. Figure 6 shows the birds eye view of the fully assembled system.

145

Fig. 5. The downstream part of the scattering chamber with the target wheel installed into the forward hemisphere of the Plastic Ball.

Fig. 6. A view of the full system as installed at the Bevalac, looking downstream.

6. Signal processing

6.1 GENERAL LAYOUT

Figure 7 schematically shows the layout of the electronics for the experiment. The event trigger is derived from the beam start counter and from the inner Plastic Wall detectors. Presently, the Plastic Ball is not involved in the trigger decision. Therefore, the signals from all 815 Ball modules are delayed via 70 m cables (RG58) and fed into the Ball Box (see fig. 4) to allow time for the trigger decision, whereas the other counters have the shortest possible cables (RG 58) with a length of only 20 m.

A passive split generates three input signals for the E, ΔE_s, and ΔE ADCs and one for an amplifier. The first three signals are fed via 25 fold 50-ohm ribbon cables to LRS 2282B ADCs. The fourth amplified signal is given to a discriminator and via twisted pair cables to LRS 4291 TDCs with a 10 μs range. The Wall signals are split in the octal constant fraction discriminators (Wall Box), and both analogue and logical outputs are given via 50 m long 8 fold 50 ohm ribbon cables to

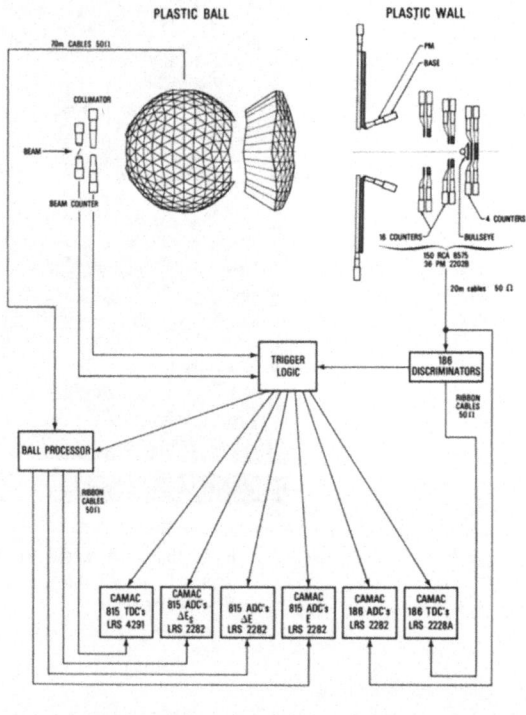

XBL 8110-1504

Fig. 7.   Electronics layout.

ADCs and 100 ns full-scale TDCs. The Inner Wall discriminator outputs are also used
in the trigger logic described below. The Wall discriminator signals to the TDCs
are regenerated just before the TDCs because of attenuation in the long cables.

6.2 TRIGGER

The time reference for the trigger signal is derived from the sum of the two
photomultiplier signals of the start counter.

Beam particles that pass through the active collimator placed 50.5 cm
downstream of the start counter veto the trigger signal. The discriminator on the
summed collimator signal is set high, so that only beam particles, and not back
scattered reaction products from the target, are vetoed.

The simplest trigger for high energy heavy ion studies is to require that the
beam particle underwent a reaction between the beam counter and the Wall counters.
This so-called "minimum bias trigger" is of great importance since all more restric-
tive trigger selections are normalized to it. After detecting a beam particle in
the start counter, one demands that the projectile loses <u>at least</u> one charge.

During the first experiments the "central trigger" used required that no beam
velocity particles of proton pulse height or higher were recorded in the inner Wall.

6.3 GATES

The live trigger signal is used to produce the different gate signals for the
ADCs—the essential part of pulse shape analysis for particle identification--and the
start signals for the TDCs and also to start the data conversion and readout. The
timing for the ADC and TDC gates is shown in fig. 4. The Ball E-ADCs are gated with
prompt 80 ns pulses to allow only the conversion of the short fast rising plastic
scintillator signal. The gates for the two ΔE-ADCs are both delayed by 240 ns and
have lengths of 700 and 1500 ns, respectively. The discriminators in the Ball
processors receive a gate of 10 μs length that is also delayed by 240 ns to suppress
the prompt signal from the plastic scintillator and to produce a signal only for a
delayed μ^+ decay. The Wall ADCs have prompt, rather broad gates (120 ns) to allow
for variations in the time of flight of the reaction products reaching the Wall.

7. On-Line data handling

The torrent of information coming from the Plastic Ball and the Plastic Wall
requires sophisticated early inspection and selection in order to restrict the data
flow to the significant data. "Smart" readout processors were applied in the Camac

branch and a microprocessor was used for other Camac modules. One PDP-11/50 processor was dedicated for data collection and writing on tape while a second PDP-11/44 was used for on-line analysis of the data. All data produced by an event are converted and read out by Camac modules. The E, ΔE, and ΔE_S analogue signals from the Plastic Ball are fed via mass terminated ribbon cables into 48 fold LRS 2282 ADC modules. The μ^+ decay times (from the $\pi^+ \to \mu^+$ decay) are measured over a range of 10 µs with 32 fold LRS 4292 TDCs, and all wall counter times are recorded with 8 fold, high resolution LRS 2228A TDCs. There are a total of 2631 ADC channels, 817 long-range TDC channels, and 188 high resolution TDCs.

Since for each event only a few hundred modules carry valuable information, a "smart" readout system is used to avoid a huge amount of useless data words. It was primarily this requirement that led to the choice of the commercially available LeCroy ADC and TDC systems.

The Camac system is interfaced to the on-line computer with a BIRA microprogrammed branch driver (MBD).

8. Calibration and performance

8.1 π^+ CALIBRATION AT LAMPF

To study the geometrical properties of the modules and to gain information on how well the particle identification scheme would work, an experiment was performed at the Low Energy Pion Beam (LEP) at LAMPF.[11])

An assembly of 13 Plastic Ball modules was placed in the beam defined by two 2 x 2 cm scintillation counters. Different kinds of particles with the same momentum (pions and protons) were selected by setting appropriate windows on the time-of-flight spectrum between those two counters. Thirteen modules were used and arranged such that the central module is completely surrounded by all possible neighbours as shown in the front view in fig. 8. This configuration specifically permits the study of the two dominant effects determining the detection efficiency: a) the scattering out of particles during the slowing down process and b) the detection probability for the positrons stemming from the $\pi^+ \to \mu^+$ decay. (These positrons are emitted isotropically with a maximum energy of 53 MeV and have a high probability of leaving the module). Both effects make it necessary to take information in the neighbouring modules into account in order to reconstruct the event. By having particles impinge on the module in the three different points (center point, side of two modules and corner of six modules), the dependence of the reconstruction efficiency on the geometrical entry point can be determined.

A comparison of the detection efficiencies measured for the three different entry points into the module assembly yields the gratifying result that nearly all scattering out can be reconstructed.

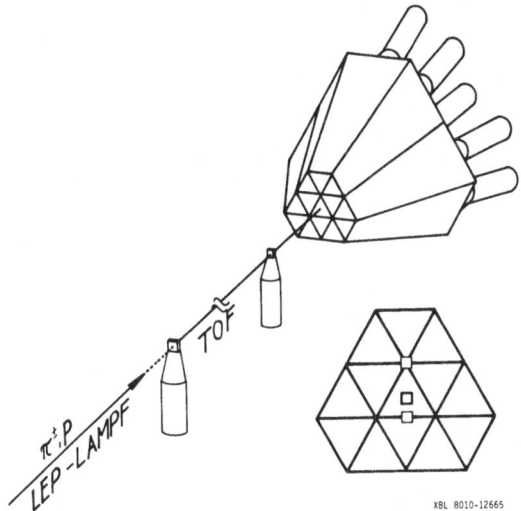

XBL 8010-12665

Fig. 8.   Experimetal setup at the low Energy Pion Channel at LAMPF. The squares in
the righthand front view indicate the three different entry points studied.

A very important task of the test was to obtain energy calibration curves for
pions and protons. For the energies covered by the measurements at LAMPF (low
energy part for protons and high energy part for pions) the light output is
surprisingly linear with energy. This was not expected, since the geometrical light
collection along the particle path as measured on the surface with an electron
source is not constant, the light collection near the CaF_2 crystal being about a
factor of 2 better than near the phototube.

The measured detection efficiency for pions is given in fig. 9. The solid
curve indicates the probability of measuring a decay time in the range between 200
ns and 10 μs. Theoretically, this probability is 90% and is independent of energy.
Due to the positron discriminator threshold at about 4 MeV and the fact that the
continuous positron spectrum extends below that energy, only 80% of the pions can be
tagged at low energies. (The final discriminator setting is at ~1.5 MeV thus
improving this number.) However, this number decreases at higher energies because
a) more pions are lost due to reactions, b) the stop point of the pions is closer to
the end of the scintillator where the light collection is lower (due to the geometry
of the detector) and the probability for positrons escaping the counter system
without giving a detectable signal increases. The dashed line in fig. 9 shows the
measured probability of detecting a pion within the same decay time and also with an
energy within 10% of the incident energy. Both efficiency curves have been
corrected for accidental stops in the TDCs. The number of accidentals can be
determined as the flat background under the exponential decay curve.

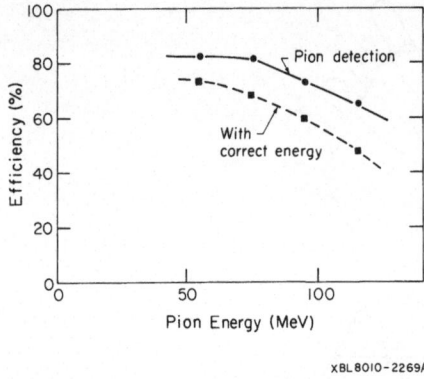

XBL 8010-2269A

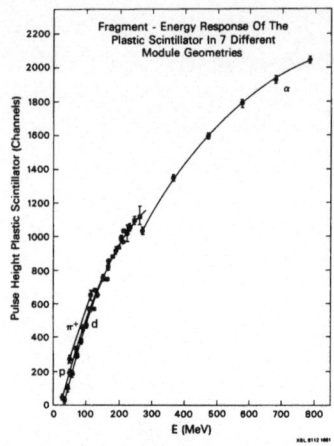

Fig. 9. π^+ detection efficiency
(solid curve) and detection efficiency
with correctly (within 10%) measured
energy (dashed curve).

Fig. 10. Response of the plastic
scintillator in the Plastic Ball
detector geometry to the energy of
π^+, p, d, and ^{4}He.

8.2 CALIBRATION AT THE 184" CYCLOTRON

After assembly each individual module of the Plastic Ball was calibrated at the
Berkeley 184" cyclotron with 800 MeV and 400 MeV α beams. This procedure allowed
the determination of the proper high voltage for each individual photomultiplier and
of an important constant for each module, the ratio of E to ΔE pulse heights at a
given energy. In addition, the energy calibration for the hydrogen and helium
isotopes was obtained by observing the fragmentation products of an 800 MeV α beam
hitting a thick target and by measuring the time of flight of the products in front
of the module. The E calibration for α-particles was obtained by varying the energy
of the incident α beam with moderators in steps of 100 MeV.

The relation between measured pulse-height and deposited energy is shown in
fig. 10 for p, d, ^{4}He, and positive pions. Immediately after the irradiation each
module was cross-calibrated with a calibrated Xe-flasher. After the assembly of the
Plastic Ball each module was calibrated again with the same flasher, and thus all
information from the calibration measurements could be related to the actual
experiment.

8.3 PERFORMANCE

8.3.1 Plastic Ball

The tests performed at the LAMPF low energy pion beam line and at the Berkeley 184" cyclotron with 800 MeV α particles showed that the energy resolution of a single module is sufficient to achieve the desired particle identification. The following energy resolutions (FWHM) were measured: for 75 MeV protons the plastic scintillator (E signal) had a resolution of 5% and the CaF_2 crystal (ΔE signal) of 12% respectively. For the 800 MeV α beam the figures were 2% for the plastic scintillator and 10% for the CaF_2.

The first Bevalac experiment was performed in June 1981 with a 800 MeV/u Ne beam on a Pb target and ^{40}Ca beams at 400 MeV/u and 1.05 GeV/u on a calcium target. From the actual data and from the calibration measurements, correction factors for the ΔE and the E pulse heights could be derived in order to achieve the proper gain matching for all modules. The quality of the particle identification for the 655 modules between 30° and 160° is shown in fig. 11. A cut in the ΔE-E plane perpendicular to the particle identification lines is selected with ΔE and E projected on this cut in a certain energy range. The integrated yield is shown as a function of this projection in fig. 12. The dashed curve represents the raw data after gain-matching, whereas the solid line shows the particle separation after all scattered out particles have been reconstructed by taking into account up to 12 neighboring modules that surround a center module and that contain only E but no ΔE information. Windows were selected along the valleys of those distributions allowing one to assign a mass and charge value with high confidence. For the particular case of 800 MeV/u Ne on Pb and a central trigger configuration, 35% of all detected tracks fall into a particle identification window, 22% of those being stopped in the plastic scintillator and 13% being high energy particles that do not stop in the detector and give only Z information. 13% of all particles have very low energy and stop in the 4 mm thick CaF_2 crystal, where only energy can be measured but no mass or charge. In 47% of all detected tracks only an E but no ΔE signal is present. These are presumably neutral particles (mostly neutrons) interacting in the scintillator; however, scattered out particles unaccounted for in our reconstruction algorithm may also give the same signature. For 5% of the detected tracks a ΔE and an E signal are obtained, but an assignment to a defined particle is not possible. These cases are due to double hits and reaction losses in the scintillators.

In the ΔE-E contour plot (fig. 11) the pion branch coincides partly with the punch through hydrogen branch and pions can only be identified by taking information from the decay measurement into account. Figure 13 shows the decay curve for the identified positive pions. The solid curve is a fit of the experimental decay with the decay constant of 2.2 μs and a constant background that indicates that only 3% of all pions are misidentified.

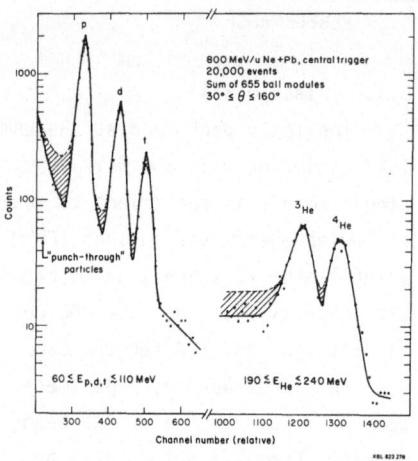

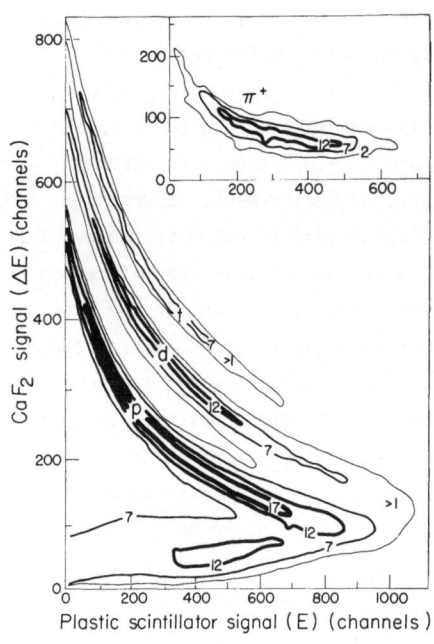

Fig. 11. ΔE-E contour diagram of 655
modules after gain matching and
scattering out reconstruction of
approximately one million reaction
products. The contour lines are
labeled as to their relative height in
arbitrary units. The upper contour
diagram for π^+ is obtained by
requiring a delayed decay signal
measured in the 10 μs TDCs.

Fig. 12. Particle identification
spectrum for 655 modules after
gain-matching and with and without
scattering out reconstruction.

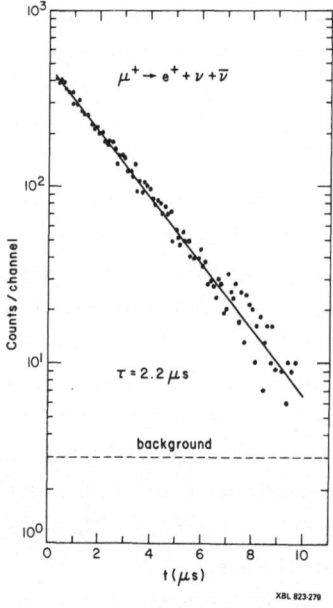

Fig. 13. Time spectrum of delayed
coincidences of 655 detectors. The
straight line is a best fit yielding a
proper decay time of 2.2 μs with a 3%
background.

8.3.2 Plastic Wall

Since the Plastic Wall can detect beam velocity particles the time calibration is taken from each individual experiment. In the 800 MeV/u ^{20}Ne on Pb experiment a time resolution of ~350 ps was observed between the inner Wall detectors and the beam counter. The pulse height resolution of the Inner Wall was good enough to have Z-identification up to the projectile Z of ten. For the Outer Wall, where dominantly hydrogen and helium fragments are detected, a pulse height resolution of ~20-25% was good for an excellent separation of hydrogen, helium, and lithium.

These performance figures of the Ball and the Wall allow one to draw a figure for the system's response as a function of fragment momentum and rapidity. Figure 14 shows the response of the Ball and the Wall for protons or α-particles. The curves for deuteron, tritons and ^{3}He are slightly different, with deuterons and tritons lying at smaller $p_\perp$ values and ^{3}He at larger $p_\parallel$ values than for the protons.

9. Physics with 4π data

The beauty in 4π data is that nearly the whole physics is contained in each event. But as with so many (sleeping) beauties, thorny hedges surround them and huge efforts are needed to set them free.

Most of the early attempts with AgCl and emuslion detectors used the 4π feature to extract angular distributions for specific event selections. The streamer chamber groups went one step further and studied π⁻–π⁻ (two particle) correlations to learn something about the size of the reaction volume. However, visual detectors have their strength in showing sparkling stars that strongly stand out from the average.

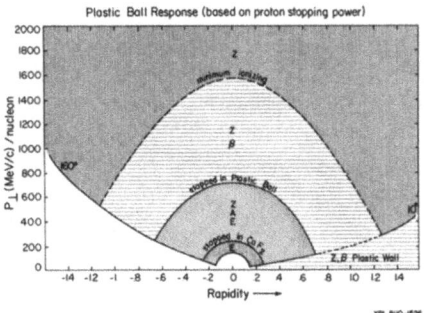

Fig. 14. Plastic Ball and Plastic Wall response in the transverse momentum $p_\perp$ versus rapidity y plane. The limits labeled "stopped in CaF$_2$" and "stopped in Plastic Ball" refer to protons and ^{4}He fragments; those for d, t are lower but somewhat higher for ^{3}He fragments.

Here, using the Plastic Ball data, a few attempts will be discussed to study one of the most fundamental question of relativistic heavy ion physics, "<u>Do relativistic nuclear collisions allow to study the equation of state of nuclear matter?</u>"

Because in single particle inclusive data the averaging occurs over impact parameter and multiplicity, all viewpoints on the reaction mechanism extracted so far are questioned until found in agreement with these new 4π data.

We will focus onto three subjects and their beauties (in parentheses)

1) Cluster formation (Entropy)
2) Thermalization   (Temperature)
3) Flow   (Density)

9.1 CLUSTER FORMATION

Recently it was pointed out that the entropy production can be related to the cluster production, especially to the ratio of deuteron-to-proton yield[12]), i.e., the larger this ratio the smaller the entropy produced in the reaction. This has stimulated a large dispute in the theory community[13-18]) of whether this is at all possible or how to construct the right observable from cluster production cross sections. If one really could measure in this way the entropy then one could look for the predicted change in entropy as the dense nuclear matter undergoes, as it were, a phase transition.

This focuses onto the mechanism of cluster production. In one of our early theoretical adventures we formulated the nuclear coalescence model to describe the observed cluster production.[19]) Basically, it states that two nucleons have to be close in phase space to have a chance to coalesce. In the first formulation the nucleons had to be close only in momentum space but there is of course also a requirement that they are close in configuration space to interact in one short time interval. With $\vec{p}_1$ and $\vec{p}_2$ the momenta of particles 1 and 2 and $\vec{r}_1$ and $\vec{r}_2$ the coordinates in configuration space, one requires for coalescence

$$|\vec{p}_1 - \vec{p}_2| < p_0 \quad \text{and--usually neglected!--} \quad |\vec{r}_1 - \vec{r}_2| < 2r_0$$

$$|t_1 - t_2| < \Delta t_0$$

Then $\dfrac{\sigma_{deuteron}}{\sigma_{proton}^2} = \dfrac{4\pi}{3} p_0^{\,3}$.

If we assume a thermal equilibrium (as in a fireball) then Mekjian[20]) has shown that the volume V of the emitting source can be obtained via

$$\frac{4\pi}{3} p_0^{\,3} = \frac{8}{\gamma} \frac{(2\pi)^3}{V}$$ where γ is the Lorentz factor of the emitted particle in the center of mass frame of the fireball (at the energies discussed $\gamma \approx 1$). From single particle inclusive data[21]) this volume seems to increase as the mass of the target

increases but, e.g., for ^{20}Ne + Pb does not change with bombarding energy. We can now employ for the first time 4π data in this context. Because of the Plastic Ball's particle identifying feature, one can study the cluster production as a function of event multiplicity in each event. Furthermore, a comparison is possible of the deuteron to proton ratio versus the deuteron-like to proton-like ratio as suggested in ref. 5. with:

$$n_{deuteron-like} = n_d + \frac{3}{2} (n_t + n_{3_{He}}) + 3 n_{4_{He}}$$

$$n_{proton-like} = n_p + n_d + n_t + 2 (n_{3_{He}} + n_{4_{He}})$$

For the investigation of entropy it is of importance to compare protons and clusters originating from the same volume of phase space. It has been reported[22] that in the ^{4}He spectrum (e.g., 2.1 GeV/u ^{20}Ne + Au), there is a large cross section of up to 13b at very low α-energies due to an α-particle production at a very late state in the reaction when the fast particles have left the reaction zone. It is our opinion that this late state has to be excluded in this study and to do this in an adequate way a low-energy threshold has to be applied to the data. We have chosen for all particles a threshold of E/A = 40 MeV. Such a low energy cutoff should be incorporated into theoretical comparisons since values obtained this way for ratios of deuteron-like/proton-like are lower than those extracted using the full spectrum. Furthermore, this study limits data to emission angles of $9° \leq \theta \leq 160°$ ignoring the small yield of particles going into 0° to 9°. This forward region contains dominantly projectile-spectator residues for large impact parameter reactions. As we will stress the high multiplicity events, this region is not affecting the results.

Figure 15a shows the deuteron-to-proton ratio for Ne + Pb at 800 MeV/u as a function of the charged particle multiplicity M_c in the event. There is a steep rise of d/p up to a value of 0.3. These data point out that the deuteron production increases the more nucleons are involved in the reaction. Figures 15b and 15c show for the same reaction the ratio of deuteron-like to proton-like particles. Notice the larger values compared with d/p and a more pronounced saturation at high M_c. Figures 16a and 16b represent these ratios for the reaction of 400 MeV/u ^{40}Ca on Ca.

Can the increase in deuteron production with increasing multiplicity be understood in terms of the current models for cluster production? In the thermal model assuming chemical equilibrium[23] the ratio of d/p was independent of the size of the thermal source, and d/p as a function of M was expected to be constant. In the original coalescence model[19] the size of the participant volume was neglected and the coalescence volume in momentum space was taken as proportional to the inverse of the size of the deuteron. This predicted that d/p^2 as a function of multiplicity (not d/p) was expected to be constant. Jennings et al.[24] have shown the practical equivalence of the coalescence and thermal models. If one

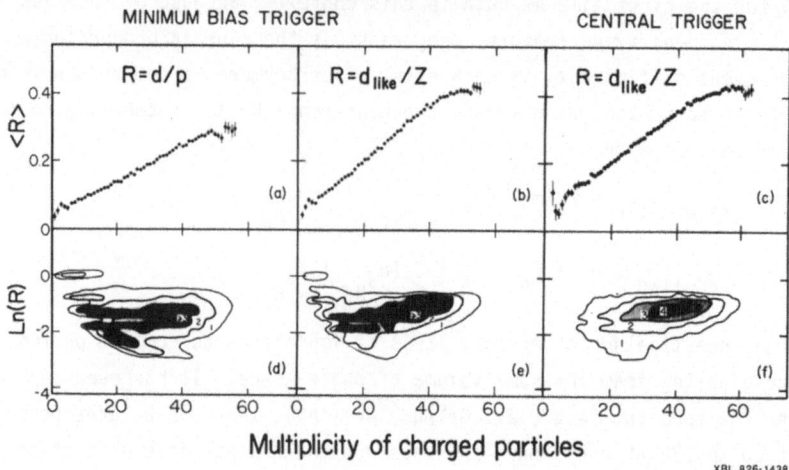

Ne+Pb E/A=800 MeV

MINIMUM BIAS TRIGGER

CENTRAL TRIGGER

$R = d/p$

$R = d_{like}/Z$

$R = d_{like}/Z$

Multiplicity of charged particles

XBL 826-1438

Fig. 15. a,b,c: Ratio of deuteron to proton production (a) and ratio of deuteron-like to proton-like particles with two different trigger conditions (b,c) as a function of the total observed charged particle multiplicity for the reaction 800 MeV/nucleon Ne + Pb
 d,e,f: Corresponding event by event contour plots of the logarithm of the ratio versus the charged particle multiplicity. Relative intensities are indicated by the contour lines.

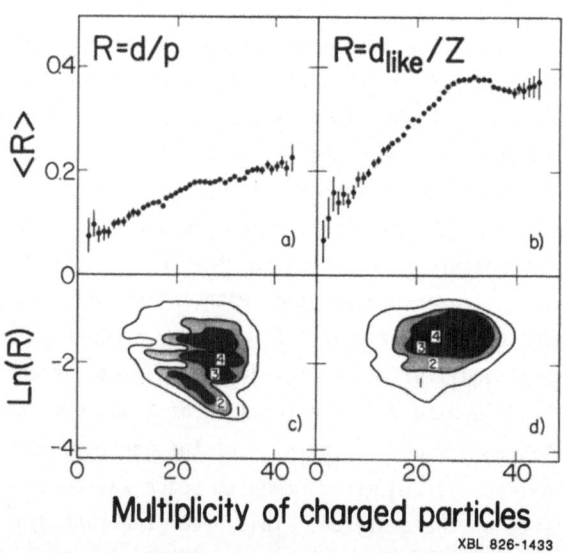

Ca+Ca E/A=400 MeV
CENTRAL TRIGGER

$R = d/p$

$R = d_{like}/Z$

Multiplicity of charged particles

XBL 826-1433

Fig. 16. Same as fig. 15 but for the reaction 400 MeV/nucleon Ca + Ca.

introduces into the coalescence model the volume of the deuteron in configuration space and integrates over the number of particles, this produces a term proportional to the volume of the participants as in the thermal model. Recently, Sato and Yazaki[25]) have formulated a model taking into account both the size of the deuteron and the volume of the participants. The coalescence radius p_0 in momentum space is related to the deuteron radius r_d and that of the participant volume r_p via

$$\frac{d_{like}}{p^2_{like}} = \frac{4\pi}{3} p_0^3 \propto \frac{1}{(r_d^2+r_p^2)^{3/2}} \quad .$$

The deuteron radius r_d is an average over the clusters used in the definition of d_{like}. The radius r_p of the participant volume can be related to the observed charges by

$$r_p = r_0 (2 p_{like})^{1/3}$$

with r_0 a free parameter (since the participants are most probably not in a compact sphere). For the d_{like} to p_{like} ratio one finds

$$\frac{d_{like}}{p_{like}} \propto \frac{p_{like}}{(r_d^2+r_0^2(2p_{like})^{2/3})^{3/2}} \quad .$$

Figure 17 shows the ratio d_{like}/p_{like} as a function of p_{like}. The solid line is a fit to the data drawn with $r_d/r_0 = 2.6$. By neglecting one or the other of the terms in the denominator one can obtain both limiting behaviors described above. If the volume of the deuteron were neglected the line would be horizontal; on the other hand, if the volume of the participants were neglected, the line would be linear through the origin. Therefore, to extract valuable information on entropy one has to know the variation with multiplicity. Entropy values from single particle inclusive data may be meaningless because they integrate over all impact parameters.

Data with a central trigger are shown in fig. 15c for comparison with the minimum bias data (fig. 15b). They practically overlap proving that the central trigger is not producing a special bias in the cluster production but only an enhancement of high multiplicity events.

Figures 15d,e,f and 16c,d show the ratio of d/p and of d_{like}/p_{like} in each event as a contour plot. Each plot contains approximately 140 000 events. Focusing on the central trigger data (Figs. 15f and 16d), one observes a depletion of low multiplicity data as expected. A comparison of the contours in d/p versus those in d_{like}/p_{like} shows a narrower distribution for d_{like}/p_{like}, supporting the choice of these variables in the analysis. The largest ratios are almost as large as those that would be calculated by Bertsch and Cugnon[15]) for zero impact parameter collisions, but it must be remembered that the 40 MeV/u cutoff has not been put in the theoretical calculation. However, these authors predict a decrease in entropy with decreasing impact parameter, which in turn would predict an increase

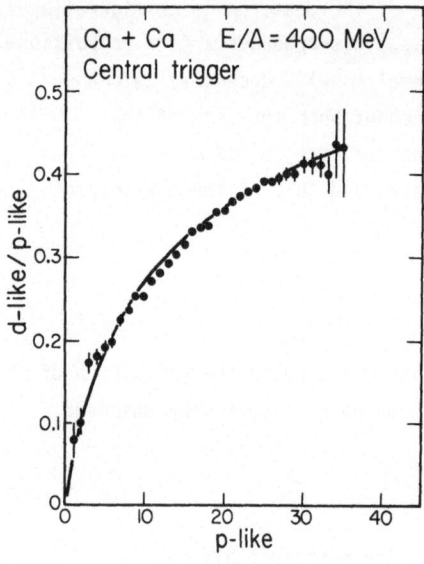

XBL 829 – 1162

Fig. 17. Ratio of the number of deuteron-like to proton-like particles as a function of the number of proton-like particles. The solid curve represents a fit to the data.

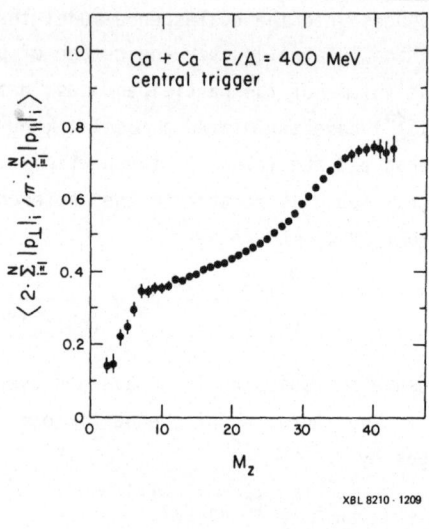

XBL 8210 · 1209

Fig. 18. Ratio of $2 \cdot \sum |p_\perp|_i \Big/ \pi \cdot \sum |p_\parallel|_i$ as a function of nucleonic charge multiplicity averaged over 50 000 events. A value of 1 would correspond to a completely thermalized event.

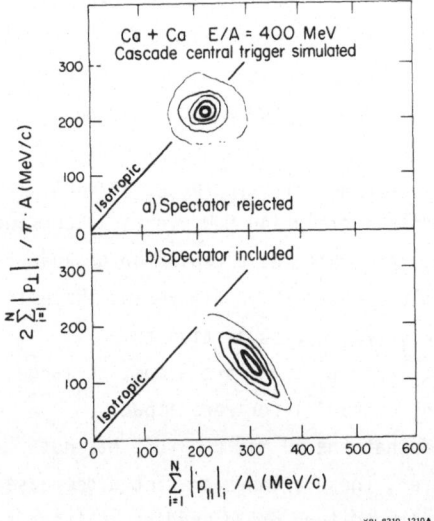

XBL 8210 1210A

Fig. 19. Contour plot of $<p_\perp/A>$ versus $<p_\parallel/A>$ for a cascade calculation Ca on Ca at 400 MeV/u with the Plastic Ball/Wall acceptance and a central trigger simulated. The diagonal straight line indicates where fully thermalized events are expected. In a) the spectator matter is rejected, while it is included in b).

in cluster production with increasing multiplicity, as observed in our data. In ref. 5 this increase of entropy is explained by the smearing of the participant volume by the mean free path of the nucleons, which may be equivalent to including the finite size of the deuteron. In fact, Biro et al.[17] found they had to include the volume of the deuteron in their thermal model.

9.2 THERMALIZATION

One huge step towards the equation of state would be a clear signal of thermalization in the primary reaction zone. In the late state of the reaction an apparent temperature of ~20 MeV seems to be clearly established and is discussed in association with the boiling or condensation point of nuclear matter.[22] The nuclear fireball was introduced with the assumption of total thermalization among the participants whereas the hydrodynamical model[26] assumed local thermal equilibrium. A crucial question is that of the mean free path of nucleons in these regions of high particle densities. Thermalization in the reaction is characterized by the effect that the originally longitudinal energy is randomized over longitudinal and transverse degrees of freedom. The degree of thermalization can be expressed by the ratio $2 \cdot \sum_{i=1}^{N} |p_\perp|_i \bigg/ \pi \cdot \sum_{i=1}^{N} |p_\parallel|_i$. This ratio is shown in fig. 18 as a function of multiplicity for the reaction of 400 MeV/u ^{40}Ca on Ca. The ratio is increasing with multiplicity but at least at this energy and target-projectile combination never reaches the value 1, which would be a necessary criterion for thermalization. Cascade calculations with the code of Yariv and Fraenkel[27], however, show that even a few spectator fragments deform a spherical event in momentum space into a prolate spheroid (fig. 19). From that it is obvious that it is necessary to find a reliable method that is able to distinguish between participants and spectator matter.

9.3 GLOBAL ANALYSIS

The global analysis methods (sphericity, thrust) were developed as a tool to detect and distinguish predicted two jet events at high energy e^+e^- storage rings from events with spherically symmetric emission patterns[28]. Both methods define a jet axis, the sphericity by minimizing $\sum_i p_{i\perp}^2|$ and the thrust by maximizing $\sum |p_{i\parallel}|$ relative to this axis. The sphericity is calculated analytically by diagonalizing the sphericity tensor. One obtains the orientation of the sphericity axis, e.g. relative to the beam direction and three eigenvalues which define an ellipsoid that describes the shape of the event. The thrust analysis yields in addition to the orientation only the magnitude of the thrust, a quantitative measure distinguishing between isotropic and back-to-back emission.

The use of global methods to analyze the more complex events from heavy ion collisions was proposed by several authors[29,30]). Sphericity (p^2) overweights leading particles and gives two nucleons a different weight from a deuteron with the same energy per nucleon. Corrections for these shortcomings have been proposed[30]), e.g. the flow analysis[31]). Since a global analysis has to be performed for each event, statistical fluctuations due to finite number effects and to limited experimental acceptance and efficiency are expected. Experimental data have to be compared with results from an analysis of theoretically calculated events, which have been filtered for experimental acceptance and efficiency. Most theoretical models have not yet reached the sophistication of the experimental equipment in the sense that they are not able to calculate all the measured quantities. Cascade codes do not include composite particles and hydrodynamical codes do not produce event-to-event fluctuations. This makes the comparison between experiment and theory difficult. Complete events generated with a statistical model calculation by Randrup and Fai[32]) will be extremely useful to study the effect of finite number fluctuations and experimental biases. As global analysis allows to determine the shape of the event in phase spaces and takes into account all the measured correlations, it should be well suited to distinguish between emission patterns as predicted by the hydrodynamical and the cascade model. Figure 20 shows the result of a flow analysis of 400 MeV/u Ca on Ca data. The angle of the main axis of the flow ellipsoid is plotted versus the square of the ratio of the largest to the smallest axis. A comparison of those results with theoretical predictions can be only done after the calculated events have been filtered for the experimental acceptance. This procedure is presently being developed for cascade events.

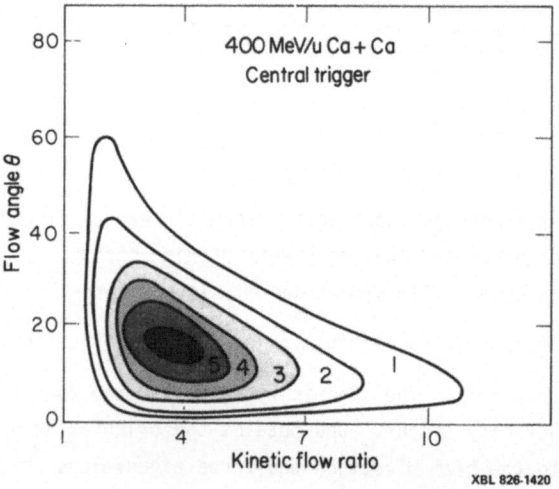

Fig. 20. Flow plot for 400 Mev/u Ca on Ca.

10. Conclusions

Through the available 4π detector data we have shown that the deuteron to proton ratio increases with increasing charge particle multiplicity. This can be explained by phase space considerations taking both the size of the participant volume and the finite size of the deuteron cluster into account. These findings should explain the target-projectile mass dependence of d/p values but shed serious doubts on entropy discussions based on single particle inclusive data which average over impact parameter. From observables such as the ratio of transverse to longitudinal momentum the degree of thermalization can be measured. If equilibrium is found, the concept of temperature can be applied and compared to predictions of various models. Temperature, together with a better understanding of the entropy values, could lead to a big step forward towards the establishment of the equation of state of nuclear matter.

This work was supported in part by the Director, Office of Energy Research, Division of Nuclear Physics of the Office of High Energy and Nuclear Physics of the U.S. Department of Energy under Contract DE-AC03-76SF00098.

References

1) A. Sandoval, H.H. Gutbrod, W.G. Meyer, A.M. Poskanzer, R. Stock, J. Gosset, J.-C. Jourdain, C.H. King, G. King, Ch. Lukner, Nguyen Van Sen, G.D. Westfall, and K.L. Wolf, Phys. Rev. C21 (1980) 1321.
W.G. Meyer, H.H. Gutbrod, Ch. Lukner, A. Sandoval, Phys. Rev. C22 (1980) 179.
A.I. Warwick, A. Baden, H.H. Gutbrod, M.R. Maier, H.G. Ritter, H. Stelzer, F. Weik, H.H. Wieman, S.B. Kaufman, B.D. Wilkins, E.P. Steinberg, J. Peter, LBL-13831, to be published.
2) J.C. Tompkins, SLAC Report No. 224 and SLAC Report No. 578 (1980) unpublished
E.D. Blum, Proceed. of the 9th International Symposium on Lepton and Photon Interactions at High Energies, Batavia, 1979, edited by T. Kirk and H. Abarbanel, Fermilab, 1980
M. Oreglia, Ph.D. thesis, SLAC Report No. 236 (1980), unpublished
3) D.H. Wilkinson, Rev. of Scient. Inst. 23 (1952) 414
D. Bodansky and S.F. Eccles, Rev. of Scient. Inst. 28 (1957) 464
4) M. Mayhugh of Harshaw Chem. Co., Solon, Ohio, proposed the use of CaF_2(Eu)
5) C.I. Industries, Tokorozawa, Japan, plastic scintillator similar to NE114.
6) PM 2202 AMPEREX
7) Crofon 1410 (Dupont) light fiber
8) J. Peter, LBL internal report 1979, unpublished
9) R. Maday and W. Schimmerling, private communications
10) Silicon PIN photo diode, SGD-100A, EG&G Electro optics, 35 Congress St., Salem, MA 10970
11) H.H. Gutbrod, M.E. Maier, H.G. Ritter, A.I. Warwick, F. Weik, H. Wieman, and K.L. Wolf, IEEE Transactions on Nuclear Science, Vol. NS-28, No. 1, February 1981
12) P. Siemens and J. Kapusta, Phys. Rev. Lett. 43 (1979) 1486
13) H. Stöcker, LBL-12302 (April 1981)
14) J. Knoll, L. Münchow, G. Röpke, and J. Schulz, Phys. Lett. 112B (1982) 13
15) G. Bertsch and J. Cugnon, Phys. Rev. C24 (1981) 2514; and J. Cugnon, Proc. of ICOSAHIR, Nucl. Phys. A387 (1982) 191c
16) S. DasGupta, B.R. Jennings, and J.I. Kapusta, Phys. Rev. C26 (1982) 274

17) T. Biró, H.W. Barz, B. Lukács, and F. Zimányi, to be published
18) R.K. Tripathi, Phys. Rev. C25 (1982) 1114
19) H.H. Gutbrod, A. Sandoval, P.J. Johansen, A.M. Poskanzer, J. Gosset, W.G.
    Meyer, G.D. Westfall, and R. Stock, Phys. Rev. Lett. 37 (1976) 667
20) A.Z. Mekjian, Phys. Rev. C17 (1978) 1051
21) S. Nagamiya, M.C. Lemaire, E. Moeller, S. Schnetzer, G. Shapiro, H. Steiner,
    and I. Tanihata, Phys. Rev. C24 (1982) 971
22) H.H. Gutbrod, A.I. Warwick, and H. Wieman, Proc. of Maria Workshop, Banff,
    Canada, and Proc. of ICOSAHIR, Nucl. Phys. A387 (1982) 177c
23) J. Kapusta, Phys. Rev. C21 (1980) 1301
24) B.R. Jennings, S. DasGupta, and N. Mobed, Phys. Rev. C25 (1982) 278
25) H. Sato and K. Yazaki, Phys. Lett. 98B (1981) 153
26) H. Stöcker, W. Greiner, and W. Scheid, Z. Physik A286 (1978) 121
    H. Stöcker, J.A. Maruhn, and W. Greiner, Phys. Lett. 81B (1979) 303
27) Y. Yariv and Z. Fraenkel, Phys. Rev. C20 (1979) 2227
28) S. Brandt and H. Dahmen, Z. Physik C1 (1979) 61
29) J. Kapusta and D. Strottman, Phys. Lett. 106B (1981) 33
30) J. Knoll, Proceedings 5th High Energy Heavy Ion Study, LBL-12652 (1981)
31) M. Gyulassy, K.A. Frankel, and H. Stöcker, Phys. Lett. 110B (1982) 185
32) J. Randrup and G. Fai, Phys. Lett. 115B (1982) 281

THE DARMSTADT-HEIDELBERG-CRYSTAL-BALL

V. Metag

II. Physikalisches Institut, Universität Gießen

D. Habs, K. Helmer, U. v. Helmolt, H.W. Heyng, B. Kolb,

D. Pelte, D. Schwalm

Physikalisches Institut, Universität Heidelberg

W. Hennerici, H.J. Hennrich, G. Himmele, E. Jaeschke,

R. Repnow, W. Wahl

Max-Planck-Institut für Kernphysik, Heidelberg

R.S. Simon, R. Albrecht

Gesellschaft für Schwerionenforschung, Darmstadt

In this talk the properties of the crystal-ball spectrometer installed at the Heidelberg tandem-postaccelerator facility are described. This detector system is a joint project of Gesellschaft für Schwerionenforschung, Darmstadt, Physikalisches Institut Universität Heidelberg, and Max-Planck-Institut für Kernphysik, Heidelberg. The spectrometer went into operation in spring 1982 after a three years' design, test, and installation period.

The main aim of this new instrument is to study nuclei at high excitation energies and angular momenta by detecting all electromagnetic radiation emitted in the collision process. New insight into the internal structure of such nuclei is expected since this novel spectrometer combines two rather successful experimental approaches, the measurement of the total energy and multiplicity of the emitted γ-radiation. Furthermore, it allows for the first time the simultaneous measurement on an event-by-event basis of all parameters characterizing a γ-ray cascade. From the scientific program discussed at the end of the talk it becomes evident that the applicability of this instrument is not limited to nuclear structure investigations. In combination with particle detectors it opens up new ways of studying nuclear reaction mechanisms and also allows the investigation of special problems in atomic physics, e.g. studies of the radiation from superheavy atoms.

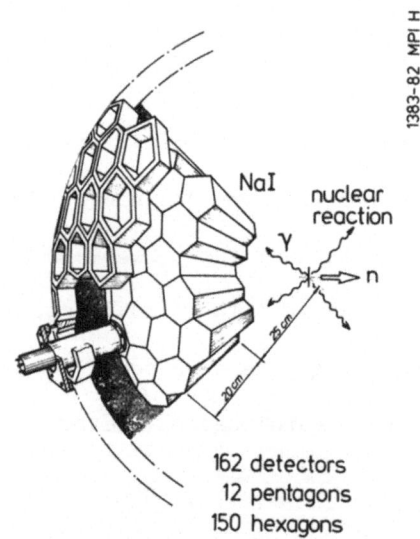

162 detectors
12 pentagons
150 hexagons

Fig. 1: Schematic diagram of a section of the crystal-ball showing the geometrical configuration of the modular NaI-shell

The configuration of the crystal-ball is shown in fig. 1. The γ-radiation emitted from a source in the center, mostly a nucleus highly excited in a nuclear reaction, is absorbed in a spherical shell of NaI-detectors which have a thickness of 20 cm. This allows the measurement of the sum energy which is related to the total excitation energy of the nucleus at the beginning of the cascade. In addition we are interested in the multiplicity, i.e. the number of γ-rays per event, in the energy and angular distribution of the individual γ-rays as well as in their correlation in time. To obtain this information the spherical shell of NaI is subdivided into 162 independent detectors of equal solid angle, 12 pentagons, and 150 hexagons.

The crystal-ball can be operated in a stand alone mode but also in coincidence with additional detectors. Any individual NaI module can be removed and replaced by e.g. high resolution Ge-detectors. Particle counters can be mounted in a spherical scattering chamber of 48 cm in diameter inside the NaI shell. The photograph in fig. 2 shows a side view of the crystal-ball and its support structure. To allow access to the target the spectrometer is divided into two hemispheres which can be moved appart vertically to the beam direction.

fig. 2: Side view of the crystal-ball
and the support structure

The number of detector modules is a compromise between desirable resolution in multiplicity and financial constraints. The fact that it is precisely 162 is related to the interesting mathematical problem of dividing a sphere into a large number of segments with equal solid angle and as few as possible geometrical shapes. Fig. 3 illustrates the underlying construction principle[1]. Starting from a dodecahedron the pentagonal surfaces are divided into triangles, and the resulting triangles into even smaller ones. In contrast to the design of the plastic ball[2] the triangles are again recombined to pentagons and hexagons. This procedure is subsequently repeated several times and one obtains a progressively finer subdivision of the full solid angle. Such a construction principle can only lead to polyhedra with

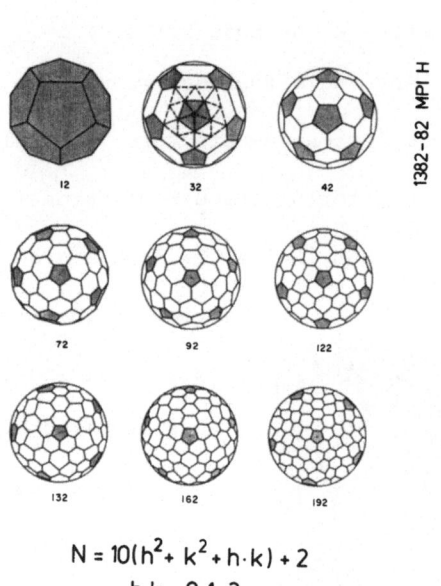

$$N = 10(h^2 + k^2 + h \cdot k) + 2$$
$$h,k = 0, 1, 2 \dots$$

fig. 3: Polyhedra with N surfaces derived from a pentagonal dodecahedron by subsequent application of the construction principle described in the text

specific numbers of surfaces given by the formula in fig.3. Since a number of modules between 100 and 200 appears to be optimal under the given constraints, as will be shown next, a configuration with 162 detectors has been chosen.

The parameters to be measured with the crystal-ball and the corresponding resolutions are listed in table I. The energies and times of the
individual γ-rays are measured with resolutions close to what is techni-
cally feasible. A good time resolution is essential to establish which
γ-rays belong to a given cascade and to disting-uish neutrons from γ-
rays by time of flight

table I:
parameters and resolutions (FWHM)

E_i	7.8 % at 662 keV 5.5 % at 1332 keV
t_i	2.8 ns
$W(\theta_i)$	14°
$\langle n_{isol}\rangle$	6.7 for $M_\gamma \simeq 20$
E_{tot}	18 - 22 % for $M_\gamma = 20$
M_γ	25 - 30 % for $M_\gamma = 20$

using a pulsed beam. This separation is illustrated in fig. 4 which
shows time spectra for detectors at 3 distinguished angles with different neutron yields, as expected from the kinematics of a fusion reaction.
The accuracy with which angles can be defined is $\pm 7°$, determined
by the size of the detectors.

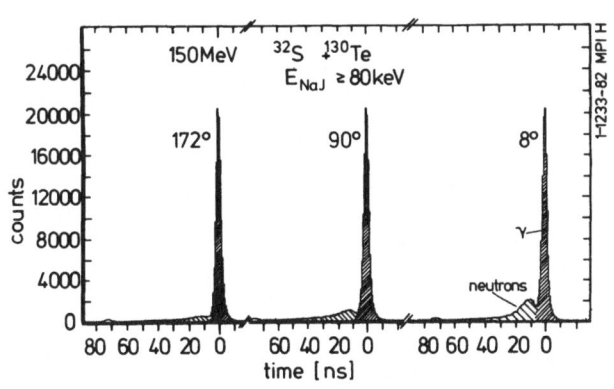

fig. 4: Time distribution of events relative
to the beam pulse, recorded with NaI detectors at 3 selected angles

The crystal-ball can be operated as a 162-fold anti-Compton spectrometer by selecting events where certain detectors but none of their neighbours have fired. Fig. 5 demonstrates that for a total of 162 detectors one observes on the average 6.7 isolated hits of this kind for a γ-ray cascade with 25 responding modules. Combinatorial considerations[1,3] and Monte Carlo simulations show that the average number of isolated hits decreases rapidly when the number of segments is reduced. It becomes ≤ 1.5 for a detector system with only 70 elements. This illustrates the advantages of a crystal-ball with 162 modules for the study of γ-ray energy correlations which has been shown to be a powerful technique[4] to reveal changes in nuclear structure with increasing rotational frequency.

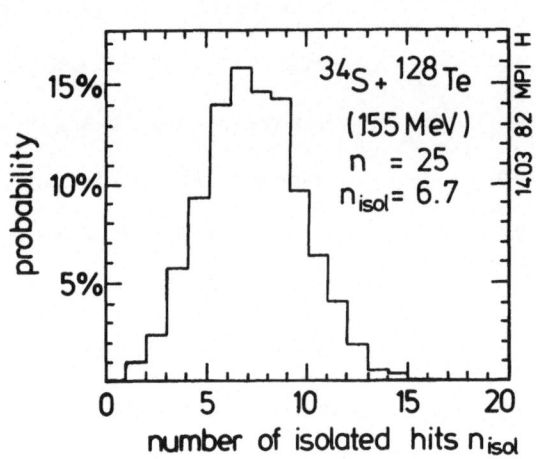

Fig. 5: Probability of isolated hits for γ-ray cascades with 25 responding detectors, observed in the ^{34}S + ^{128}Te reaction at 155 MeV

The improvement in the spectral response of the NaI detectors by using the anti-Compton option is illustrated in fig. 6 which shows the line shapes for standard two line γ-ray sources. The photopeak of one γ-ray is detected in a Ge- counter and the second γ-ray is registered in the crystal ball. After appropriate gain matching the spectra of all individual NaI modules are sorted into one spectrum. The peak-to-total ratios deduced from the response functions in fig. 6 are summarized in fig. 7. As illustrated, the peak-to-total ratio can be further improved by selecting events where one and only one detector has fired or by adding up the energies in all detectors.

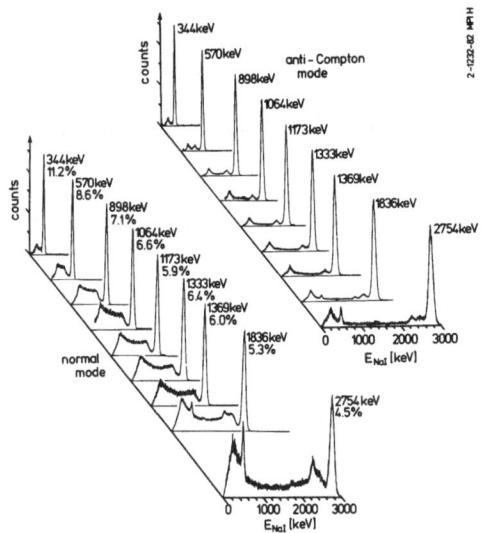

fig. 6: Response function of the individual NaI detectors operated in normal and anti-Compton mode, respectively.

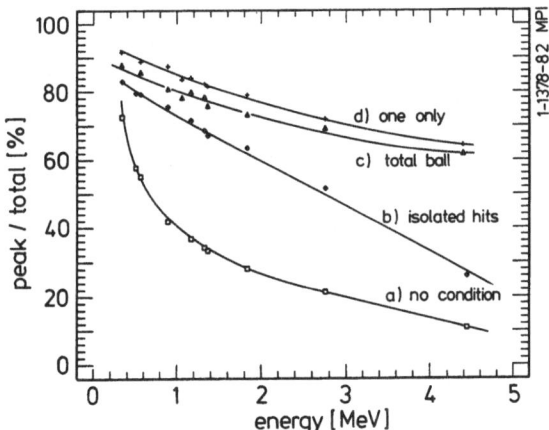

fig. 7: Peak-to-total ratios in four different modes: a.) no condition on pattern of responding detectors, b.) only isolated hits are accepted, c.) energy deposited in all crystals is added up, d.) only events with one and only one detector responding are accepted.

The most important parameters measured by the crystal-ball are total energy and multiplicity. The resolution of the total energy is determined by the probability to absorb the γ-radiation in the NaI-shell. The probability that at least one module of the crystal ball responds when a γ-ray is emitted from a source in the center is plotted in fig. 8a as a function of the γ-ray energy. This probability has a maximum of 99 % at 300 keV and drops to 94 % at 2 MeV. The experimental efficiencies are 0.5-1.0 % smaller then expected from the probabilities of transmission of the γ-rays through the detectors of given geometry, measured in a separate experiment. This difference is attributed to inactive gaps inbetween the detectors. The average energy $<E_{CB}>$ registered in the crystal-ball relative to the energy E_{γ} of the incoming γ-ray is plotted in fig. 8 as a function of E_{γ}.

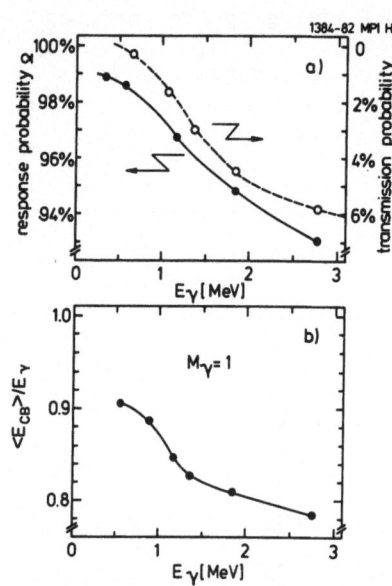

fig. 8: a.) The probability that at least one detector of the crystal-ball fires (threshold > 80keV) when a γ-ray is emitted from the center, and the probability that a γ-ray penetrates a NaI module without interaction, plotted versus the γ-ray energy. b.) the average energy registered in the crystal-ball relative to the energy of the incoming γ-ray as a function of the γ-ray energy.

The resolution of the total energy for higher multiplicity events has been determined by combining in the off-line analysis multiplicity 1 events to large events of any given multiplicity[5]. The resulting total energy resolution, defined as the FWHM of the response curve relative to the average energy registered in the crystal-ball is given in Fig. 9 as a function of the γ-ray multiplicity. Depending on the energy of the γ-radiation the resolution of the total energy varies between 12 % and 23 % for typical multiplicities of M_γ = 20-30.

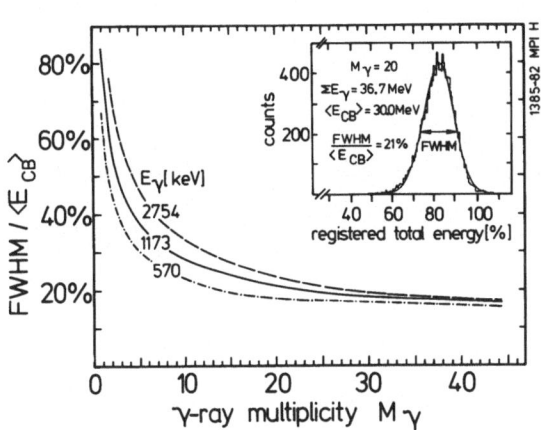

Fig. 9: The FWHM of the total energy distribution divided by the average energy registered in the crystal-ball plotted as a function of the multiplicity for events with M_γ γ-rays of 2754, 1173, and 570 keV, respectively. Inset: Distribution of the total energy registered in the crystal-ball in percent of the incoming energy for events with 20 γ-rays of 1836 keV.

The multiplicty resolution is mainly determined by two opposing effects. If the number of modules in the crystal-ball is not appreciably larger than the γ-multiplicity M_γ of the event the number of responding detectors will be considerably smaller than M_γ because of multiple hits, i.e. two or more γ-rays enter one and the same detector. This loss can be reduced by increasing the number of modules. On the other hand, the number of responding detectors becomes larger than M_γ due to Compton cross-talk between the detectors, i.e. one γ-ray triggers not only one detector but after Compton scattering also a neighbouring one. The distortion due to Compton scattering depends on the geometrical shape and size of the detectors. By choosing hexagon and pentagon shapes there are always only 3 detectors next to each other instead of 6 in case of triangular shapes, which automatically reduces the cross-talk. The probability to completely absorb a γ-ray in the same detector after a first Compton scattering is the lower the smaller the detector; i.e. for a given volume of NaI scintillator material the subdivision into independent elements should not be too fine, the number of detectors not too high. After studying these two opposing effects in the design phase[1,3] of the project a number of elements between 100 and 200 turned out to be optimal. Together with the number of isolated hits per cascade and the geometrical constraints discussed above this result led to a configuration with 162 elements.

The number of responding detectors per γ-ray is shown in fig. 10a. Due to Compton cross-talk up to 1.5 detectors fire on the average for one γ-ray emitted, depending on the γ-ray energy. This number is reduced for increasing multiplicities because of multiple hits. Fig. 10b shows the actual number of responding detectors versus the true γ-ray multiplicity for 3 different γ-ray energies. While for γ-rays of 2754 keV the number of responding detectors exceeds the true multiplicity by ≈ 30% this deviation is about 20% for γ-rays of approximately 1.2 MeV which are typical transition energies in rapidly rotating deformed nuclei. For 570 keV this excess amounts to ≈ 10%. In fig. 11 the multiplicity resolution,

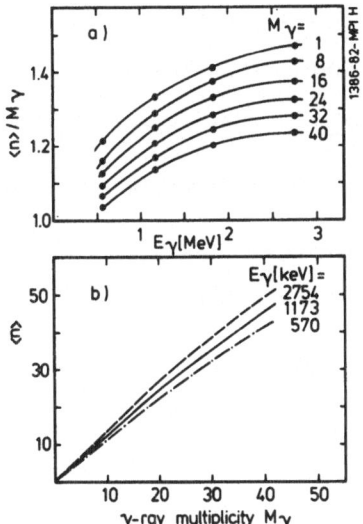

defined as the FWHM of the response curve (see inset) divided by the average number of firing detectors, is plotted as a function of the true multiplicity. For typical multiplicities of M_γ=20-30 a resolution of 20-30 % is obtained. Events with low energy γ-rays show a better multiplicity resolution because of the smaller probability for Compton cross-talk.

fig. 10: a.) Average number of responding detectors divided by the true γ-
ray multiplicity plotted as a function
of the γ-ray energy for selected multiplicities. b.) Average number of responding detectors as a function of the
true γ-ray multiplicity for 3 typical
γ-ray energies

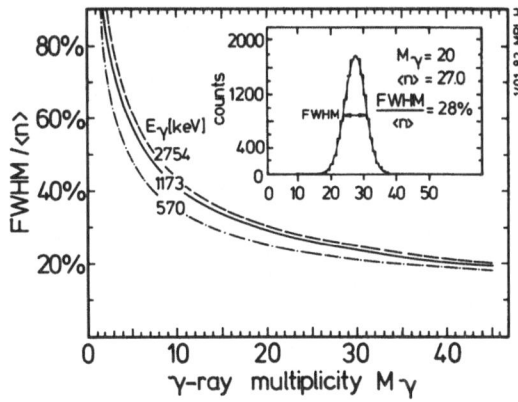

fig. 11: The FWHM of the multiplicity distribution divided by the
average number of responding detectors plotted as a function of the
multiplicity for events with M_γ γ-rays of 570, 1173, and 2754 keV,
respectively. Inset: Distribution of the number of responding detectors for events with 20 γ-rays of 1836 keV.

It should be noted that with the crystal-ball it is now possible to determine the multiplicity of any given event within the quoted resolution while previously only moments of the multiplicity distribution could be deduced by analyzing a large statistical ensemble of events. Since rapidly rotating deformed nuclei decay predominantly by long cascades of stretched E2 transitions also the spin of a nucleus at the beginning of the cascade can be determined from the observed multiplicity.

The nuclear spin can, however, not only be determined in its magnitude but also in its direction by analyzing the angular distribution of the γ-rays which is measured for each event. This is illustrated in fig. 12. The probability for emitting E2-radiation is largest in a plane vertical to the spin direction. This plane is determined experimentally for each event by requiring a sufficiently large number of responding detectors to lie within a narrow ring

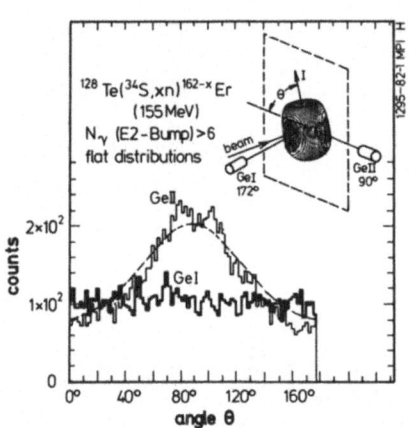

fig. 12: Determination of the nuclear spin axis by exploiting the characteristic angular correlation of stretched E2-transitions. The orientation of the spin is taken to be vertical to the ring zone containing most of the responding detectors. The number of coincidences between the crystal-ball and detectors at 90° and 172° to the beam direction, respectively, are plotted as a function of the angle θ of the spin axis with respect to the horizontal plane.

containing the beam axis. The spin axis is taken to be vertical to
this ring. This procedure has been tested for fusion reactions where
the nuclear spins are known to lie in a plane perpendicular to the
beam. If one considers only coincidences of the crystal-ball with
a detector close to the beam axis (GeI) all spin directions (angles
θ with respect to the horizontal plane) should be equally likely, as
experimentally found in fig. 12. For a detector (GeII) at 90° to the
beam direction, however, the probability to detect an E2 γ-ray depends
on the orientation of the nuclear spin and goes through a maximum if
the nuclear spin is oriented perpendicularly to the plane defined by
the detector and the beam axis. This opens up the possibility to plot
angular distributions of γ-rays with respect to the spin direction
rather than with respect to the beam, as usually done, leading to anisotropies which are typically a factor 2 larger.

Exploiting the features of the crystal ball described above a series
of experiments has been performed. In view of the short operation
time of the detector system most of the results are still preliminary
but, nevertheless, demonstrate the new experimental possibilities
opened up with the use of a crystal-ball. More detailed descriptions
of the experiments are found in contributions to other conferences[6-8].

The current status of the experiments can briefly be summarized as
follows: Using the option of the crystal-ball to select events within
a certain total energy and multiplicity range it has been possible to
considerably reduce the background in high resolution γ-ray spectra

taken with Ge-detectors in coincidence with the crystal-ball. In the reaction ^{128}Te $(^{34}$S, 4n$)^{158}$Er transitions up to spins of 38^+ with a possible candidate for a 40^+ state have been observed[6]. The results can be improved by using Ge-detectors with larger efficiencies.

In the same reaction giant resonance transitions in the energy range between 10 and 20 MeV have been observed. The remarkable feature is that the resonance energy seems to shift[7] from 15 MeV at low multiplicities (spins ≈ 15 $\hbar$) to 11 MeV for high multiplicities (spins ≈ 60 $\hbar$). This shift is much larger than expected theoretically for a pure giant dipole resonance based on yrast states. The interpretation as giant dipole resonances is also difficult to reconcile with the observed angular anisotropies which are rather characteristic of E2 radiation, at least for the lowest multiplicities.

Using light ion induced reactions in the actinide region vibrational excitations of fission isomers in ^{240}Pu and ^{238}U have been identified[8]. These studies supplement and extend previous electron spectroscopy experiments.

In a first attempt to exploit the features of the crystal-ball for studying nuclear reaction mechanisms resonances in the inelastic scattering of ^{12}C on ^{12}C have been investigated. The high detection efficiency of the crystal-ball has been used to search for collective rotational transitions between quasimolecular states in the composite nucleus ^{24}Mg, possibly formed in the scattering. An upper limit of

$2 \cdot 10^{-6}$ has been derived for the γ-decay of the resonance at E_{cm} = 25.7 MeV which makes the interpretation of the observed resonances in terms of quasimolecular configurations very unlikely.

Theoretical attempts to unify all fundamental forces have led to the prediction of a new elementary particle[9], the axion, a light neutral pseudoscalar boson with $m_a c^2 \leq 1$ MeV. A search for the decay of the $^3S_1$ state in positronium into an axion and one γ-ray has been started In this experiment the normal decay of the $^3S_1$ state into 3 γ-rays is vetoed by the crystal-ball and only events with one and only one γ-ray in 4π are accepted. An upper limit on the branching ratio of these two decay modes of $7 \cdot 10^{-7}$ to $2 \cdot 10^{-6}$ has been derived, depending on the mass of the axion. This limit excludes partly theoretical predictions[10] for this branching ratio. It appear possible to improve the sensitivity of the experiment by a factor 5.

This list of current experiments illustrates that the application of the crystal-ball is not limited to nuclear spectroscopy. Further experiments also in atomic physics are planned. More detailed investigations of nuclear reaction mechanisms in deeply inelastic collisions or studies of multi-phonon excitations in Coulomb excitation require to either install the crystal-ball at the UNILAC accelerator at GSI Darmstadt or to upgrade the Heidelberg booster facility.

References:

1. D. Habs, F.S. Stephens, and R. M. Diamond,
   Lawrence Berkeley Laboratory PUB-5020, March 1979

2. H. Gutbrod: Plastic ball and wall, contribution to this conference

3. R.S. Simon, J. Physique C 10 (1980) 281

4. O. Andersen, J.D. Garret, G.B. Hagemann, B. Kerskind, D.L. Hillis,
   and L.L. Riedinger, Phys. Rev. Lett. 43 (1979) 687

5. M. Jääskeläinen, D.G. Sarantites, R. Woodward, F.A. Dilmanian,
   J.T. Hood, R. Jääskeläinen, D.C. Hensley, M.L. Halbert, and
   J.H. Barker, to be published in Nucl. Instr. Meth (1982)

6. R.S. Simon, R. Albrecht, D. Habs, K. Helmer, U. v. Helmolt,
   H.W. Heyng, B. Kolb, D. Pelte, D. Schwalm, V. Metag, H. Gräf,
   W. Hennerici, H.J. Hennrich, E. Jaeschke, R. Repnow, and W. Wahl,
   Int. Symp. on Dynamics of Nuclear Collective Motion, July 1982,
   Mt. Fuji, Japan

7. W. Hennerici, V. Metag, H.J. Hennrich, R. Repnow, W. Wahl, D. Habs,
   K. Helmer, U. v. Helmolt, H.W. Heyng, B. Kolb, D. Pelte, D. Schwalm
   R.S. Simon, and R. Albrecht, EPS meeting, Amsterdam, Aug. 1982
   and to be published in Nucl. Phys. C

8. D. Habs, U. v. Helmolt, H.W. Heyng, R. Kroth, B. Kolb, D. Pelte,
   D. Schwalm, H.J. Specht, W. Hennerici, H.J. Hennrich, G. Himmele,
   R. Repnow, W. Wahl, R.S. Simon, R. Albrecht, and V. Metag,
   Nordic Meeting on Nuclear Physics, Fuglsø, Denmark, Aug. 1982,
   and to be published in Physics Scripta.

9. S. Weinberg, Phys. Rev. Lett. 40 (1978) 223
   F. Wilczek, Phys. Rev. Lett. 40 (1978) 279

10. W. Bernreuther and O. Nachtmann, Z. Phys. C 11 (1981) 235

A 4π NEUTRON MULTIPLICITY DETECTOR FOR HEAVY-ION EXPERIMENTS

U. Jahnke, G. Ingold, D. Hilscher, H. Orf
Hahn-Meitner-Institut Berlin, D-1000 Berlin 39, Germany

E.A. Koop, G. Feige, R. Brandt
Kernchemie, FB. 14, Philipps-Universität, D-355 Marburg, Germany

I. Introduction

Our contribution to this symposium concerns the rediscovery of one of the oldest types of detectors in nuclear physics research, neutron scintillator tanks, for modern heavy-ion reaction studies. As you may know, scintillator tanks have been developed long before the invention of heavy-ion accelerators and solid state detectors some 30 years ago in Los Alamos, namely by the ingenious work of B.C. Diven, J. Terrell (Di 55) and F. Reines (Re 54). Since then until very recently they have very successfully been applied in many different fields, for instance as γ-ray detectors in

i)   the systematic investigation of neutron capture cross sections throughout the periodic table (Di 60, Gu 78)

and, much more frequently, for neutron detection in

ii)   studies of neutron multiplicities following spontaneous (Hi 56, Ho 80) and neutron induced (Ma 64, So 69) fission

iii)   (n, xn) x=1,2,3 cross section measurements (Fr 76, Ve 77)

iv)   photonuclear cross section experiments (Kn 75, Le 81)

and finally in

v)   the search for superheavy elements (Ch 72, Br 80)

So far these detectors have not been used on-line with a charged particle beam. Our efforts to accommodate the detector to heavy-ion experiments have essentially been stimulated by two considerations: the one is, that heavy-ion reactions are marked by a large amount of incident kinetic energy being dissipated into internal excitation which in turn the reactants rid themselves off dominantly by evaporating large numbers of neutrons. The multiplicity distribution of these neutrons is expected to provide closer insight into the reaction and energy dissipation mechanisms, but yet has not been investigated.

The concept of the scintillator tanks for this purpose - and this is the second reason - is still unique, in the sense that so far no other technique is known that could compete with its principle when neutron multiplicity distributions - not only average values - are to be recorded.

Compared to the long history of the scintillator tank our own acquaintance with it is still rather short. Thus, instead of giving a review of its merits in the past we will concentrate on our experience only. About one and a half years ago the 1m-diameter spherical tank with 500 l of scintillating liquid from Marburg University (Be 79) was transported to Berlin, where we adopted it to single and coincidence experiments with the heavy-ion beam of the VICKSI-accelerator of the Hahn-Meitner-Institut. In the first half of this article we will discuss the more technical aspects when operating the detector with the heavy-ion beam for a much higher range of neutron multiplicities of up to 20 or 30 than what it has been used for before. In the second half we will report some preliminary results of recent experiments with it in order to illustrate what the use of the new experimental parameter "neutron number" can be in heavy-ion reactions.

II. Technical Aspects of the Multiplicity Detector

II.1 Principle of Gamma and Neutron Detection

Scintillator tanks basically are γ-ray detectors. Because of the low atomic numbers of the scintillator material (toluene, C_7H_8) practically all interactions of γ-rays are due to the Compton effect; only below 100 keV photoabsorption and above 10 MeV pair creation become important. Furthermore, the large volume of the tank ensures that the major fraction of the total γ-ray energy is absorbed. Spectra of γ-ray sources thus show a total absorption sum peak and a low energy Compton tail. The energy resolution, however, is rather poor because of insufficient light collection from different parts of the large volume and the low electron statistics on the photocathodes which cover only about 3 % of the interior surface, the rest being coated by white reflecting paint (we are referring to our present detector, a schematic set-up of which for heavy-ion experiments is shown in fig. 1). Pulse-height spectra for ^{137}Cs-, ^{54}Mn-, ^{60}Co-sources, for laboratory background and for fission neutrons measured with our detector some 6 years ago (Be 79) when it was rather new are shown in fig. 2. Now either the scintillator or the reflecting paint probably has deteriorated so that an energy calibration is hardly possible.

The scintillator tanks become very efficient neutron detectors when a loading material, usually 0.3 or up to 0.5 % Gadolinium by weight, is added to the liquid. Gd is chosen because two of its isotopes, ^{155}Gd and ^{157}Gd with 15 and 16 % abundance respectively, have very high thermal neutron capture cross sections of $6.1 \cdot 10^4$ and $2.5 \cdot 10^5$ barn.

The neutron detection process then is as follows: neutrons originating from the source or the target in the middle of the tank loose most of their energy within a few tens of nanoseconds by elastic scattering from the hydrogen atoms in the liquid giving rise to the prompt proton recoil signal (in conventional neutron time-of-flight detectors this signal indicates the arrival of a neutron and is most often made subject to n/γ-pulse shape discrimination - not so here). They then diffuse in the tank for some microseconds finally in thermal equilibrium with the scintillator molecules and are eventually being captured by the Gd nuclei.

The Gd (n,γ)-capture γ-rays, three on the average with a total energy of about 8 MeV, are detected with a very high efficiency of about 98 % (Po 74). So that for the neutron capture event a twofold coincidence signal form the 12 phototubes can be asked for without lowering sensibly the neutron detection efficiency but substantially reducing the noise background.

Already here one of the most unfavourable disadvantages of the scintillator tank becomes apparent: its extreme sensitivity to laboratory and cosmic background radiation. Because of this exponentially growing background (for decreasing energy) lower thresholds have to be set on the phototube signals; they determine the neutron detection efficiency of the tank (s. II.5).

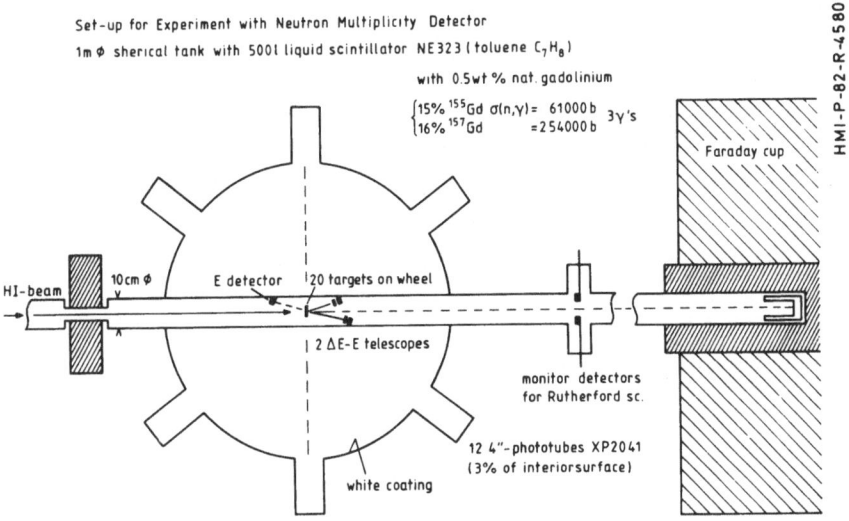

Fig. 1: Schematic set-up for heavy-ion experiments with the scintillator tank.

The essential trick of this neutron multiplicity counter then is, that neutrons which are simultaneously emitted in a reaction event are dispersed in time and only after an average storage period of some 10 to 12 μs are being counted one by one by the fast photomultiplier and electronic recording system.

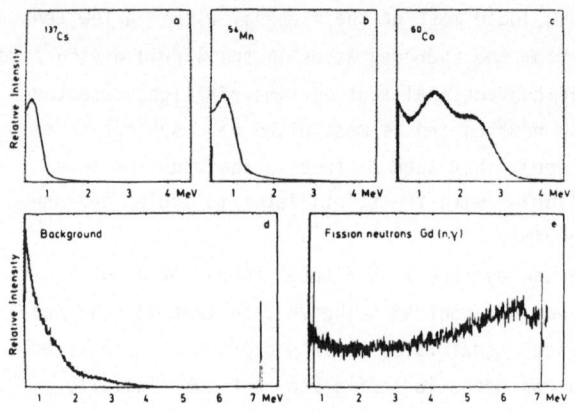

Fig. 2: Pulse-height spectra from the tank for ^{137}Cs, ^{54}Mn, ^{60}Co, laboratory and cosmic background and fission neutrons from ^{252}Cf. From (Be 79).

This storage duration on the other hand makes it a rather slow detector, because about three times this average storage time has to be allowed for the detector in order to collect all neutrons before the next reaction can be initiated.

II.2 Neutron Capture Time Distribution

The capture time distribution for our detector is shown in fig. 3 as measured with a ^{252}Cf fission source in the middle of the tank. The fission fragments detected in a solid state detector provide the start signal, the stop signal is from a twofold coincidence of any of the 12 phototubes. The prompt peak is due to fission γ-rays and recoil protons. The most probable capture time is after about 7 μs. Within the intervall of 1 to 35 μs after prompt 98 % of all neutrons are being captured. So in the experiment the heavy-ion beam should be on the target for about 1 μs or less and off during the subsequent 35 μs counting gate - and on again after about 40 to 50 μs.
This cycle is indicated at the bottom of fig. 3. With our accelerator these conditions are most easily met by a low-energy RF-pulsing system which deflects the beam from the Van-de-Graaff injector before the cyclotron during the pauses.
The experimental capture time distribution can be approximated with a simple 2-parameter exponential formula (Pa 68)

$$h(t) \approx e^{-\lambda t}\{t(\beta-\lambda)-1\} + e^{-\beta t}$$

where the parameter λ characterizes the moderation properties of the scintillator and thus is proportional to the proton density in the liquid; whereas β describes the capture strength of the scintillator and is proportional to the density of the Gd- nuclei.The fit of this formula to the experimental distribution in fig. 3 indicates that the present loading of our scintillator is rather closer to 0.4 wt % than to 0.5 wt %

which it nominally had when first put into operation some 9 years ago.

In order to explore the detector properties for different loading concentrations we collected in table 1 the mean life time of a neutron in the tank $\langle t \rangle = \beta^{-1} + 2\lambda^{-1}$ and the corresponding counting gate lengths for 98 % of the neutrons to be captured.

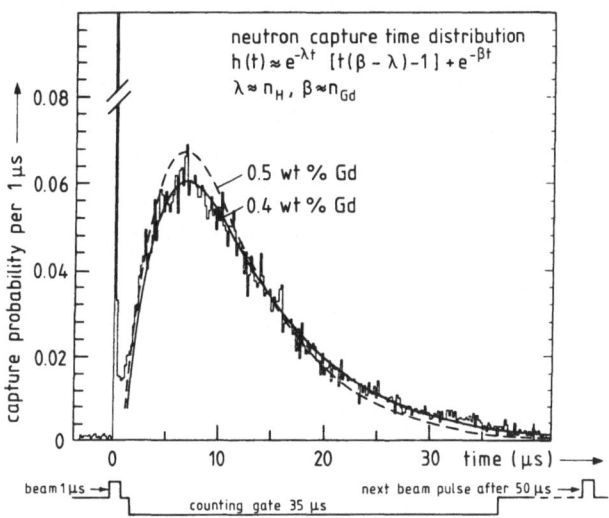

Fig. 3: Neutron capture time distribution and irradiation cycle (bottom).

Table 1: Mean storage times and 98 %-counting gate lengths versus Gd-loading.

Gd-loading [wt %]	mean storage time $\langle t \rangle$ [μs]	98 %-gate length [μs]
0	180	600
0.1	41.1	149
0.3	16.4	53
0.5	11.5	34
1.0	7.7	20

The capture time characteristic sensitively depends on the amount of loading: Without Gd the neutrons predominantly would be captured by hydrogen to form deuteron and emit the 2.2 MeV capture γ-ray. The 98 %-gate length then has to be as long as 600 μs. For the most often used concentrations of 0.3 and 0.5 wt % the 98 %-gate length is 53 and 34 μs, respectively. The detector could be made faster with still higher Gd-concentrations. The chemical stability of the solution, its pulse-height and optical transmission properties may however limit the loading.

II.3 Limitation of the Reaction Rates

The reason why it is desirable to shorten the counting gate length and thus increase the duty cycle of the beam on the target comes from an indispensable condition for experiments with the accelerator. That is, that the neutrons in the counting gate have to originate from a single reaction only, because it is nearly impossible to correct for multiple reaction events which would simulate enhanced neutron multiplicities. So the average number of neutron creating reactions per beam burst has to be kept so low that the probability for multiple reaction events is negligible.
As a rule of thumb only 3 % of the beam bursts should initiate a reaction because then the ratio of multiple to single events is of the order of 0.01 only. Together with the duty cycle this sets the limit of the maximum permissible reaction rate to:

0.03 reactions per burst x 1 burst in 35 µs ≈ 1000 reactions per second.

This limitation can be released only by shortening the counting gate length.

As an illustration, we consider the example of a 500 µg/cm^2 medium heavy (A=100) target and an assumed cross section of 500 mb for neutron producing reactions: the average beam current then has to be kept below about 100 particle pA. This current is roughly of the same order of magnitude of what can be expected from present heavy-ion accelerators considering the necessary duty cycle. Also for coincidence experiments this limitation is not too severe: The 4π-solid angle of the tank roughly compensates the reduction due to the duty cycle and the reaction rate limitation.

The duty cycle and the rate restriction on the other hand make the experiments with the neutron tank ideally suited for parasitic beams.

II.4 Time Response and Coincidence Counting Losses

For high neutron multiplicities of up to 20 or 30 or, when stronger Gd-concentrations are taken into consideration, neutron counting losses due to insufficient time resolution of the phototubes or dead time of the electronic recording system become very important.
If there are only two neutron signals within the counting gate of length T the probability that one is lost due to overlap within the resolution or dead time Δτ is

$$P_2^1 = 2\Delta\tau \int_0^T h^2(t)dt$$

where h(t) is the capture time distribution. For T=35 µs and 0.5 wt % loading this probability is $P_2^1 = 2 \cdot \Delta\tau \cdot 0.05 \ \mu s^{-1}$, or less than 1 % for Δτ < 100 ns.

For higher multiplicities, however, the counting losses drastically increase. As an illustration we calculated these losses (table 2) in a Monte-Carlo simulation for multiplicities 2, 20 and 30 and a set of dead times $\Delta\tau$ ranging from 20 to 150 ns:

Table 2: Average number of signals lost due to overlap within the dead time $\Delta\tau$ - calculated for 0.5 wt % Gd-loading.

dead time [ns]	number of events within the 35 µs counting gate		
	2	20	30
150	0.02	2.4	5.1
100	0.01	1.7	3.6
50	0.005	0.9	1.9
20	0.002	0.4	0.9

From table 2 it is obvious that in order for the counting losses not to become too overwhelming (though principally they can be corrected for in a similar way as the detection efficiency, s. II.6) the time resolution for high multiplicity experiments should not exceed 20 to 50 ns. This can be achieved even for such a large detector volume when fast phototubes and constant-fraction timing discriminators are used in each branch. In fig. 4 the prompt time response of the tank as measured relative to the RF-signal from the cyclotron is shown. 10 beam pulses with 72 ns spacing in time have been let on the target every 100 µs. The width of the peaks is about 6 ns (FWHM), very close to the basic limit given by the scattering of the light path lengths in the tank volume.

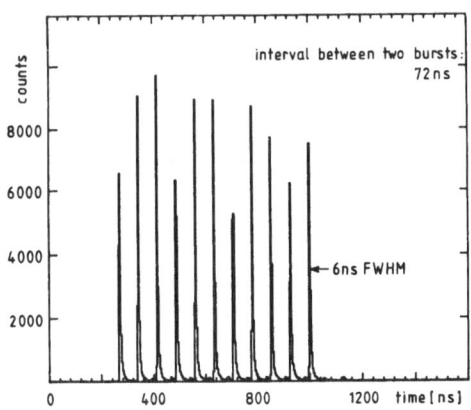

Fig. 4: Prompt time response of the neutron tank to 10 beam bursts on the target. The time intervall between two bursts is 72 ns. A time resolution of about 6 ns (FWHM) is achieved by the use of fast phototubes of the typ XP 2041 (Valvo).

We see that the timing requirements by itself easily can be met. At this very fast

timing, however, the phenomenon of after-pulsing of the phototubes which is not completely eliminated by the coincidence circuit between any two of the 12 tubes tends to
become a problem. The after-pulsing would give rise to anomalous multiplicity distributions that cannot be corrected for. The after-pulsing can be identified by a deviation of the background multiplicity distribution from Poisson statistics or by an enhanced width of the ^{252}Cf-calibration spectrum.

II.5 Single-Neutron Detection Efficiency

The most prominent feature of the detector is its very high neutron detection efficiency. As mentioned before (II.1) in practice the efficiency is determined by the
lower threshold setting on the phototube signals and these thresholds in turn determine the background rate. The efficiency-to-background relation, shown in fig. 5 for
our detector, thus is the characteristic quantity of these detectors. The efficiency
with our present detector can be set as high as 85 % at the expense of about 12 %
background rate or it has to be reduced to 70 % when background up to 5 %, that is 5
events per 100 gate openings, is tolerable only.

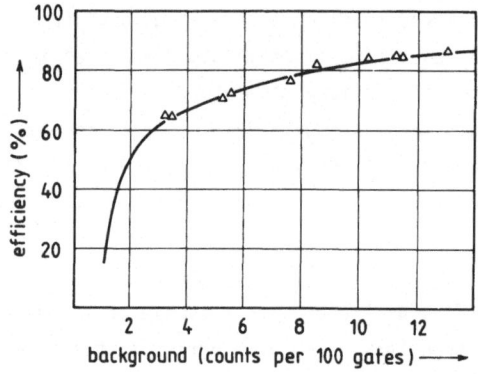

Fig. 5: Efficiency-to-background characteristic for our neutron detector,
measured with a ^{252}Cf-fission source
in its middle.

In the experiments we simultaneously use two different threshold settings for about
75 and 85 % efficiency, first of all in order to allow for a reliability check for
the efficiency corrections (II.6), but also, because single and coincidence experiments favour different settings. The background rates are measured with identical experimental conditions, the target however being replaced by an empty target frame.When
the beam is properly adjusted there is little difference to the background without
beam.

The efficiency is determined and constantly monitored during the experiment with a
^{252}Cf-fission source ($\nu=3.78 \pm 0.04$ %) close to the target position. The spontaneous
fission sources, however, have the disadvantage of rather low average multiplicities

as compared to what is expected in the experiment. Heavy-ion reactions by themselves
in this respect may provide much better calibrations with multiplicities ranging up
to 8 or 10.

These calibrations, of course, provide a detector efficiency as weighted with the
neutron energy distribution of the source or reaction that has been used, since in
spite of the large dimensions of the tank its efficiency is indeed energy dependent.
High energy neutrons have a chance to escape from the tank without being captured. As
an example for this energy dependence fig. 6 shows the results of a rather advanced
Monte-Carlo calculation (Po 74). For a 100 cm-∅-tank and a threshold at 500 keV γ-ray
energy the efficiency drops from about 90 % for slow neutrons to about 70 % at 8 MeV
neutron energy (fig. 6 shows the relative efficiency decrease only). The efficiency
could be made less energy dependent by still increasing the tank dimensions. Since,
however, for energies above 1 MeV the mean free path of the neutron in the liquid increases roughly linearly with energy, table 3, this encounters technical as well as

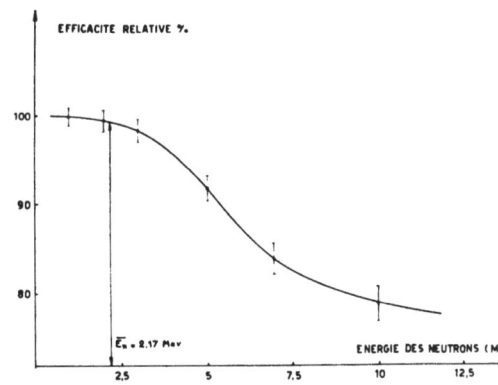

Fig. 6: Relative efficiency versus
neutron energy, calculated in a Monte-
Carlo simulation (Po 74) for a 100 cm-
diameter tank. The average energy
E_n=2.17 MeV of neutrons from a
^{252}Cf-source is indicated. From
(Be 69).

Table 3: Mean free path of a neutron in the scintillator liquid (Re 54)

neutron energy (MeV)	mean free path in the scintillator (cm)
1	4
5	11
10	20
20	41
50	120

economical limits. Instead, it seems more reasonable to provide a rather thick outer
tank wall from iron (Le 81) which on the one hand would act as a very good reflector
for high energy neutrons and on the other hand provides an optimal shielding against
laboratory and especially cosmic radiation since a tight shielding does not tend to
convert distant cosmic radiation into neutron showers which is often the problem with
spacy shieldings.

II.6 Detection Efficiency for High Neutron Multiplicities

The detection efficiency discussed so far, is the efficiency for a single neutron
emitted from the middle of the tank with no preference in direction. Now, if ν neutrons are released simultaneously from a reaction, the detector will recognize this
event with a probability

$$P_{k,\nu} = \binom{\nu}{k} \varepsilon^k (1-\varepsilon)^{\nu-k} \quad ; \quad k = 0,1, \ldots \nu$$

as an event with $k < \nu$ neutrons. The probabilities $P_{k,\nu}$ form the recognition pattern
for the real multiplicity ν. As an example, fig. 7 exhibits these patterns for multiplicities up to 20 and 25 and detector efficiencies $\varepsilon=70$ to 90 %.

We will try to make you familiar with these patterns: for 70 %-efficiency and an event
with multiplicity $\nu=8$ the probability to detect all neutrons is 6 %; with 22 % probability the event will be recognized as $k = 7$. With a maximum probability of about 30 %
it will be observed as $k = 6$. The width of this pattern, that is the multiplicity resolution, is about 3.5 units.
This - very roughly speaking - have been the conditions in the early stages of the
neutron tanks, when neutron multiplicities up to 8 or 9 from spontaneous or induced
fission were observed with about 70 % efficiency.
Now, for 80 % efficiency the $\nu=12$ multiplicity exhibits a very similar pattern concerning the chance $P_{\nu,\nu}$ to observe all emitted neutrons, the maximum and the width
of the pattern.
For 90 % efficiency a similar shaped pattern already belongs to a multiplicity $\nu=25$.
The conditions concerning the data taking as well as the subsequent unfolding procedure - this is the point we want to make - are very similar when detecting multiplicities up to 8 or 9 with some 70 % efficiency like in the past or when going up to $\nu =$
20 or 25 with efficiencies of 85 or 90 %.

When reconstructing the real multiplicity distribution from the experimental one, the
inverse problem has to be solved, taking the experimental uncertainties into account.

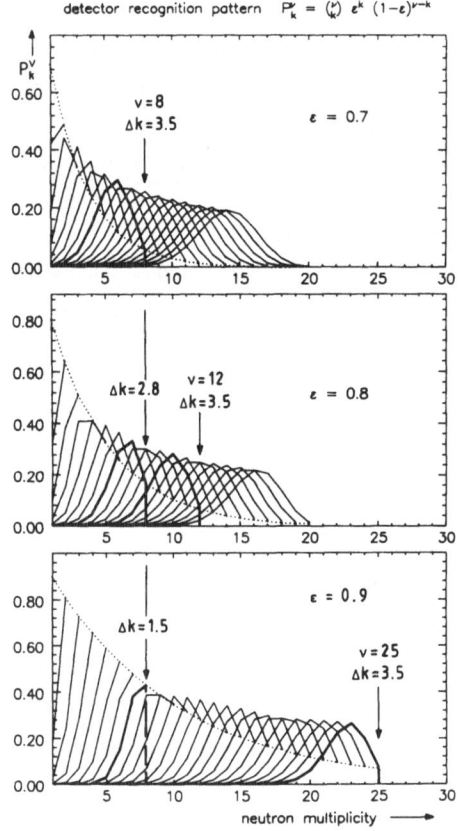

detector recognition pattern $P_k^\nu = \binom{\nu}{k} \varepsilon^k (1-\varepsilon)^{\nu-k}$

Fig. 7: Detector recognition patterns $P_{k,\nu}$ for single-neutron efficiencies ε=0.7, 0.8 and 0.9. The patterns give the probability to observe k (k<ν) neutrons, when ν neutrons have been emitted in a reaction. The dotted lines connect the probabilities $P_{\nu,\nu}$ to detect all ν emitted neutrons. The patterns for fixed multiplicity ν=8 have been selected (lines intensified) for the three efficiencies in order to show the gain in resolution from Δk=3.5 to 1.5 with increasing efficiency. The patterns for ν=8 (ε=0.7), ν=12 (ε=0.8) and ν=25 (ε=0.9) have identical shape with Δk=3.5 and $P_{max}\approx$30 %. They demonstrate the growing range of multiplicities accessible under similar experimental conditions when the efficiency can be increased from 70 to 90 %.

The recognition patterns constitute the column elements of the matrix which converts the real multiplicity vector into the measured one.

As is well known from other instrumental inverse problems for instance in photonuclear cross section work with bremsstrahlungs x-rays or in neutron spectroscopy with the proton-recoil-technique, the direct solution of the problem often tends to give false or oscillating results. For the multiplicity problem as a rule of thumb the three selected recognition patterns for ν=8 (ε=70 %), ν=12 (ε=80 %), ν=25 (ε=90 %) indicate the highest multiplicities up to which the simple direct unfolding procedure (Di 55) will give satisfactory results.

For higher multiplicities or less sufficient experimental statistics more elaborate mathematical techniques have to be applied (Da 74).

As an example for the unfolding procedure, fig. 8 shows the experimental neutron multiplicity distribution for evaporation residues from 220 MeV ^{20}Ne+^{165}Ho, observed at an angle of Θ=13^0. The dot-dashed line denotes the reconstructed real distribution with the proper recognition patterns at the bottom, which add up to the experimental distribution. The efficiency correction thus tends to preserve the shape of the distribution, it shifts, however, the maximum by $1/\varepsilon$.

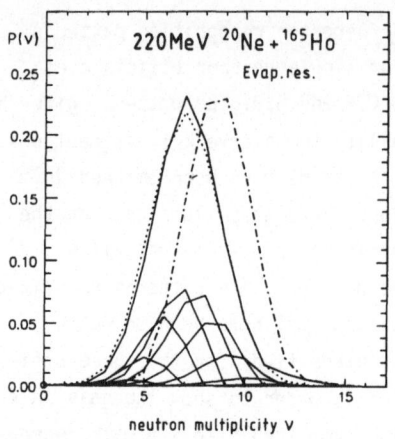

Fig. 8: An example for the effect of the efficiency correction on the multiplicity distribution: neutrons detected (with 83 % efficiency) in coincidence with heavy evaporation residues from 220 MeV ^{20}Ne + ^{165}Ho. The dotdashed line gives the reconstructed real distribution with the proper recognition pattern at the bottom, which very closely add up (dotted line) to the observed distribution (full line).

II.7 Summary of the Technical Aspects of the Multiplicity Detector, Comparison to Alternative Techniques and Side Remarks

Two alternative detectors have to be considered before summarizing the technical aspects of our present investigation: BF$_3$- or ^{3}He-filled proportional counters embedded in paraffin or plastic material (Te 81) and a neutron ball assembled from a large number of individual neutron detectors, for instance the ones presently used for time-of-flight experiments. Both types of detectors provide additional advantages. Much less γ-ray sensivity, if at all, on the one hand, energy and angular information on the other hand. Both types, however, cannot compete with the neutron detection efficiency of the scintillator tank, it is 60 % at best for the proportional counter array and some 30 to 40 % for a neutron ball neglecting cross talk and other problems inherent to detector assemblies (Me 82).

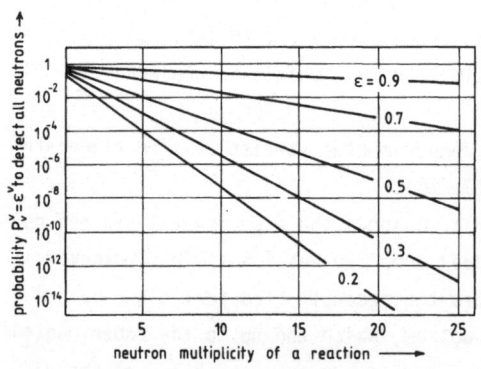

Fig. 9: Probability $P_{\nu,\nu}=\varepsilon^{\nu}$ to detect all emitted neutrons versus neutron multiplicity for a set of efficiencies ranging from $\varepsilon=0.2$ to 0.9.

Fig. 9 demonstrates the simple exponential relation $P_{\nu,\nu}=\varepsilon^{\nu}$ for the probability to observe all neutrons from a reaction. With an efficiency of 40 or 60 % it would be practically impossible to observe all emitted neutrons when there are more than 6 or 12, respectively. For 20 neutrons the chance is 10^{-9} or 10^{-5}. Also the multiplicity resolution is drastically worsened.

In particular heavy-ion coincidence experiments, which rely on an event-by-event analysis, often do not allow for an efficiency correction (when, for instance, particle or γ-ray energy spectra are made subject to the simultaneous occurrence of a definite number of neutrons). We thus consider a detection efficiency superior to 80 or 85 % as a conditio sine qua non.

With our present detector, so far, we did not pay attention to the prompt response signal, because of its desolate energy resolution. Principally, however, it measures the total kinetic energy converted into internal excitation, all the more as an energy resolution of some 20 % or better seems to be feasible. Since the neutron emission is measured separately by the delayed signals, the prompt response thus might provide very similar information as total γ-ray energy spectrometers (Me 82) - at least, when charged particle evaporation is negligible.

The geometrical disadvantage of our present detector could be overcome rather easily: with a modest increase in the outer diameter (up to 110 or 120 cm) it could incorporate a small reaction chamber of about 25 to 35 cm in diameter in its middle for all kinds of charged particle detectors, thereby still preserving or even increasing the neutron detection efficiency.

In summary:

Neutron scintillator tanks seem to be very well suited for heavy-ion experiments where neutron multiplicities up to 20 or 30 are to be expected.

- they count the number of neutrons released in each single reaction event with an efficiency of the order of 80 or 90 % in 4 π.
- the total number of emitted neutrons is detected, unless the reaction offers the favourable condition of a clear kinematical separation between neutrons disposed off by the one or the other fragment (s. III.4).
- they provide no information on the neutron energy or angular distribution
- the heavy-ion beam has to be pulsed: it can be on the target for about 1μs and has to be switched off for the subsequent counting gate of $\approx$ 35 μs duration.
- the average rate of neutron producing reactions is limited to roughly 1000/s.

III. Perspectives of the Multiplicity Detector for Heavy-Ion Experiments

In this section we will by means of very preliminary results of some recent experiments try to outline what the use of the number of ejected neutrons might be in heavy ion experiments. According to our present understanding the neutron number may provide a threefold insight:

i)   The neutron number - strongly simplified - is proportional to the energy dissipated into thermal excitation of the reactants. It thus measures the degree of inelasticity of a reaction and allows to select different reactions according to the energy conversion.

ii) Since the neutron emission tends to reduce the angular momentum brought into a reaction only rather modestly, the neutron number might be used to select certain bands of angular momenta. It thus may provide an angular momentum filter - complementary to the sum energy γ-ray spectrometers (Me 82) which are more sensitive to the rotational part of the excitation energy.

iii) The neutron number information may be used to select specific reaction channels or residual nuclei. This may be helpful when investigating rare decay modes or the production of exotic nuclei.

So far we examined and will discuss the first point only.
The experiments have been somewhat limited by the narrow geometry inside the neutron tank (fig. 1). The 10 cm-∅ beam pipe transversing the detector had to accommodate a 20-position target wheel and three solid-state detector telescopes restricted to either fixed foreward or backward angles. About 5 m behind the target the beam was stopped in a heavily shielded Faraday-cup from carbon. It was not felt necessary to shield the tank itself against stray radiation from the accelerator or the beam line in front of it, except for some lead bricks screening irradiated quarzes.

III.1 Inclusive Neutron Multiplicity Spectra

Inclusive neutron multiplicity spectra for 220 MeV ^{20}Ne on a sequence of targets ranging from ^{100}Mo to ^{209}Bi are displayed in fig. 10.

In a rigorous description, these spectra may be taken for internal excitation energy spectra. The highest multiplicities originate from the most inelastic, i.e. the compound nucleus reaction. The maximum dominating all spectra then corresponds to the

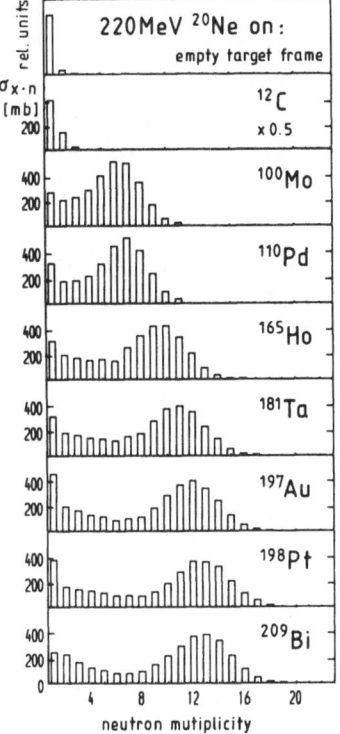

rel. units

$\sigma_{x \cdot n}$ [mb]

220 MeV ^{20}Ne on:

empty target frame

^{12}C x 0.5

^{100}Mo

^{110}Pd

^{165}Ho

^{181}Ta

^{197}Au

^{198}Pt

^{209}Bi

neutron mutiplicity

Fig. 10: Inclusive neutron multiplicity spectra for 220 MeV ^{20}Ne on a variety of targets ranging from ^{100}Mo to ^{209}Bi. For comparison a spectrum for the background (measured with an empty target frame) and for target impurity (^{12}C) is also shown.

weighted average neutron emission from the compound nucleus decay. It increases from 6.5 for ^{100}Mo to 13 for ^{209}Bi, the reason being twofold: with increasing N/Z of the compound system the charged particle emission becomes less competitive relative to the neutron emission and also for heavier systems the fission with slightly higher neutron release (III.2) as compared to compound nucleus evaporation becomes dominant. The lower neutron multiplicities in the spectra are due to more peripheral reactions where less energy is dissipated.

Simultaneously to the neutron measurements the Rutherford scattering of the projectiles was observed with a monitor detector at very forward angles (fig. 1). The relative neutron emission probabilities therefore can be converted to partial cross sections $\sigma(^{20}$Ne, i•n) on an absolute scale.

The partial cross sections now can be added up to give - to a good approximation - the total reaction cross section:

$$\sigma \text{ Reaction} = \sum_i \sigma \, (^{20}\text{Ne, i•n})$$

Certainly there are some restrictions to be made: all reactions with only γ-rays or charged particles as products are not included in the sum. In particular all those are omitted, where less than the neutron binding energy is dissipated.

For heavy-ion reactions at higher energies and not too neutron deficient reactants this omission probably is marginal. Furthermore, from exclusive multiplicity experiments (III.3) a fair estimate of the amount of non-neutron releasing reactions might be derived.

Very recently total photonuclear cross sections up to the pion threshold E_γ=140 MeV have been deduced in a similar way by the Saclay group (Le 81). Also the 20 year old total cross section measurements by Viola and Sikkeland (Vi 62) with highly fissionable targets of ^{238}U bear some resemblence.

Presently the light element contamination of the targets (see for instance ^{12}C in fig. 10) prevents us from explicitely quoting reaction cross sections. During the last experiments this difficulty has not been properly taken care of.

The inclusive neutron multiplicity spectra provide an overview on the reaction characteristics concerning the relative strength of contributions according to their intrinsic energy deposit (this property also may be used for an event-by-event distinction between central and peripheral collisions, for instance.)

The method offers a new alternative to total cross section measurements with all its experimental ease in favour: the inclusive spectra in fig. 10 each have been accumulated in about 15 min of time.

III.2 Neutron Multiplicity Spectra from Compound Nucleus Decay

In the following sections we will discuss exclusive neutron multiplicity spectra. However, before going into detail we show in fig. 11 as an illustration the whole variety of neutron number spectra for all kinds of charged reaction residues observed at 13°, close to the grazing angle for 290 MeV [20]Ne on [197]Au.

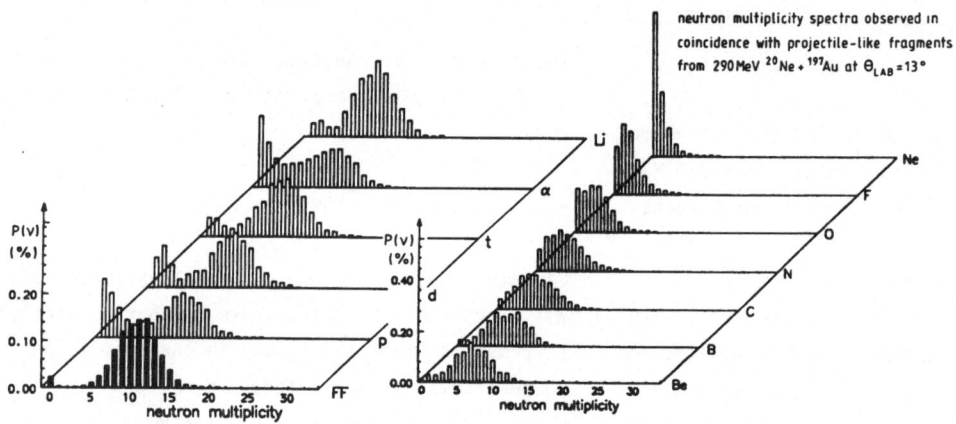

Fig. 11: Neutron multiplicity spectra as observed during the experiment, 290 MeV [20]Ne + [197]Au, in coincidence with fission fragments, light ions and projectile-like fragments detected at $\theta_{lab} = 13°$. The spectra are not corrected for background, efficiency etc.

Neutron multiplicity spectra observed in coincidence with compound nucleus decay products, e.g. fission fragments, compound nucleus evaporation residues, α-particles and protons for 220 MeV [20]Ne + [165]Ho as an example are shown in fig. 12. They all exhibit a pronounced maximum at high neutron numbers and there are practically no low multiplicities. The peak is centered at 12 for fission and is about 3 units lower for heavy and light evaporation residues. Protons and α-particles in this experiment have been observed at far backward angles (160⁰) so that they can originate from the compound nucleus decay only. The evaporation residues, however, were detected at 16⁰ and therefore could not clearly be separated from target-like recoils from damped reactions. The corresponding neutron spectrum thus on its low multiplicity side is contaminated from these reactions. We estimate that the pure compound nucleus neutron spectrum would be shifted by about 1 unit towards higher neutron numbers and its width somewhat reduced as compared to the one shown in fig. 12.

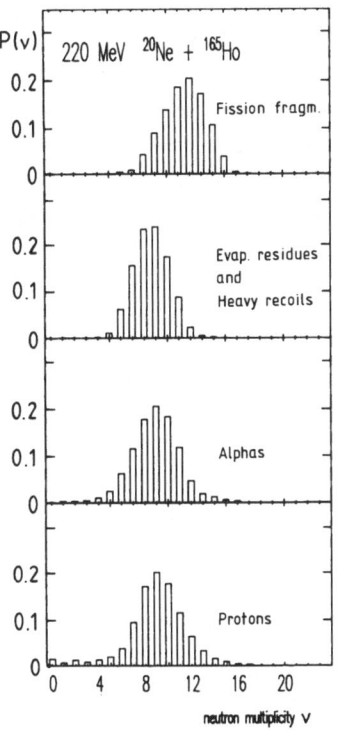

The shift of fission neutron numbers relative to the ones from the evaporation process is due to the additional gain in the fragment excitation energy from the fission mechanism itself. The enhanced neutron width observed for fission (probably more evident in other experimental results not shown here) reflects the large spread in the fragment mass and the kinetic energy release.

We expect a closer insight into the fission process in particular at high excitation and strong deformation from the neutron multiplicity spectra when both fragments are observed simultaneously. For low energy fission the heavy-ion in-beam experiments offer the possibility to extend the systematic of the average fission neutron yield $\bar{\nu}$ to heavier and less stable nuclei than could be reached before.

For the compound nucleus the neutron multiplicity distribution provides a new means to investigate its evaporation properties and thereby tests the reliability of statistical model calculations. The multiplicity spectra can be measured in coincidence with the heavy residues as well as all kinds of light evaporation particles and thus provide a link for the combinatorial reconstruction of the decay chains.

Fig. 12: Exclusive neutron multiplicity spectra for all kinds of compound nucleus decay products from 220 MeV ^{20}Ne + ^{165}Ho: fission fragments, heavy and light (p,α) residues.

It will be interesting to persue the neutron multiplicities from compound nucleus evaporation and fission up to the highest excitation energies in order to find out if there is a limit for the maximum energy transfer and, if so, if it is the same for both processes.

In fig. 13 we give another example for the determination of cross sections from neutron multiplicity measurements. The neutron number spectrum observed in coincidence with fission fragments at 16^0 (the neutron spectra for two other fission fragment emission angles have identical shapes) from 220 MeV ^{20}Ne + ^{197}Au has been fitted into the corresponding inclusive multiplicity distribution. The fission neutron spectrum fits very well to the broad maximum in the single spectrum and completely exhausts it. From this fit the partial cross sections σ (i·n) in the inclusive spectrum originating from fission can be added up to give the total fission cross section.

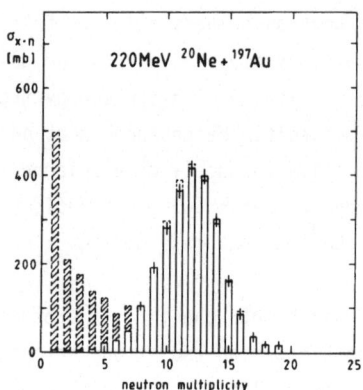

Fig. 13: Inclusive neutron multiplicity spectrum (hached bars) from 220 MeV ^{20}Ne + ^{197}Au and exclusive spectrum due to fission neutrons (blank bars) fitted into it. The error bars indicate the relative uncertainty from experimental statistics and the unfolding procedure.

The restrictions mentioned above (III.1) for the determination of total reaction cross sections do not apply here. Since the fission process rather exclusively proceeds with the emission of a large number of neutrons it does not escape from observation. Also target contamination from light elements would not affect this new method since their neutron multiplicities are restricted to very small numbers.

III.3 Reconstruction of the Distribution of Primary Reaction Products

It is a common situation in heavy-ion experiments that the observed distributions in mass, charge or energy differ considerably from the outcome of the primary reactions; this as the result of sequential decay of the highly excited primary fragments.
As an example for the use of the neutron multiplicity in the reconstruction of the initial distributions, we discuss preliminary results of a recent study (Bü 82) of quasielastic collisions of 290 MeV ^{20}Ne on ^{197}Au.
At the intermediate energy between 10 and 20 MeV/nucleon the quasielastic cross section for the formation of a projectile-like fragment is generally believed to be composed of two constituents: a transfer component and a direct or sequential break-up component. In the inclusive energy spectra of projectile-like fragments, a

sample of which is shown in fig. 14, both components are intimately mixed, since both processes give rise to ejectile energies roughly corresponding to the beam velocity. The target nucleus excitations, however, and therefore the number of emitted neutrons is quite different for transfer and break-up and provides a gratifying means of distinction: while in the break-up a third particle, in our example generally an α-particle because the respective binding energies are the lowest, is set free and moves on with beam velocity, in the transfer it deposits its share in the projectiles kinetic energy into the target nucleus giving rise to higher intrinsic excitation as compared to the break-up.

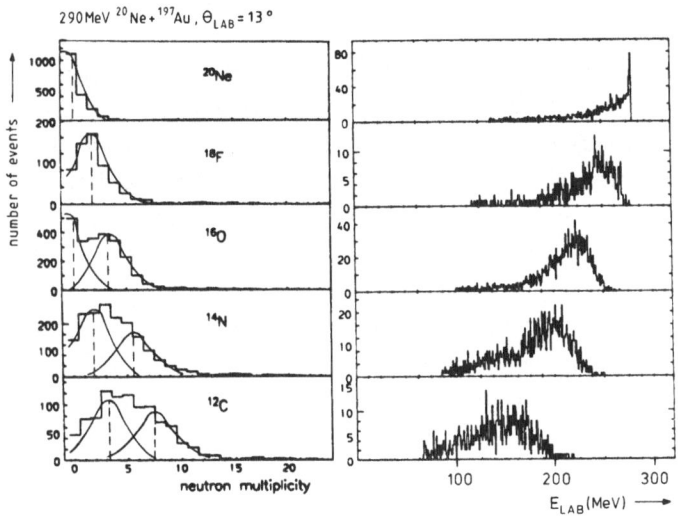

Fig. 14: Energy spectra (right) and neutron distributions (left) for projectile-like fragments ^{20}Ne, ^{18}F, ^{16}O, ^{14}N and ^{12}C observed at $\theta_{lab} = 13^0$ from quasielastic collisions of 290 MeV ^{20}Ne and ^{197}Au. Handdrawn Gaussians in the distributions of neutrons show their provenance from transfer (high neutron numbers) or break-up (low neutron numbers). The neutron spectra are not corrected for detector efficiency.

The neutron multiplicity spectra for the five isotopes ^{20}Ne to ^{12}C on the left side of fig. 14 exhibit these two components. The transfer component shifts monotonically towards higher neutron numbers, its centroid coinciding roughly with the number of nucleons missing relative to the incident projectile, it is close to 0 for ^{20}Ne and close to 8 for ^{12}C: very roughly speaking each transferred nucleon deposits 15 MeV of excitation, allowing for the emission of one neutron.

From ^{16}O on downwards there is a second component centered at about 4 neutrons lower than the transfer for the same observed isotope and in shape and position very simi-

lar to the transfer component of an ejectile with 4 more nucleons: this is the break-up component. The primary fragment sequentially has emitted an α-particle. For ^{16}O, ^{14}N and ^{12}C one may conclude that both the primary transfer and the secondary break-up contribute with roughly equal strength to the quasielastic cross section - indistinguishable, when only inclusive energy spectra are available.
Certainly the results for the other observed isotopes will have to confirm our preliminary interpretation.

III.4 Energy Partition in Deep Inelastic Collisions, a Future Experiment

The last experiment in this experimental vista has stimulated our mind since the beginning of this work because of its particular challenge from the experimental side as well as from the expected insight into the process of dissipation and thermalization of energy. The idea to it is due to Hilscher (Hi 79) and to Morrissey and Moretto (Mo 81), but it has already had its analogue in spontaneous fission studies (Ni 69).

From many experiments it is rather well established, that in strongly damped reactions the total excitation energy is shared between the fragments in relation to their masses. The statistical nature of the energy equilibration causing this partition is expected to manifest itself in fluctuations about the average energy division, which so far could not be investigated. The energy partition and the fluctuations, when observed for different degrees of inelasticity or damping could allow an estimate of the time scale for the thermalization of energy.

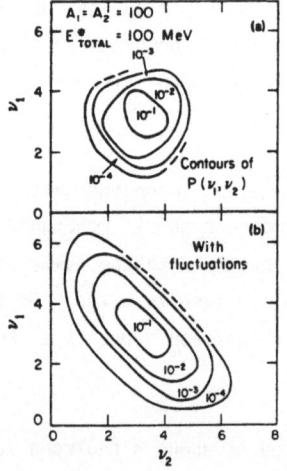

Fig. 15: Monte-Carlo calculation of the numbers ν_1 and ν_2 of neutrons evaporated from the correlated fragments 1 and 2 of a deep-inelastic collision. Top: for fixed partition of excitation energy, bottom: for statistical equilibrium. From (Mo 81).

Now, when the width of the neutron multiplicity distribution for one of the fragments is considered as the experimental observable, it has to be kept in mind, that the evaporation process itself is of statistical nature. The number n of released neutrons is not strictly related to the excitation energy E^*, $E^* = E_{rotation} + \langle E_n + B_n \rangle$ n, where E_n and B_n are the kinetic and binding energies, respectively, for the neutrons. This natural neutron width has to be unfolded from the observed one in order to obtain the spread in excitation induced by the primary reaction.

Instead, it is more promising to examine the anticorrelation in the neutron number emitted from the one or the other fragment - which, however, requires two neutron multiplicity detectors. In a Monte-Carlo simulation (Mo 81), fig. 15, the experimental outcome has been anticipated, when the two neutron numbers v_1 and v_2 in the case of two symmetric fragments with mass A = 100 and a total excitation energy E^* = 100 MeV are observed. The upper part of the figure exhibits the probability contours $P(v_1, v_2)$ for fixed energy division, i.e. the natural width. In the lower part of the figure fluctuations with a width $\sigma^2 = T^3$ $a_1 \cdot a_2 / a_1 + a_2$ (T = temperature of the composite system, a_1 and a_2 level density parameters of the binary fragments) are turned on, thereby introducing a strong anticorrelation in the neutron numbers v_1 and v_2.

In the experiment the strong focussing of the neutron emission into the direction of the diverging fragments has to be made use of, as has been done in many neutron experiments before (Hi 79). Also a compromise has to be found between maximum kinematical separation of the respective neutron emission cones (for asymmetrical fragmentation) and strongest fluctuations favouring symmetrical fragmentations.

The essential advantage of the two-detector experiment is its strict separation between the natural and experimentally induced (by insufficient separation of the neutron emission cones, detector cross talk etc.) neutron width along the $v_1 = v_2$ diagonal in fig. 15 and the effect of fluctuations in excitation energy division increasing the width only along the perpendicular diagonal $v_1 + v_2$ = const.

We hope to be able to begin this experiment within near future.

E. Baltrusch and B. Mertesacker deserve our thanks for the difficult mechanical set-up inside the neutron tank, also the help of L. Schmidt and H. Jungclas during the early phases of this work is gratefully acknowledged. We are indepted to Prof. K. H. Lindenberger for his continuous interest and support of the present investigation.

References

Be 69   H. Beil, R. Bergère, A. Veyssière, Nucl. Instr. 67, 293 (1969)

Be 79   H.-J. Becker, R. Brandt, H. Jungclas, T. Lund, and D. Molzahn,
         Nucl. Instr. 159, 75 (1979)

Br 80   R. Brandt, T. Lund, D. Molzahn, P. Vater, H. Jungclas, and A. Marinov,
         Nucl. Instr. 173, 121 (1980)

Bü 82   these experiments have been performed in collaboration with M. Bürgel, Ch.
         Egelhaaf, H. Fuchs, and H. Homeyer; U. Jahnke et al. to be published.

Ch 72   E. Cheifetz, R.C. Jared, E.R. Giusti, and S.G. Thompson,
         Phys. Rev. C 6, 1348 (1972)

Da 74   M. Dakowski, Yu. A. Lazarev, V.F. Turchin, and L.S. Turovtseva,
         Nucl. Instr. 113, 195 (1973)

Di 55   B.C. Diven, H.C. Martin, R.F. Taschek, J. Terrell, Phys. Rev. 101, 1012
         (1956)

Di 60   B.C. Diven, J. Terrell, and A. Hemmendinger, Phys. Rev. 120, 556 (1960)

Fr 76   J. Frehaut, Nucl. Instr. 135, 511 (1976)

Gu 78   S.K. Gupta, J. Frehaut, and R. Bois, Nucl. Instr. 148, 77 (1978)

Hi 56   D.A. Hicks, J. Ise jr., and R.V. Pyle, Phys. Rev. 101, 1016 (1956)

Hi 79   D. Hilscher, J.R. Birkelund, A.D. Hoover, W.U. Schröder, W.W. Wilcke, J.R.
         Huizenga, A.C. Mignerey, K.L. Wolf, H.F. Breuer and V.E. Viola jr. in
         proceedings of the Symposium on Deep-Inelastic and Fusion Reactions with
         Heavy Ions, Berlin 1979. Lecture Notes in Physics, Springer Verlag 1980

Ho 80   D.C. Hoffman, G.P. Ford, J.P. Balaga, and L.R. Veeser,
         Phys. Rev. C 21, 637 (1980)

Kn 75   U. Kneissl, G. Kuhl, H.-H. Leister, and A. Weller, Nucl. Phys. A247, 91
         (1975)

Le 81   A. Leprêtre, H. Beil, R. Bergère, P. Carlos, J. Fagot, A. de Miniac, and A.
         Veyssière, Nucl. Phys. A367, 237 (1981)

Ma 64   D.S. Mather, P. Fieldhouse, and A. Moat, Phys. Rev. 133, 1403 (1964)

Me 82   V. Metag et al., crystal ball, contribution to this conference.

Mo 81   D.J. Morrissey and L.G. Moretto, Phys. Rev. C 23, 1835 (1981)

Ni 69   H. Nifenecker, J. Frehaut, M. Soleilhac, 2. IAEA-Symposium on the physics and
         chemistry of fission, Vienna 1969, proceedings p. 491

Pa 68   J.B. Parker, P. Fieldhouse, L.M. Harrison, D.S. Mather, Nucl. Instr. 60, 7
         (1968)

Po 74   J. Poitou, C. Signarbieux, Nucl. Instr. 114, 113 (1974)

Re 54   F. Reines, C.L. Cowan jr., F.B. Harrison, and D.S. Carter,
Rev. Sci. Instr. 25, 1061 (1954)

So 69   M. Soleilhac, J. Frehaut, and J. Gauriau, J. Nucl. Energ. 23, 257 (1969)

Te 81   G.M. Ter-Akopian et al., Nucl. Instr. 190, 119 (1981)

Ve 77   L.R. Veeser, E.D. Arthur, and P.G. Young, Phys. Rev. C 16, 1792 (1977)

Vi 62   V.E. Viola jr., and T. Sikkeland, Phys. Rev. 128, 767 (1962)

DIOGENE : A 4π PICTORIAL DRIFT CHAMBER

J. Poitou

DPh-N/MF, CEN Saclay, 91191 Gif-sur-Yvette Cedex, France

In the last few years, many efforts have been made both on the theoretical and experimental sides to investigate high energy heavy ion collisions [1]. The first experiments, the only ones which had been made when we decided to build Diogène, were inclusive experiments. On the theoretical side, different approaches were investigated. In one type of approach, the nuclei were considered as interacting like fluids in hydrodynamics calculations. A completely different point of view was to assume that the interaction occurs through single nucleon collisions as investigated in the various intranuclear cascade calculations. Both types of calculations were equally good in reproducing the inclusive data, which could even be reproduced by simple thermal models like the fireball. In order to check what kind of model best describes the physics, it was thus necessary to make exclusive measurements. In order to get a clear signal, one should indeed measure all ejectiles from the heavy ion collision. This can be achieved only with a detector covering as close as possible to 4π sr with the possibility of measuring high ejectile multiplicities. With such a detector it should then be possible to measure completely heavy ion reactions on an event by event basis. The distinction between the various models would then be possible for instance by computing suitable global variables of the event like sphericity, thrust,energy flow... [2]. Another advantage of such a detector is the possibility to measure correlations between the emitted particles. For instance if the resolution were high enough, one should be able to measure the size of the participant region from the correlations between two pions measuring the Hanbury-Brown-Twiss effect [3]. High compression and high excitation energy may be reached in central collisions ; it is important to measure the behaviour of nuclear matter in this special regime. Depending on the kind of physics one is interested in, it should be possible to measure preferentially in the forward or in the backward direction. So we decided to build a versatile large solid angle detector : Diogène (this is the name of our 4π solid angle detector) is being developed by a collaboration [4] between physicists from Saclay, Strasbourg and Clermont-Ferrand, to be installed at the Saturne Synchrotron in Saclay.

MOTIVATIONS FOR THE CHOICE OF A PICTORIAL DRIFT CHAMBER

The Saturne Synchrotron at Saclay will deliver beams of heavy ions up to Ar at energies up to 1.2 GeV/nucleon. The following characteristics of the detector should be achieved in order to make useful measurements with such beams : a very large solid angle, as close as possible to 4π sr ; the possibility of handling multiplicities of light particles greater than 40 ; the momentum of all particles whatever their type, energy and direction should be measured within an accuracy of 10 to 20 % up to momentum values of 1.5 GeV/c ; the different types of charged particles ($\pi^{\pm}$, p, d, t ...) should be clearly identified ; it should be possible to trigger the data acquisition on central collision events.

In principle all these requirements are met by streamer chambers. However the full tridimensional analysis of the events from the stereoscopic photographs is a huge problem especially for high multiplicity events. This problem is not yet solved satisfactorily in an automatic way. Hence the number of events which can be fully analyzed with this kind of detector is strongly limited.

Another possibility would have been to build an absorption detector like for instance the plastic ball [5], where the particles are identified by the combination of an energy loss and their total energy. We rejected this solution because of its limited energy range, and the quasi impossibility to measure π^{-}.

Another type of detectors can be used for high multiplicity event measurements, and have already been developed for several high energy physics experiments : they are based upon drift chambers located in a magnetic field and allow full three dimensional recording of points along the trajectories. The particles are identified by their energy loss and the curvature of their trajectories which is a measure of their momentum. Diogène is an adaptation of these huge high energy physics detectors to the mass and energy range of the species to be detected at Saturne. In particular its internal detector is a pictorial drift chamber derived from the "Jet chamber" of the Jade detector [6] installed at Petra.

PRINCIPLE OF THE PICTORIAL DRIFT CHAMBER

Let us consider a particle passing through an ionization chamber which is filled with an appropriate gas (Fig. 1). An uniform electric field drives the ionization electrons towards the anode wires. In the vicinity of a wire, the electron multiplication occurs. The electrons are collected, giving rise to an electric pulse at each end of the wire. With these two signals, it is thus possible to get a three dimensional recording of the location of the track in each drift cell : the x coordinate is simply given by the wire location : the Z coordinate (parallel to the wire) is obtained from charge division at both ends of the wire : the drift time is representative of the distance y between the track and the wire. There is however an ambiguity on the y coordinate : since the measured quantity is a drift time it is not a priori possible to decide whether the track is on the left hand side or on the right hand side of the wire plane. The way we solve this ambiguity will be explained later.

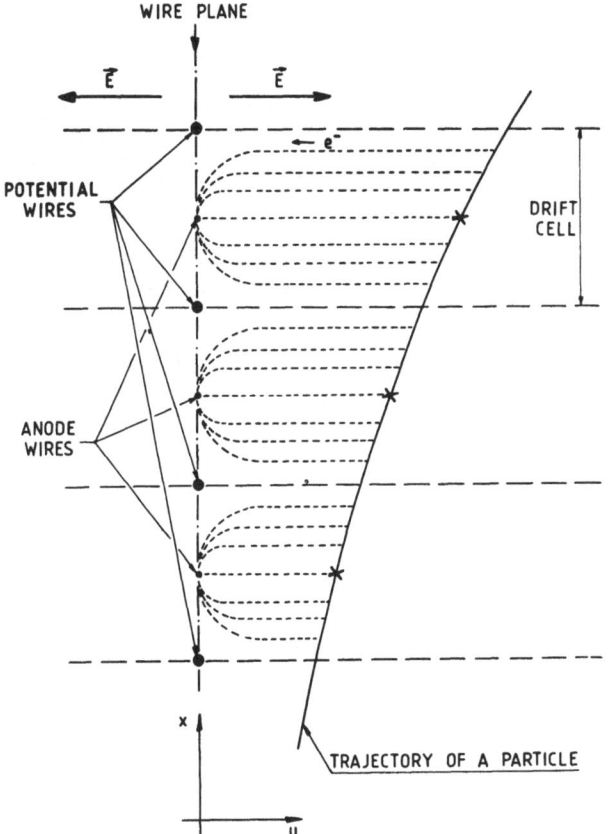

Fig. 1 - Principle of a pictorial drift chamber.

In order to measure high multiplicity events, one has to measure many tracks in the chamber. This can be done by having small drift cells so that the probability for two tracks from the same event to go through the same cell is negligible. Another possibility is to have larger cells, with an electronics which is suited for the handling of multiple hits. This latter choice has the advantage of reducing considerably the number of wires and preamplifiers. This was indeed our choice following in that the "jet chamber" of JADE [7]. I have the pleasure to use this opportunity to express our thanks to all the people involved in the development of this detector and especially to Professor J. Heintze for their permission to copy their electronics and for their kind advices at the various stages of the development of our detector.

So far we have the information about the geometry of the trajectories. From the measured amplitudes on each wire we have also a sampling of the energy loss of each track in the gas. In order to achieve the particle identification, one measures also the momentum of the particle by putting the chamber in an uniform magnetic field. The curvature of the trajectories is a measure of the momentum.

It must be decided which direction to give to the electric and magnetic fields with respect to the beam direction. In order to preserve the cylindrical symmetry around the beam we chose to have the magnetic field parallel to the beam axis. The electric field could then be either parallel to both the beam direction and the magnetic field, or perpendicular to both. If the magnetic and electric fields are parallel, the electrons drift also parallel to the fields. If both fields are not parallel, then the drift direction is tilted by an angle which depends on the fields and the drift velocity. However, in the case of the fields perpendicular to each other, the accuracy on the curvature of the trajectory and hence on the momentum of the particle is just the accuracy on the drift time measurement, which can be fairly good. So this was our choice following again in that respect the jet chamber of Jade.

THE "DIOGENE" DETECTOR

The configuration of Diogène is shown on Fig. 2. A solenoid made with aluminium coils and surrounded on top and bottom by an iron yoke provides a magnetic field of 1 Tesla. This field is homogeneous within 1 % in the whole active volume, 80 cm in length and 70 cm in diameter. Both left and right hand sides of the coil are left available for future detection of neutral particles.

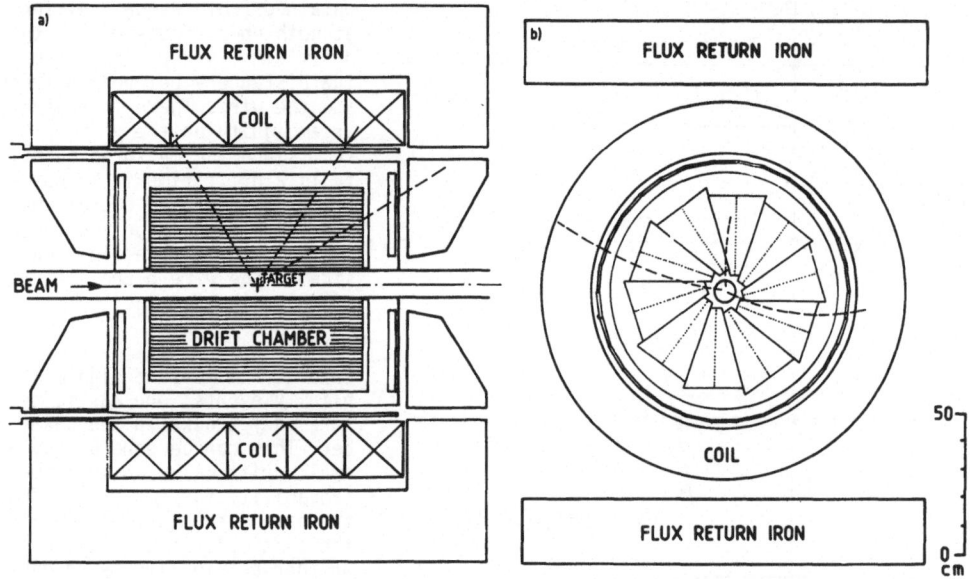

Fig. 2 - Two views of the Diogène detector : a) radial cut in the vertical plane ; b) cut through the target perpendicular to the beam axis.

The drift chamber consists of ten sectors, each of them subtending 36 degrees of azimuthal angle around the beam. In each sector there are 16 sense wires regularly spaced along 26 cm. These wires are made of Ni-Cr, 30 μm in diameter. This configuration leads to an average momentum resolution of about 10-15 % for particles leaving the active volume laterally after hitting 16 wires (Fig. 3). The chamber is filled with an argon (86 %) and propane (14 %) mixture at a pressure that can be varied between 1 and 4 atmospheres depending upon the incident beam energy. This enables us to make a compromise between a good momentum resolution which requires low pressure in order to minimize multiple scattering and a good energy loss resolution which requires high pressure. The argon propane mixture was chosen to fulfil the following requirements : the drift angle should be kept to a reasonably small value ; this implies a small enough value of the drift velocity, which should saturate at the used electric field ; finally the electric field should not be too high in order to avoid high voltage problems. With our argon-propane mixture, the drift velocity saturates at about 40 mm/μs for an electric field of around 1.5 kV/cm leading to a drift angle slightly below 25 degrees. The electric field is fixed by the potential wires (Fig. 1), printed circuits at the top and bottom of each sector and strips of coppered kapton between the sectors. In order to stand the 4 atmospheres pressure, the chamber is put in a stainless steel vessel, 4 mm thick. The gas is separated from the beam by a 1.3 mm thick stainless steel pipe 10 cm in diameter which gives at 90° a lower cut-off of 26 MeV for protons, 12 MeV for pions and 120 MeV for α particles. This pipe might be replaced in near future by a carbon fiber one of the same thickness reducing thus the cut-off to 12 MeV, 5 MeV and 60 MeV for protons, pions and α particles respectively.

In order to help in solving the left-right ambiguity between tracks on the right hand side or the left hand side of the wire plane, two different solutions are used in Diogène. Like in ma-

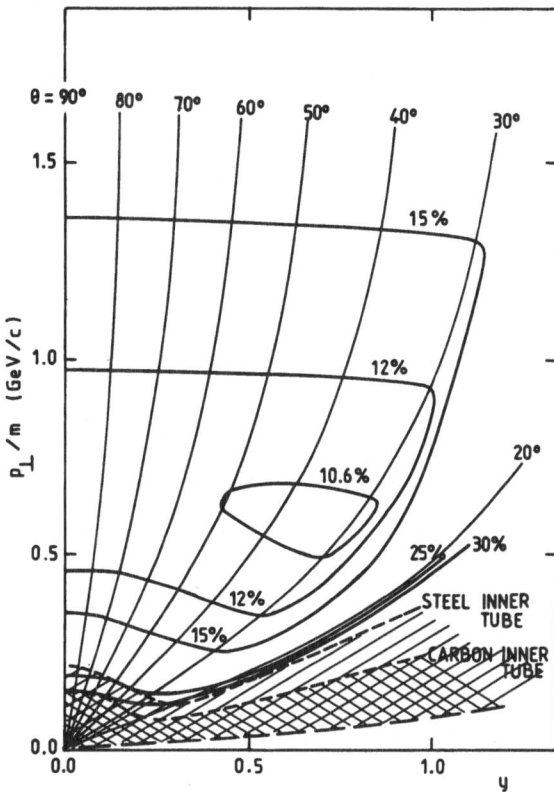

Fig. 3 - Momentum resolution of Diogène for protons detected in the forward hemisphere. The target is located 20 cm upstream of the geometrical centre of Diogène. The cuts due to the inner tube are shown for the actual steel tube and the future carbon fiber one. The shaded area is a dead area for Diogène. The region below the shaded area is to be detected with the plastic wall.

ny other similar drift chambers the sense wires are alternately staggered by ± 200 μm from the average wire plane. If one takes this into account, the χ^2 of the fit of the true track should be significantly smaller than the χ^2 of the wrong track, provided that the number of points is large enough since there is always an uncertainty on the measured coordinates. This usual way can be used wherever the particles come from. However Diogène will be used with a target so that we know where a trajectory is to come from. We decided to shift all the wire planes by about 2 cm from the axis of the chamber. The wrong track symmetric with respect to the wire plane of the true trajectory can then not originate from the target, which makes the solution of the left-right ambiguity very easy.

The trigger of Diogène consists of three parts. The drift chamber is surrounded by a barrel of 30 scintillators (hence the name "Diogène") which measures the multiplicity of particles emitted between 45 and 135° (when the target is at the geometrical center of the detector). Two multiwire proportional chambers are placed upstream and downstream of the drift chamber in order to measure the multiplicity of particles emitted forwards and backwards. A plastic wall is also being built to detect, and identify by means of time and flight and energy loss measurements, all the particles emitted in the forward direction and going through the hole of the iron yoke. It has been shown [8] in the framework of cascade calculations, that the multiplicity of emitted particles is a good trigger on the centrality of the events.

STATUS

The whole detector, that is the ten sectors of the drift chamber, the pressure vessel and the magnet are completed. The upstream and downstream wire chambers are being tested. The plastic wall is being designed.

Tests of one prototype sector were performed with secondary beams of protons and pions from the Synchrotron Saturne, to study its response to the particles. The measured resolutions : 400 μm (full width at half maximum) for the drift distance, and 1.5 cm for the coordinate along the wires, agreed with the expected values.

Tests of one sector and then of four sectors in the magnetic field, with beams from Saturne hitting a Cu target were performed in the first half of this year, confirming the resolution values from the tests of the prototype. The particle identification is very easy as can be seen from Fig. 4. The separation between protons and pions becomes difficult above 600 MeV/c, where we expect nearly no more pions at Saturne beam energies. The true resolution should be even better since several corrections to the measured energy losses have still to be applied.

The full detector is just being tested during this conference.

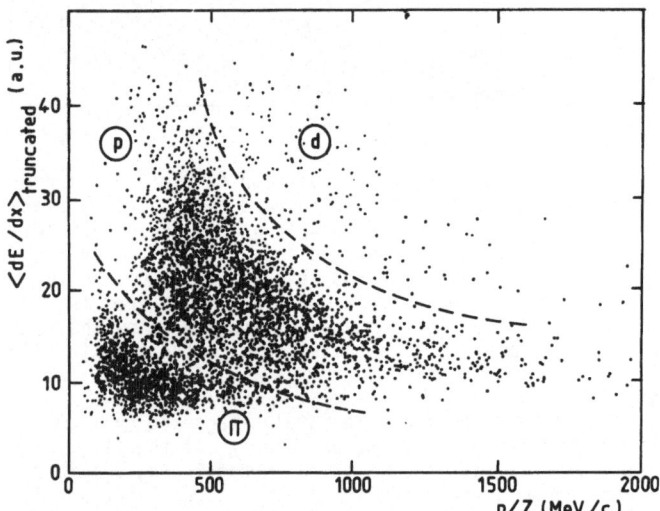

Fig. 4 - Energy loss versus momentum scatter plot from a test run shooting 2.5 GeV protons in a copper target. The separation between $\pi^{\pm}$, p, d is obvious on this figure.

A very important task in such a project is the track reconstruction program. The problem is to put together points belonging to the same track, and at the same time to solve the left-right ambiguity. This has to be done for all the tracks of the event. None of the programs existing for large detectors of the same kind used by high energy physicists could be applied to Diogène since the detector size is much smaller (typically the number of wires is one order of magnitude smaller). The program which is running takes advantage of the fact that we know the vertex position. Since the z resolution (coordinate along the wires) is much worse, by a factor of 40, than the (x, y) resolution, the program more or less decouples the treatment in the (r,z) plane and in the (x,y) plane. In the (r,z) plane, a track should be nearly a straight line. Thus the program looks sector by sector for peaks in θ histogram, θ being the polar angle. One peak should correspond to one or a few tracks. In the (x,y) plane, a track should be a circle. The program makes some clever choice of two points inside a θ peak, one of them being measured on an outer wire, the other one being measured on an intermediate wire. These two points together with the target define a circle in the (x,y) plane. The program looks for points close to this circle, and, if a large enough number of points is selected, fits a track through these points. If some criteria on

the quality of the fit are fulfilled, the program assumes it is a true track. Otherwise it selects another couple of points, etc... The left-right ambiguity is easily
solved by the fact that the wrong track cannot stand the tests of coming from the target, due to the shift of the wire plane with respect to the beam axis. Another advantage of this shift is that no track, even with the highest rigidity can stay along
the wire plane where there is non linearity in the drift path measurement. It also
cannot stay along the frontier of one sector where there are important edge effects.

The program has been first tested on simulated events. Since January 1982 where there
was the first test run with the magnetic field, it is being steadily used and improved with real data. An example is shown in Fig. 5.

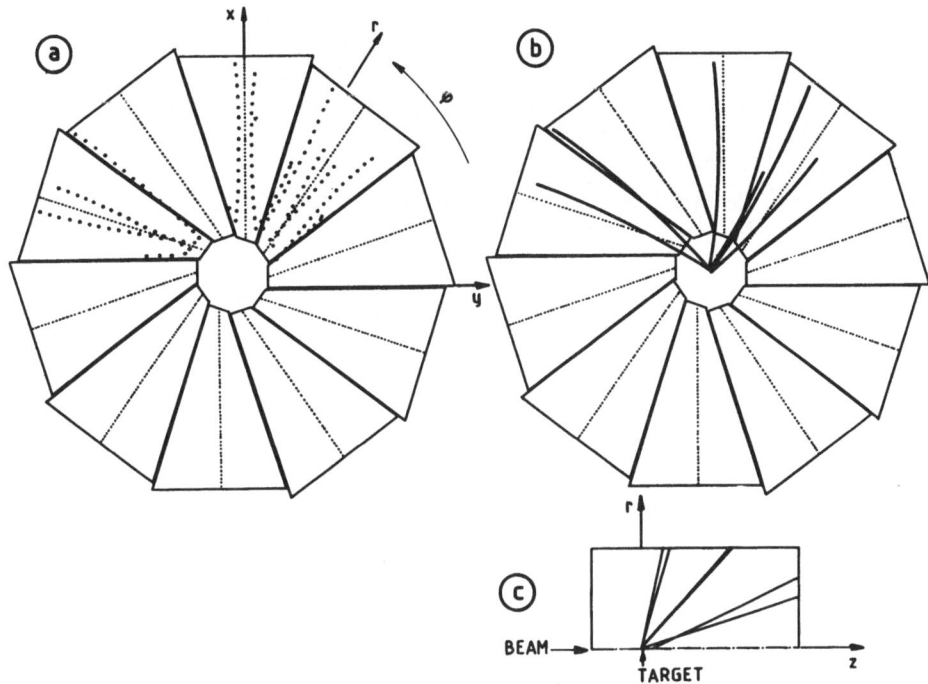

Fig. 5 - Track reconstruction of one event from the test run of four sectors with a
800 MeV/A α-particle beam hitting a Cu target : a) (r,φ) projection of all measured
points with the two possibilities due to the left-right ambiguity ; b) (r,φ) projection of the event after reconstruction by the program. c) (r,z) projection of the
reconstructed tracks.

CONCLUSION

With Diogène, we have a detector which can handle charged particle multiplicities up
to at least 40. The maximum of counts which can be recorded per wire is 8, limiting
theoretically the possible multiplicity to 80. The actual limit is certainly smaller
due to double track resolution and the difficulty to reconstruct the pattern, which
increases exponentially with the density of tracks. The drift chamber covers about
85 % of the full 4π solid angle, with polar angles ranging from 20° to 155°. However
the momentum resolution becomes poor at very forward and backward angles when the number of hitted wires becomes small. The following resolutions are expected 10-15 % for
the momentum, 1° to 2° for the polar angle and less than 1° for the azimuth φ. The various particles, $\pi^{\pm}$, p, d, t ... can be easily identified. The measurable energy range
for light particles is very large as can be seen from Fig. 3. A drawback of Diogène
is that it does not measure neutral particles, thus leading to some lack of information

about the events. The situation is however more or less the same with other large solid angle detectors. The facts that our coils are in aluminium (and not in copper)
and that the yoke is only located at the top and the bottom of the chamber will allow
us in future to put detectors for neutral particles on the sides of Diogène.

REFERENCES

[1] See for instance the proceedings of the Workshop on Future Relativistic Heavy
    Ion Experiments, GSI Darmstadt, October 1980, GSI 81-6 and the proceedings of the
    5th High Energy Heavy Ion Study, Berkeley, May 1981, LBL-12652.
[2] R. Stock, 5th High Energy Heavy Ion Study, LBL-12652 (1981) p. 284.
[3] S.E. Koonin, Phys. Lett. 78B (1978) 556.
[4] J.P. Alard, J. Augerat, R. Babinet, F. Brochard, Y. Cassagnou, J.P. Costilhes,
    L. Fraysse, J. Girard, P. Gorodetzky, J. Gosset, J. Julien, C. Laspalles, M.-C.
    Lemaire, D. L'Hôte, B. Lucas, G. Montarou, A.P. Papineau, M.J. Parizet, J. Poitou,
    C. Racca, M. Suffert, J.C. Tamain and Y. Terrien, Letter of intent n°8 and Propo-
    sal n°43, Laboratoire National Saturne.
[5] H. Gutbrod, this conference.
[6] J. Heintze, Nucl. Instr. Meth. 156 (1978) 227.
[7] W. Farr and J. Heintze, Nucl. Instr. Meth. 156 (1978) 301.
[8] J. Cugnon and D. L'Hôte, submitted for publication.

THE MUNICH RF-RECOIL SPECTROMETER

S.J. Skorka, K.E.G. Löbner, R. Pengo, U. Quade, K. Rudolph, I. Weidl
University of Munich, 8046 Garching, Germany

Abstract

Design and recent applications of the Munich RF-recoil spectrometer
are described. This device allows the identification of heavy ion
reaction products at 0° within a large acceptance angle. Beam projectiles are suppressed by a crossed field radio-frequency velocity selector. Velocity, energy, and energy loss of the recoils of interest are
determined by a high resolution time-of-flight measurement and with
a multiparameter ionization chamber. The performance of the instrument is demonstrated by describing its application in the search for
heavy ion radiative capture, in subcoulomb fusion measurements and in
the investigation of high angular momentum states in ^{27}Al.

Introduction

Recoil spectrometers allow the detection of nuclear reaction
products recoiling from the target independent of the high intensity
of the projectile beam.

Their functions are: 1) The suppression of the projectiles leaving the target more or less in their original direction and with their
original energy: 2) The detection and identification of the recoiling
reaction products.

It is useful to distinguish between two basically different ver-
sions of these instruments. In one version (often called recoil mass
spectrometer) combinations of magnetic and electric fields (including
higher order magnetic fields) are used for both suppression of projectiles and identification of recoils. Their features are mass dispersion with high mass resolving power (500 and more), double focusing of recoils of given mass, and narrow charge window. In some designs isochronism is sacrificed. They are large in size and relatively
expensive. An instrument of this kind is in use at Rochester[1];
others are under construction or are being designed at Osaka[2], Daresbury[3] and Legnaro[4].

In the second version electromagnetic fields are employed only
in order to suppress the projectiles. Identification of recoils is
achieved by nuclear detection devices like time-, ΔE-, and E-detectors,
position sensitive detectors or detectors allowing the observation of

the recoil decay properties (e.g. α-decay). These spectrometers offer no mass dispersion and have a limited mass resolving power. Generally the recoil transport is highly isochronous making high velocity resolution possible. Based on this property, these instruments might be called recoil velocity spectrometers. Their limited ion optical performance allows a large charge window and they can be relatively small and inexpensive. Examples are: The SHIP (GSI) at Darmstadt[5], the Brookhaven/MIT separator[6] (similar: the MIT/ORNL recoil mass selector) and the instrument described in this article.

Typically, recoil spectrometers are employed at a fixed reaction angle of $0°$ with respect to the incident beam. However, some can be varied in position within a limited angular range around $0°$. Desirable properties of recoil spectrometers are: 1) large acceptance angle, 2) high transmission of recoils, 3) large integral suppression of projectiles (the term integral refers to the full range of projectile velocities), 4) good separation of the recoils of interest from other recoils,and 5) low background of falsely registered and falsely identified particles in the recoil spectra of interest. Sufficient suppression can be achieved by a combination of two (or more) stages of proper filters. A special feature of the Munich recoil spectrometer is the restriction to only one stage and, at the same time, the use of an electric radio-frequency field in a Wien-filter type velocity selector. In this combination an integral suppression of at least 10^8 is possible, together with relative simplicity, high transmission and low cost. In the following section a short description of the Munich recoil spectrometer is presented. More details can be found in a forthcoming article.[7]

The Munich recoil spectrometer

Design and performance of the spectrometer are described with the help of fig. 1. A pulsed projectile beam is employed. Projectiles and recoils leave the target with different velocities. The flight path from the target to the deflector and the frequency of the electric field are selected in such a way that the recoils pass the deflector near a phase angle of π , while the projectiles - generally faster - arrive during the period of reversed field direction. Magnetic fields guide the recoils back onto the optical axis and a lens is provided to focus the recoils into the detector.

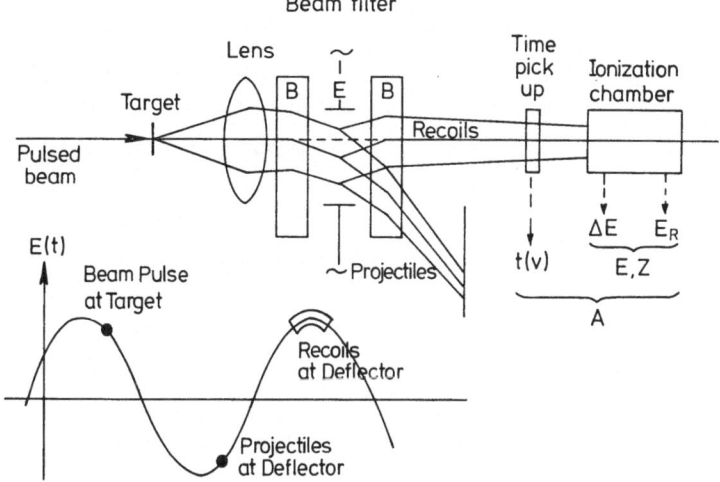

Fig.1: Schematic set-up of the Munich recoil spectrometer. The electric field oscillates in phase
with the beam pulsing. The indicated arrival
times of the projectiles (insert) are possible
examples and depend on the particular experiment.

Background counts in the detector consist of projectiles suffering large energy losses in the target, of beam halo particles hitting the target frame and of projectiles scattered at the edges of the
entrance aperture as well as (once or several times) at other mechanical components. Also, very rare projectiles with extremely low charge
states might contribute to the background. The background particles
cover a large range of velocities below that of the primary beam. The
large deflection angle of the projectiles as a function of their velocity as shown in fig. 2 (in comparison to a dc-filter) is of particular importance in order to understand the satisfactory suppression
achieved with just one Rf-filter. The additional windows at low velocities are very narrow and of no practical importance. Due to the
straight geometry the magnetic quadrupole lens is the only charge dispersive component in the system. Yet the relatively short distance
from lens to detector (approx. 2 m) and the large entrance windows
of the detectors (60-70 mm) allow a charge window of typically
30-40 % (FWHM). The distance from target to deflector plates is, depending on the particular experiment, 1.5 to 2.5 m. The components
can be moved on precision rails without losing the alignment. Rf-fre-
quencies of 2.5 and 10 MHz have been used so far in different experiments.

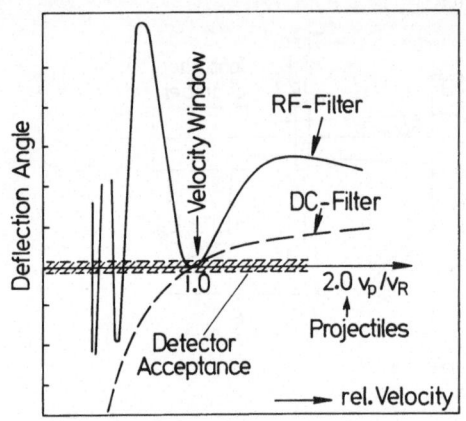

Fig.2: Deflection angle as a function of relative velocity v_p/v_R for a typical case of identical target and projectile masses. The rf-filter shows a larger deflection than the dc-filter particularly for the significant velocity region above the velocity window.

The time detector consists of a grid supported carbon foil of 70 mm diameter placed on the beam axis and inclined by 45°. Secondary electrons from the foil are accelerated by 1.5 kV and detected by a pair of large channel plates connected in series. If necessary a second system can be added in a roof-like geometry to compensate the time errors due to the inclination. With this system a time resolution of about 200 ps is obtained. Since the flight time of the recoils is measured relative to the phase of the beam pulsing system, the beam pulse width generally defines the total time resolution. With a recently installed rebuncher a pulse width of 300 ps is obtained for ^{32}S.

Important parameters of the ionization chamber are the total length of 21 cm, the polypropylene (50 µg/cm^2) entrance window of 64 mm Ø, the Z-resolution of 30-50 (for particles of more than 1-2 MeV/u) and the energy resolution of 0.9 to 1.4 %. The electric field extends transversally, the energy loss in the ΔE part is typically about 60 % of E. Methane of 99.95 % purity flows at a rate of about 100 mg/min at a pressure of 100 mb depending on the expected range of the recoils to be detected.

The total detection efficiency is a product of the acceptance of the recoils through the entrance aperture and the transmission from this aperture into the ionization chamber. The former depends on the kinematic angular spread of the recoils in relation to the entrance angle. The transmission is a function of the atomic charge distribution, of the geometrical dimensions, and of the angular distribution of the recoils within the accepted solid angle. Various supporting grids and those defining the accelerating field of the time detectors limit the transmission, too.

For a particular nuclear reaction the efficiency can be determined by a Monte-Carlo calculation which includes all important parameters of influence. In a recent fusion experiment near Coulomb energy (e.g. ^{32}S+^{104}Ru) the acceptance of the 32 mrad entrance aperture (corresponding to a solid angle of 3 msr) was approximately 50 %, the transmission roughly 40 %. The resulting high total counting efficiency of about 20 % shows the advantage of $0°$-detection in heavy ion nuclear reactions utilizing the kinematic concentration of the flux in the forward direction.

The efficiency has been carefully checked by several experiments. In one measurement the yields of special radioisotopes from fusion reactions collected directly behind the entrance aperture were observed and compared to the corresponding yields behind the window of the ionization chamber. These measurements agreed with the results of the Monte-Carlo simulations within 15 % or less. Another method of experimental determination of the transmission employed a light particle coincidence as explained in fig. 3. Each α-particle observed by a ring detector at $180°$ from the reaction ^{16}O(^{12}C,α)^{24}Mg* populating a particular level in ^{24}Mg indicates the injection of a ^{24}Mg recoil of defined energy and solid angle into the recoil spectrometer. Consequently, the coincidence rate is a measure of the transmission. Again, the agreement between measurement and Monte Carlo simulation is very good.

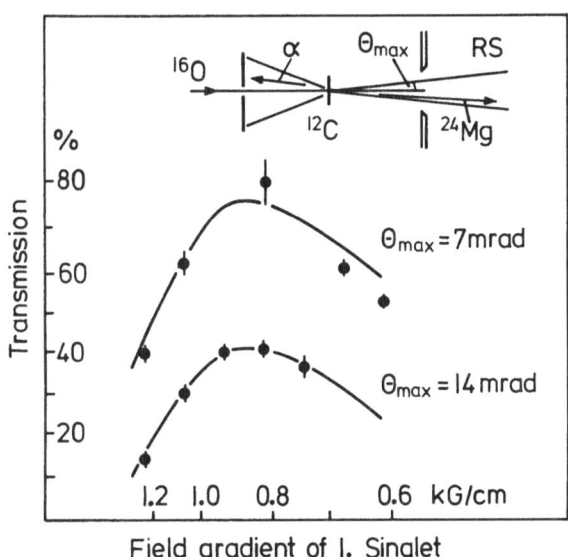

Field gradient of I. Singlet

Fig. 3: Comparison of measured (points) and Monte Carlo simulated transmission (line) through the spectrometer as a function of the lens setting. Experiment: ^{12}C+^{16}O → ^{24}Mg+α . The ^{24}Mg-nucleus is detected within the recoil spectrometer in coincidence with the α-particle which is detected in a ring-detector. The angle θ_{max} of the recoils is determined by the distance target to ring detector.

Applications of the recoil spectrometer

In this section a few examples of recent applications of the Rf-spectrometer are briefly described with the aim of illustrating the possibilities of this instrument.

a) Radiative capture of heavy ions.[8]

This kind of measurement represents the 'classic' application of 0°-recoil spectrometers because of the small kinematic angular spread

$$\Delta\alpha \simeq \frac{E_{\gamma}}{\sqrt{2 \ Mc^2 E}}$$

of the recoiling capture products with mass M and energy E. In the case of the reaction $^{12}C(^{12}C,^{24}Mg)\gamma$ at the E_{CM} = 19.3 MeV resonance this spread is as small as $\simeq$ 20 mrad, even assuming a γ-ray of 20 MeV being involved in the deexcitation. This spread is smaller than the acceptance angle of the recoil spectrometer. On the other hand the intense flux of Rutherford- and multiply scattered projectiles prevents any direct access to the capture products in a conventional measurement. Due to the particularly high efficiency of the recoil spectrometer in this favourable case the expected detectable cross section should be as low as 10 nb. However, background from residual target impurities of N_2 which could not be removed even with a simultaneous target bombardment with high intensity electrons, limited the sensitivity of the experiment to about 100 nb. Fig. 4 shows the velocity spectrum

<u>Fig. 4:</u> Search for radiative capture of $^{12}C+^{12}C \rightarrow ^{24}Mg$ at the 19.3 MeV resonance. Dashed line: approximate shape of velocity window (transmission as a function of recoil velocity), histogramm: Background spectrum from target impurities (see text). Points: Expected velocity distribution of ^{24}Mg recoils for a capture cross section of 200 nb according to a Monte-Carlo simulation. Conclusion: σ_{capt} < 100 nb.

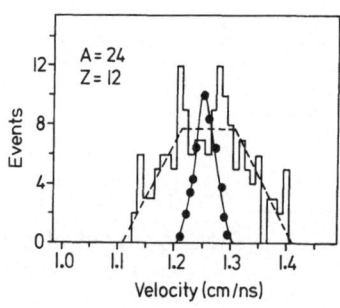

of events identified as ^{24}Mg recoils by the mass and energy loss determination. The velocity window of the recoil spectrometer displays a "white" spectrum of $^{14}N(^{12}C,np)^{24}Mg$ reaction products.

b) Subcoulomb fusion of heavy ions.[9]

Important information on the as yet not fully understood subcoulomb fusion cross section of medium mass nuclei can be obtained from the analysis of the excitation function, particularly below 1 mb. Systematic measurements with different target/projectile combinations are required. The recoil spectrometer with its high efficiency at $0°$ is again the proper instrument. In the reaction

$$^{32}S + ^{104}Ru \rightarrow ^{136}Nd^* \rightarrow \text{Evaporation Residues}$$

the average recoil energy is 27 MeV. At these low energies (and relatively high masses) the mass resolving power of the flight time/energy detection system is not sufficient to resolve individual masses of the fusion products. Knowledge of the relative intensities of the different evaporation residues is, however, necessary to determine the total counting efficiency. This information can be obtained by an analysis of the velocity spectrum (fig. 5), which offers - complementarily to the low mass resolution - a relatively high resolution.

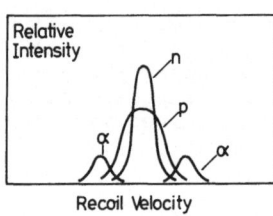

<u>Fig. 5:</u> Schematic velocity spectrum of pure n-, p(+xn)-, and α(+xn)-evaporation residues accepted by the $0°$-recoil spectrometer. The velocity distribution reflects the kinematic energy spread due to the light particle recoil effect. The central part of the α-evaporation residues is cut off by the limited angular acceptance.

To illustrate the residual background elimination fig. 6 shows a ΔE,E-plot from one of the runs $^{32}S + ^{104}Ru$ at 115 MeV. The events enclosed in the polygon contain the evaporation residues, not resolved from some of ^{32}S background events. In fig. 7 the corresponding

<u>Fig. 6:</u> Energy loss versus total energy of events from the $^{32}S + ^{104}Ru$ reaction. The fusion products partly overlap with scattered ^{32}S-projectiles. Very high energy projectiles are not stopped in the ionization chamber; some hit the potential grid between ΔE- and E-part of the chamber.

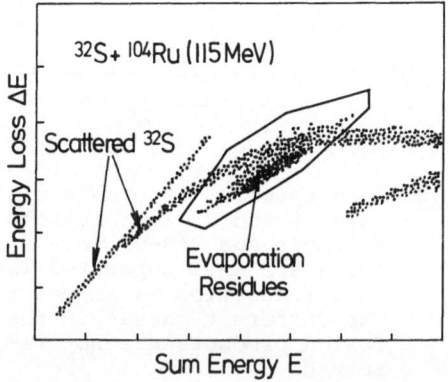

velocity spectra are displayed. Scattered ^{32}S particles are observed with apparent velocities up to the original beam velocity (2.6 cm/ns). The lower part of fig. 7 contains only those scattered projectiles which are registered within the polygon of fig. 6, well separated from the fusion products. The corresponding mass spectrum is shown in fig. 8. The mass range of interest is free of background down to very low cross sections. Finally, in fig. 9 one example of a series of cross section measurements is plotted (^{36}S + Ru isotopes). The reproducibility of the individual cross sections was found to be better than 15 % by repetitions under variable conditions.

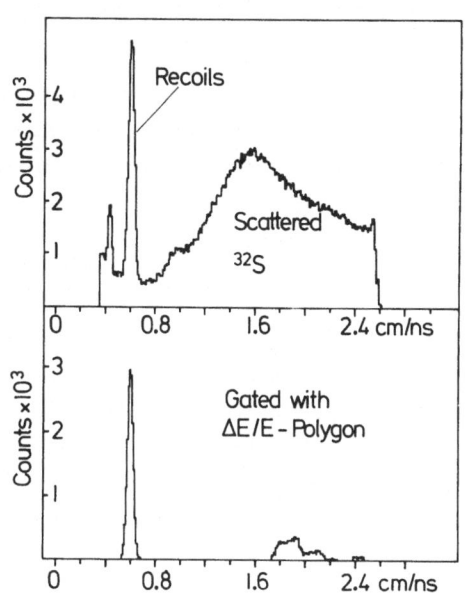

Fig.7: Events in fig. 6 as a function of their (apparent) velocity. Upper part: no gates, lower part: only events contained within the polygon of fig. 6.

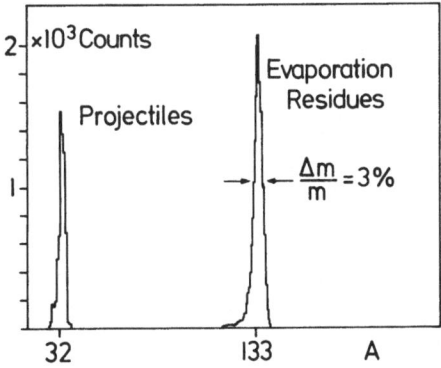

Fig.8: Mass spectrum of events of figs. 6 and 7. Evaporation residues and ^{32}S-projectiles are well separated. The mass resolution is $\Delta m/m = 3$ %. The different masses of the fusion products are not resolved.

c) Spins and decay properties of high spin states in ^{27}Al[10].

This measurement illustrates the usefulness of the recoil spectrometer for investigations of nuclear properties of pure spectroscopic nature.

States of ^{27}Al were excited by the ^{16}O(^{12}C,p)^{27}Al* reaction and the recoils (identified in the recoil spectrometer) observed in coincidence with the protons, which were discriminated against the much more intense α-particles (populating ^{24}Mg) by the time of flight to the ring detector (see fig. 10). At cer-

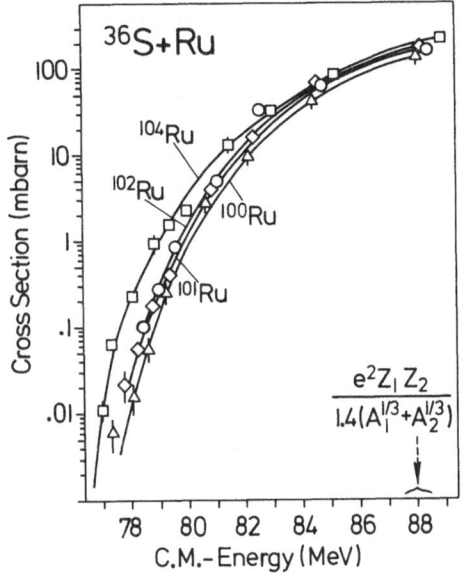

Fig.9:   Subcoulomb fusion cross
         section of ^{36}S+Ru-iso-
         topes as measured with
         the recoil spectro-
         meter. Data collection
         takes about 2 hours for
         the lowest points. The
         analysis of this and
         similar results using
         ^{32,36}S projectiles and
         Ru-, Rh-, Mo- and Pd-
         isotopes suggests form
         fluctuations and pre-
         compound neutron-trans-
         fer to be possible rea-
         sons for the unexpected
         high fusion cross sec-
         tion at sub-barrier
         energies.

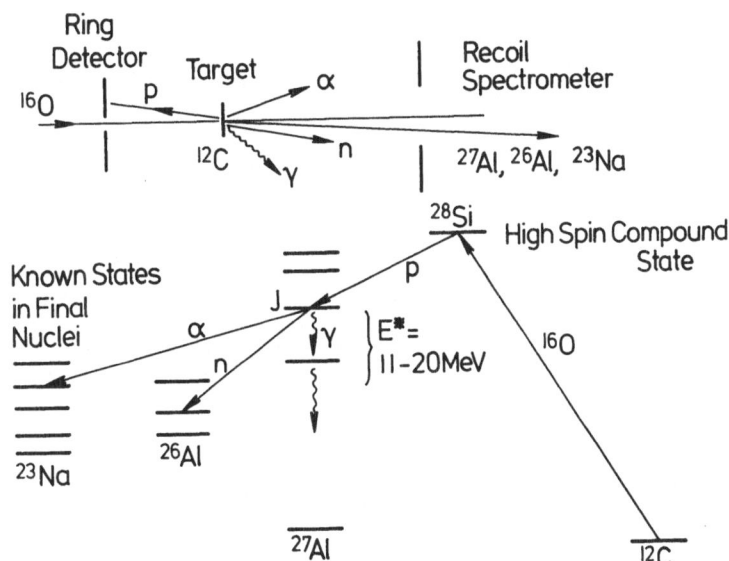

Fig. 10:   Upper part: schematic experimental arrangement for
           the exploration of high spin states in ^{27}Al. The light
           decay products of ^{27}Al are not detected. Spectroscop-
           ic information is obtained from the heavy recoils
           which peak at 0°. Lower part: Illustration of the tran-
           sitions involved. Energy levels are not to scale and
           not complete.

tain incident energies intense proton lines appear in the proton
spectrum indicating the selective population of certain high spin states near the Yrast line of ^{27}Al. The observation of the subsequent
decay residues in coincidence with these protons yields the branching ratios for the decay of the ^{27}Al-states in question:

$$\frac{\Gamma_i}{\Gamma} = \frac{N_c^{(i)}}{\varepsilon^{(i)} N_s}$$

Here the index i denotes one of the decay channels $\gamma+^{27}$Al, $n+^{26}$Al or
$\alpha+^{23}$Na. $N_c^{(i)}$ is the coincidence counting rate of a proton with the
heavy ion reaction product (e.g. ^{23}Na) populated by the unobserved
particle (e.g. α-particle) (see fig. 10). N_s is the corresponding
singles counting rate in the proton line feeding the ^{27}Al-state in
question. The efficiency $\varepsilon^{(i)}$ of the recoil spectrometer depends on
the angular distribution of the unobserved light particles and, hence,
on the spin J of the ^{27}Al state, which can be determined (to within a
few units of $\hbar$) by the condition

$$\sum \frac{\Gamma_i}{\Gamma} = 1.$$

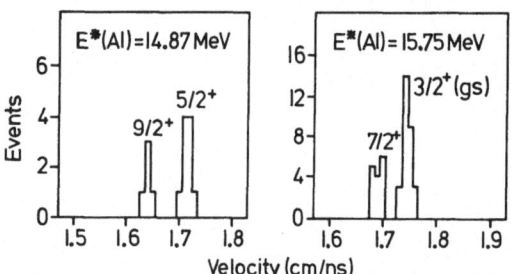

Fig. 11: Two examples of velocity spectra of ^{23}Na from the α-decay of selectively populated levels in ^{27}Al. The velocity window of the recoil spectrometer was centered at the high velocity part of the recoil spectrum (similar to situation in fig. 5). Individual levels of ^{23}Na are separated due to the different recoil energy obtained from the (unobserved) α-particle.

The particular state in the final reaction product, which is populated
by the light particle (e.g. α-particle), can be identified from the
velocity of the recoiling nucleus, which reflects the energy of the
α-particle. Fig. 11 shows two examples. The 5/2$^+$ and 9/2$^+$ states in
^{23}Na (energy difference 2.7 MeV) are well resolved. An interesting
feature of the decay of all the investigated high spin states in ^{27}Al
is that these states are not only <u>populated</u> very selectively, but they
also <u>decay</u> in a very selective manner to only a few final states,
a fact indicating the strong dependence of the decay modes on the structure of the involved states.

We thank Drs. D. Evers and P. Konrad for their valuable contributions to the early part of this work. The project is supported by the Bundesministerium für Forschung und Technologie.

References

1) T.M. Cormier and P.M. Stwertka, Nucl. Instr. and Meth. <u>184</u> (1981) 423.

2) S. Morinobu, I. Katayma and H. Nakabushi, Proc. 4th Int. Conf. on Nuclei far from Stability, Helsingør, 1982, Report CERN 81-09, p. 717.

3) R.G.P. Voss, Proc. of the III. Int. Conf. on Electrostatic Accelerator Technology, 1981, p. 3 and Daresbury Report DL/NSF/TM 38 1978.

4) C. Signorini, priv. communication.

5) G. Münzenberg, W. Faust, F.P. Hessberger, S. Hofmann, W. Reisdorf, K.H. Schmidt, W.F.W. Schneider, H. Schött, P. Armbruster, K. Güttner, B. Thuma, H. Ewald and D. Vermeulen, Nucl. Instr. and Meth. 186 (1981) 423.

6) M. Beckermann, J. Ball, H. Enge, A. Sperduto, S. Gazes, A. DiRienzo and J.D. Molitoris, Phys. Rev. <u>C23</u> (1981)1581 and H.A. Enge, Nucl. Instr. and Meth. <u>186</u> (1981) 413.

7) K. Rudolph, D. Evers, P. Konrad, K.E.G. Löbner, U. Quade, S.J. Skorka, I. Weidl, Nucl. Instr. and Meth.(1982), accepted for publication.

8) P. Konrad, D. Evers, K.E.G. Löbner, U. Quade, K. Rudolph, S.J.Skorka, I. Weidl, Zeitschr. f. Phys, to be published.

9) R. Pengo,D. Evers, K.E.G. Löbner, U. Quade, K. Rudolph,S.J.Skorka, I. Weidl, to be published.

10) I. Weidl, D. Evers, P. Konrad, K.E.G. Löbner, U. Quade, K. Rudolph and S.J. Skorka, to be published; I. Weidl, Thesis (1982), University of Munich.

PHASE-SPACE COOLING OF ION BEAMS

D. Möhl and K. Kilian

CERN, Geneva, Switzerland

1. INTRODUCTION

Cooling of proton and antiproton beams[1-5] has recently led to spectacular results. Amongst these are the following:

- the increase of the lifetime of stored beams by a factor of 50 [6-7] due to the compensation of blow-up by multiple scattering on the residual gas;

- the non-destructive observation of as few as 50 circulating particles[8] (corresponding to a beam current of 3×10^{-11} A), by reducing their frequency spread to less than 10^{-5};

- the accumulation of 2×10^{11} antiprotons from batches of about 3×10^{6} [9];

- the preparation of sharply collimated beams with a final beam size of less than a millimetre and a momentum spread of 10^{-4} at intensities of 10^{8}-10^{9} circulating protons[10].

At CERN, these new techniques have made an ambitious programme of antiproton physics[11] possible, covering a large range of energies from almost 0 to 600 GeV/c in the centre of mass and with beams of very high quality.

In the light of these achievements and prospects it is natural to consider the application of similar techniques to other rare particles like heavy-ion and/or polarized-particle beams. The use of electron cooling in this context was in fact suggested already some time ago by the Novosibirsk group[3,4]. In the present paper we try to work out in some detail the features of both stochastic and electron cooling of heavy-ion beams. The problems considered are cooling times, beam lifetime, particle loss due to electron capture, equilibrium states, and possible experimental techniques and applications.

2. ELECTRON COOLING

2.1 Principle

The principle is simple (see Fig. 1). Electrons produced in a gun travel together with the ions over part of the ion storage ring and absorb transverse and longitudinal energy deviation of the ions by Coulomb interaction. It is instructive to consider the process in the "electron rest frame", moving at the average electron

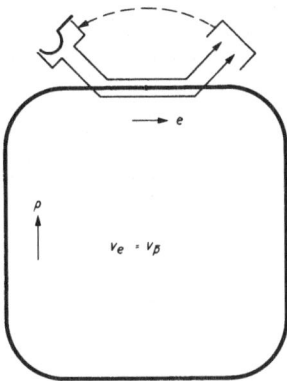

Fig. 1 The principle of an electron cooling system in a storage ring. The electrons travel together with the ions over part of the storage ring. They absorb transverse and longitudinal energy deviation of the ions via Coulomb interactions.

velocity which is equal to the desired ion equilibrium velocity $v_0 = \beta c$. In this frame, ideally, ions are "stopped" because they lose their energy deviation by friction with the electrons, similarly to the slowing down of particles in matter by virtue of their energy loss to the atomic electrons.

2.2 Cooling time and equilibrium

The cooling time can be worked out in the electron rest frame[12], taking only binary collisions into account. This picture permits us to "scale" the expressions obtained by the more general theory to the case of heavy ions of mass $A_i M_p$ and charge $Z_i e$ being cooled by particles of mass $M_c m_e$ (i.e. M_c times the electron mass) and charge $Z_c e$. One finds for the cooling rate

$$\frac{1}{\tau} = \frac{1}{k} \frac{Z_i^2}{A_i} \frac{Z_c^2}{M_c} \frac{(j_c/Z_c e) r_p r_e}{\beta^4 \gamma^5 \theta^3} n_c \ln \left[\frac{\hat{\rho}_e M_c \sqrt{M_c}}{\check{\rho}_p Z_i Z_c \sqrt{Z_c}} \right].$$ (1)

Here k depends on the ion-electron velocity distribution in the electron rest system: $k = 3/(2\sqrt{2\pi}) \simeq 0.6$ for a uniform Maxwellian distribution, $k = 1/2\pi = 0.16$ for a flattened distribution (where the longitudinal velocity spread of the electrons is negligible), $r_p = 1.54 \times 10^{-18}$ m, $r_e = 2.82 \times 10^{-15}$ m, $n_c = \ell_i/2\pi R$ is the fraction of the cooling ring circumference ($2\pi R$) over which cooling takes place.

$$\hat{\rho} = \left(\frac{v}{c} \right) \sqrt{4\pi \frac{j_c}{Z_c e} \frac{Z_c^2 e^2}{M_c m_e}} = \hat{\rho}_e \sqrt{\frac{Z_c}{M_c}}$$ (2)

is the maximum impact parameter due to Debye screening, expressed here in terms of the Debye length in the plasma of cooling particles with the current density $j_c = n_c Z_c e \beta c$. In a similar way

$$\check{\rho} = r_e \left(\frac{c}{V}\right)^2 \frac{Z_i Z_c}{M_c} = \check{\rho}_0 \frac{Z_i Z_c}{M_c} \tag{3}$$

is the minimum impact parameter entering into the Coulomb logarithm. Finally, V is. the r.m.s. of the ion-electron relative velocity and $\theta = V/\beta c = V/v_0$ is the normalized velocity spread. For transverse components of v it can be interpreted as angular spread.

For $Z_i = Z_c = 1$, $A_i = M_c = 1$ we recover the familiar results for electron cooling of protons. When all other parameters are equal (especially j_c, θ, β) then the cooling speed for heavy ions becomes slower with the mass of the "c" and "i" particles and faster with their charge:

$$\frac{1}{\tau} \propto \frac{Z_i^2 Z_c}{A_i M_c} \ln r_0 \frac{M_c^{3/2}}{Z_i Z_c^{3/2}} \, , \tag{4}$$

or neglecting the logarithm dependence

$$\frac{1}{\tau} \propto \frac{Z_i^2 Z_c}{A_i M_c} \, . \tag{5}$$

Equations (4) or (5) permit us to scale the cooling time; we recall that M_c is the mass of the cooling particles in units of the *electron* mass.

We note that the linear dependence $1/\tau \propto Z_c$ comes about because we assume a current limited cooling plasma and hence a constant current j_c of cooling particles. If, instead, one assumes a constant number density the cooling rate is proportional to Z_c^2, and thus symmetric in its dependence on Z_i and Z_c as one might expect from elementary considerations.

Another quantity of interest is the equilibrium beam velocity spread. From equalization of temperatures in the rest system $\left[(A_i m_p/2)v_i^2 = (M_c m_e/2)v_c^2\right]$, the velocity deviations of the ions and the cooling particles are related. Ideally,

$$v_i = \sqrt{\frac{M_c m_e}{A_i m_p}} \, v_c = \sqrt{\frac{M_c}{A_i}} \, \frac{v_c}{42.8}$$

or in terms of the divergence

$$\theta_i = \frac{v_i}{\beta c} = \theta_{(ep)} \sqrt{\frac{M_c}{A_i}} \; . \tag{6}$$

Here $\theta_{(ep)}$ denotes the equilibrium *proton* beam divergence with *electron* cooling. Equation (6) indicates stronger collimation for heavier ions and lighter cooling particles.

These equilibrium considerations are only true in the absence of any other heating mechanism. For intense beams the equilibrium can be considerably affected by scattering amongst the particles of the same beam ("intra beam scattering"). This leads to an intensity-dependent equilibrium spread $\Delta p/p$, as was measured at Novosibirsk[3] and CERN[8] for proton beams. As the interaction is via the Coulomb force of the particles, we expect that for ions of charge Z_i and mass number A, the interaction and hence the equilibrium spread are Z^2/A_i times the spread for the same number of protons. For a more detailed analysis we have to balance the cooling strength $1/\tau$ with the total heating rate $d(\Delta p)/dt$ or dE/dt (for momentum spread or emittance respectively), and this leads to equilibrium values for momentum spread and emittance of the order of

$$\Delta p_{eq} \simeq \frac{d(\Delta p)}{dt} \tau$$

$$E_{eq} \simeq \frac{dE}{dt} \tau \; . \tag{7}$$

Apart from intra-beam scattering, heating can be due to various other processes, for example: collisions with the residual gas, ripple of the guide fields, etc. (see Section 4.2).

2.3 Electron capture

Ions can pick up cooling electrons and get lost from the storage ring owing to the change of their charge state. At the electron densities in question the dominant mechanism is radiative capture:

$$X^{(n+1)+} + e^- = X^{n+} + h\nu \; . \tag{8}$$

This process has been well known since the late 1920's and exact calculations of the cross-section have been performed[13] for hydrogen-like ions, i.e. bare nuclei or partially stripped ions if the captured electron can be assumed to move in the Coulomb field of a (partially screened) point charge.

In connection with electron cooling of protons, recombination has been studied by Budker and Skrinsky[3] and more recently by M. and J.S. Bell[14]. The Novosibirsk Group has successfully used the neutral hydrogen formed in the cooling section to observe and optimize the cooling process.

We will now try to scale these results to heavier ion beams. Using the cross-section for radiative capture given by Seaton[15]

$$\sigma = Ax \left\{ \frac{1}{n} (\ln x) + \gamma_1 + \gamma_2 \, x^{-1/3} \right\} \tag{9}$$

where

$$x = \frac{E_b}{E} = \frac{\text{ground state binding energy}}{\text{electron energy in ion rest system}} \tag{10}$$

and

$$\gamma_1 = 0.140 \, , \quad \gamma_2 = 0.525 \, , \quad A = 2.11 \times 10^{-22} \text{ cm}^2 \, .$$

Bell and Bell[14] obtain the recombination rate $\alpha = \langle v_e \sigma(v_e) \rangle$ by averaging over a Maxwellian or a flattened velocity distribution for v_e. This distribution may be written as

$$f(\vec{v})_{\text{Maxwell}} = \left(\frac{m}{2\pi E_T} \right)^{3/2} \exp \left(-\frac{1}{2} \, mv^2 / E_T \right)$$

$$f(\vec{v})_{\text{flatt.}} = \frac{m}{2\pi E_T} \exp \left(-\frac{1}{2} \, mv^2 / E_T \right) \delta(v_{\parallel}) \, ,$$

where E_T is the typical energy spread between ions and electrons. Defining now $x = E_b / E_T$ in terms of this "temperature" E_T, one gets

$$\alpha(\text{Maxwell}) = Ac^2 \, \sqrt{\frac{2}{\pi}} \, \sqrt{x} \, \sqrt{\frac{E_b}{m_0 c^2}} \left[\gamma_3 + \gamma_4 \, x^{-1/3} + \frac{1}{2} \ln x \right]$$

$$\alpha(\text{flatt.}) = Ac^2 \, \sqrt{\frac{2}{\pi}} \, \sqrt{x} \, \sqrt{\frac{E_b}{m_0 c^2}} \left[\gamma_3 + \gamma_5 \, x^{-1/3} + \frac{1}{2} \ln 4x \right]$$

with $\gamma_3 = 0.429$, $\gamma_4 = 0.469$, $\gamma_5 = 0.334$.

The results for protons can be generalized to fully stripped ions by noting that the ground-state binding energy is

$$E_b \simeq Z^2 E_b (\text{hydrogen}) = Z^2 \cdot 13.6 \text{ eV} \, .$$

For very heavy fully stripped nuclei E_b is even larger than Z^2 times the Rydberg constant. We shall neglect this for the rest of this paper and replace E_b by $Z^2 E_b$, $x \to Z^2 x$ in Eq. (10). Numerically with E_T in eV we have from Bell and Bell

$$\alpha(\text{Maxwell}) = \frac{1.92 Z^2}{\sqrt{E_T}} \left\{ \ln \left(\frac{5.66 Z}{\sqrt{E_T}} \right) + 0.196 \left(\frac{E_T}{Z^2} \right)^{1/3} \right\} \times 10^{-13} \ \text{cm}^3/\text{s}$$

$$\alpha(\text{flatt.}) = \frac{3.02 Z^2}{\sqrt{E_T}} \left\{ \ln \left(\frac{11.32 Z}{\sqrt{E_T}} \right) + 0.14 \left(\frac{E_T}{Z^2} \right)^{1/3} \right\} \times 10^{-13} \ \text{cm}^3/\text{s} \ .$$

These are Eqs. (27) and (28) of Bell and Bell[14] generalized for $Z \neq 1$. Usually, $E_T \simeq 0.1\text{-}1$ eV so that one can neglect the $\frac{1}{3}$ power term especially for large Z. Hence, we find the scaling of α with charge state

$$\alpha \propto Z^2 \ln \left(aZ/\sqrt{E_T} \right) , \qquad a = \begin{cases} 5.66 \ \text{Maxwell} \\ 11.32 \ \text{flattened} \end{cases} \tag{11}$$

or, neglecting the logarithmic dependence,

$$\alpha \propto Z^2 \ . \tag{12}$$

The recombination rate $1/\tau_R = \alpha n_e n_c$ is obtained by multiplying α with the volume density n_e of the electron beam. For the cooling as for the recombination problem one has to include the ratio n_c (cooling length/storage ring circumference), which gives the fraction of time the ion stays inside the electron cloud; τ_R may then be interpreted as the lifetime of the ion beam against recombination. Note that an equation similar to this was given by Budker and Skrinsky[3] who write:

$$\alpha = 19.4 Z^2 r_e^2 c^2 \ \frac{\ln[Zc/(137v)]}{137 \ v_e} \ .$$

Our main result is that the recombination rate scales like $Z^2 \ln$ a $Z/\sqrt{E_T}$ or approximately $1/\tau \propto Z^2$. This is true for hydrogen-like systems, i.e. for bare nuclei. For partially stripped ions Seaton's formula (10) is no longer true as the captured electron will find itself in a shielded potential into which its orbit may partially plunge.

The difference is most pronounced for singly stripped ions. However, from the work of Bates[13] one can conclude that the recombination coefficient even for singly charged ions can be estimated to within about 40% by using formula (10) with $Z = 1$. This is explained by the fact that for the higher state orbits the ion looks more and more hydrogen like and the unexcited state of the valence orbit contributes only a fraction of 25-50% to the total capture probability (see Tables 1 and 2 taken from the work of Bates). With these reservations we will use Eqs. (10) to (12) for arbitrary ions of charge state Z_i.

Table 1

Total radiative recombination coefficients at 250 K (0.02 eV)

Ion	H^+	He^+	Li^+	C^+	N^+	O^+	Ne^+	Na^+	K^+
$\alpha(10^{-12}$ cm^3/s)	4.8	4.8	3.7	4.2	3.6	3.7	3.4	3.2	3.0

Table 2

Radiative recombination of H^+ ions (from Ref. 14)

Principle quantum number n	Rate coefficient α (in cm^3/s) for an electron temperature of:		
	1000 K (0.08 eV)	4000 K (0.35 eV)	16000 K (1.4 eV)
1	5.07×10^{-13}	2.56×10^{-13}	1.20×10^{-13}
2	2.79×10^{-13}	1.32×10^{-13}	5.63×10^{-14}
3	1.88×10^{-13}	8.44×10^{-14}	3.19×10^{-14}
.	.	.	.
.	.	.	.
.	.	.	.
Sum α_Σ	1.99×10^{-12}	7.85×10^{-13}	2.93×10^{-13}

2.4 Comparison of cooling and recombination rate

In the approximation outlined above, the beam lifetime in the presence of recombination now scales as

$$\tau_r \propto \left[Z_i^2 \ln (aZ_i/\sqrt{E_T}) \right]^{-1} , \tag{13}$$

or approximately as

$$\tau_r \propto Z_i^{-2} ,$$

whereas the electron cooling time from expression (4) varies as

$$\tau_c \propto \frac{Z^{-2}A_i}{\ln(r_0/Z_i)} .$$

The ratio τ_r/τ_c thus becomes more unfavourable (roughly as $1/A_i$) as the ion mass increases. This is illustrated in Table 3, where cooling and recombination time are scaled from protons to heavier ions.

We also see from Table 3 and Eqs. (4) and (11) that electron cooling times tend to become long for singly charged heavy ions, whereas recombination times become short for highly stripped heavy nuclei.

Table 3

Scaling of cooling times, recombination time, and equilibrium beam divergence for three different ion species singly or fully stripped. Normalization such that cooling time $\tau = 1$ s, recombination time $\tau_r = 10^5$ s, and equilibrium proton divergence $\theta_{eq} = 0.1$ mrad for electron cooling of a proton beam. Note the different dependence $\tau \propto V^3$, $\tau_r \propto V$.

Particle to be cooled			Cooling plasma	Cooling time τ	Ideal equilibrium divergence	Recombination time τ_r	τ_r/τ
Type	Mass number A	Charge state z_i		(s)	(mrad)	(s)	
p	1	1	e	1	0.1	10^5	10^5
			p	1836		-	
Ne	20	1	e	20	0.02	10^5	5×10^3
			p	3.7×10^4		-	
		10	e	0.20	0.02	580	2.9×10^3
			p	37		-	
U	238	1	e	238	0.006	10^5	4.2×10^2
			p	4.4×10^5		-	
		98	e	0.025	0.006	4	1.6×10^2
			p	46		-	

The alternative of cooling heavy ions by protons, which in turn are cooled by electrons, is also considered in Table 3. It has already been suggested by the Novosibirsk group. "Proton cooling" avoids the recombination problem but it is slower even if the high density obtainable in electron beams could be achieved in the proton beam. Certainly the need for an additional proton storage ring is a serious drawback.

One concludes that electron cooling of heavy-ion beams seems feasible. However if one wants to obtain long storage with low recombination losses one will have to reduce the electron current, which of course also reduces the cooling strength.

3. STOCHASTIC COOLING OF HEAVY ION BEAMS

3.1 Principle

This second cooling method indicated in fig. 2 uses an electronic feedback system to correct the momentum deviation and the betatron oscillations of each individual particle[2,4]. A set of pick-up electrodes senses the error of a particle; the signal is amplified in a high-gain wide-band amplifier and applied on a corrector downstream in the storage ring. Stochastic cooling may be viewed as the competition between three effects: the action of a test particle upon itself via the cooling system ("coherent cooling effect"), the perturbing action of the other particles on the test particle ("incoherent heating effect"), and the perturbing influence of the thermal noise of the amplifier (heating by amplifier noise). Owing to the finite bandwidth (W) of the cooling system, the correction signal of a particle will be present during a time $T_s = 1/2W$, where T_s is the response (i.e. decay) time of the system. All particles passing during this time interval will influence the test particle. To minimize their heating effect it is important to have large bandwidth and a small number of particles.

Another requirement is referred to as mixing: stochastic cooling only works if the test particle continuously changes its neighbours -- if it stays in the same company the perturbing effect is no longer random. After a passage through the system cooling would stop. Fortunately, owing to momentum spread, faster particles continuously overtake the slower ones and under those conditions the incoherent effect keeps the required random character.

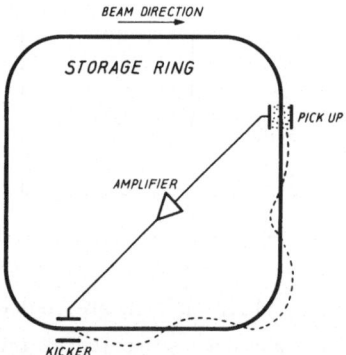

Fig. 2 The principle of (horizontal) stochastic cooling. The pick-up measures horizontal deviation; the kicker corrects angular deviation. They are spaced by a multiple plus one quarter of the betatron wavelength. A position error at the pick-up transforms into an error of angle at the kicker. This angular error is corrected.

3.2 Basic limitations

The above qualitative considerations may help to understand the cooling-rate expression, which can be written as[4]

$$\frac{1}{\tau} = \frac{W}{N} \left[2g - g^2 (M + U) \right] ,$$

(14)

where $g \leq 1$ is related to the amplifier gain, W is the system bandwidth, N the total number of particles in the coasting beam. $M \geq 1$ is the number of turns for mixing. Finally, $U \geq 0$ is the ratio of noise to signal power. In Eq. (14) the 2g term represents the coherent effect, the g^2M term the incoherent effect (including partial mixing $M > 1$), and the g^2U term the heating by amplifier noise. Note that the cooling rate (14) is fastest $1/\tau = (W/N)[1/(M+U)]$ if the optimum gain $g_0 = [1/(M+U)]$ is chosen.

We can now scale from protons to ions of charge $Z_i > 1$. For the same particle number N both the coherent and the incoherent signals will increase in the same way with Z_i; hence their balance is not affected. Mixing is unchanged, but the signal power increases with Z_i^2. Hence, for given amplifier noise

$$U = \text{noise/signal power} \rightarrow U_0/Z_i^2 \ .$$

We can thus use

$$\frac{1}{\tau} = \frac{W}{N}\left[2g - g^2\left(M + \frac{U}{Z_i^2}\right)\right] ,$$

which suggests faster cooling

$$\tau_0 = \frac{N}{W}\left(M + \frac{U}{Z_i^2}\right) \tag{15}$$

for the optimum choice $g_0 = \left[M + (U/Z_i^2)\right]^{-1}$ with the noise term reduced by Z_i^2. We thus find the same limitation with particle number and the same mixing limitations, but a more favourable noise limit than for protons.

Equation (15) is shown in Fig. 3 (from Ref. 4). The linear increase of τ with N is in the range of "large intensity" where $U \ll M$. For small intensity, where $U \gg M$, the curves level off as the noise-to-signal ratio U itself is proportional to 1/N. Together with the N outside the bracket in Eq. (15) this leads to constant cooling time independently of intensity. For multiple charged ions, the conclusion is that this noise-limited regime is "pushed downwards", thus permitting (much) shorter cooling times for (very) small particle number. Stochastic cooling obviously avoids recombination losses. There are no basic differences in the performance limits for stochastic ion and stochastic proton cooling. Figure 3 thus can be used for our scaling purposes; it also includes the working points of some CERN machines and one can see that these points fall on lines with $M \approx 5\text{-}10$.

Present-day systems work with cooling times of 1 s for 10^7 protons or antiprotons or 1 day for 10^{12}. A factor of 10 can probably be gained in the near future by pushing the bandwidth into the gigahertz region.

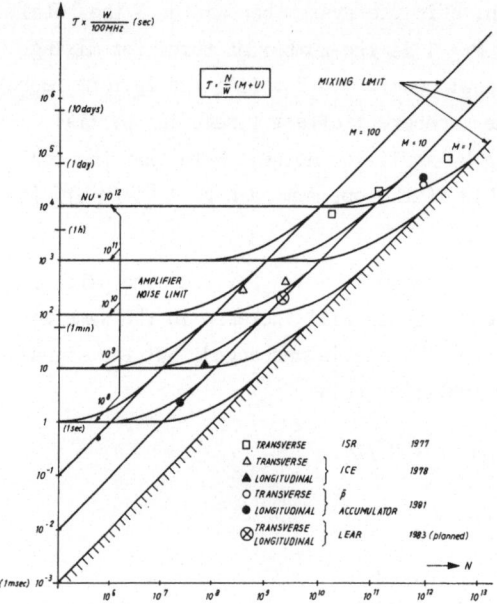

Fig. 3 Normalized cooling time versus intensity. The inclined lines represent the mixing limit. For low intensity the cooling time levels off because of noise. The points represent initial cooling in various proton and antiproton machines. During cooling noise and/or mixing become more important and the cooling time increases, i.e. the working point moves upwards. Note that the vertical scale is normalized for 100 MHz bandwidth.

Another limitation of stochastic cooling should be understandable from the preceding discussion: as the cooling of a beam proceeds, stochastic damping tends to become increasingly slower. This is because it becomes more difficult to detect and feed back the decreasing error signals from the shrinking beam and also because mixing decreases with $\Delta p/p$. This "slowing down" tendency is quite opposite to the behaviour of electron cooling, which tends to work faster for cooler beams -- at least as long as the ion-electron velocity spread is dominated by the ion beam properties.

These differences have led the scepticists to the statement that electron cooling works well when one does not need it and that stochastic cooling gets worse and worse as it proceeds. But more positively: stochastic cooling is very efficient for hot beams like antiprotons from a conversion target or ions from a large emittance source, and electron cooling is well suited to (post)-freezing and/or conditioning beams which are already cool. The combination of both techniques[16] will be used in the low-energy antiproton ring (LEAR), thus profiting from the complementarity of the two cooling techniques.

4. APPLICATION OF PHASE-SPACE COOLING

4.1 Accumulation of rare particles

In principle, all long-lived charged reaction products can be injected into a storage ring, where they can be cooled and accumulated.

One might consider collecting the rare fractions of highly stripped heavy ions which emerge from a stripper, or accumulating polarized reaction products, or rare isotopes, or stable isomers, which emerge from a reaction target. Cooling then allows preparing perfect-quality beams of these particles for new experimental applications. The particles can be used inside the storage ring on internal targets (Section 4.3), in beam-beam interactions (Section 4.4), or on external targets after fast or slow extraction. For a good duty cycle one needs extraction times longer than the accumulation time. Stochastic extraction[17] will be a means of obtaining extremely long spill-out times.

In the CERN Antiproton Accumulator[9,11] (AA) the preparation of a high-density beam of rare particles by cooling is realized on a large scale. Stochastic cooling is used to achieve stacks of several 10^{11} antiprotons from batches of 10^7 particles. These batches emerge from a conversion target where antiprotons are produced from 26 GeV protons impinging on it. Schematically the aperture of the AA vacuum chamber can be subdivided into the precooling region, which houses the fresh antiproton batch, and the stacking region, where antiprotons are accumulated.

A fast precooling system reduces the momentum spread of the injected pulse from 1.5 to 0.25% in 2.5 s before a new pulse arrives. Stochastic cooling is well adapted because the antiproton production rate is relatively small and the beam as taken from the target has extremely large emittances.

The cooled pulse is transferred into the stack region, where additional cooling systems continuously make phase space available. Momentum ($\Delta p/p$) phase space is used for stacking. At the same time, horizontal and vertical cooling (with time constants of the order of 30 min) slowly improve the transverse density as required to obtain high luminosity beams. For proton-antiproton operation of the CERN SPS, accumulation takes $\sim$ 24 hours, and during this time a factor of about 10^4 in antiproton intensity and a factor of about 10^8 in over-all phase-space density is gained in the AA by cooling. This large improvement makes antiproton physics possible with beams which have only one order of magnitude less *circulating* beam intensity (5×10^{11} particles circulating) than SPS proton beams but similar density. The repetition rate is of course much lower than for proton beams but this can be partly made up for by doing experiments with two circulating colliding beams or a circulating beam and an internal target.

4.2 Compensation of emittance increase

In LEAR[16], now in its running-in phase, at CERN, stochastic cooling will be used to compensate for the adiabatic increase of emittances. This increase occurs when a particle beam is decelerated. As is well known, the product of momentum with horizontal (E_H), vertical (E_V), and longitudinal ($E_L = \Delta p/p$ bunch length) emittance is an invariant of (ideal) particle beams as long as no cooling can be applied:

$$E_H p = \text{const} , \quad E_V p = \text{const} , \quad E_L p = \text{const} .$$

Cooling of the three emittances prior to and/or during deceleration makes it possible to reduce these invariants, and this permits slowing down the beam to small momentum without loss. Slowing down for experiments at low energy is used in LEAR because in our world antiprotons are produced in sufficient number only at relatively high energy ($\sim$ 3 GeV if 26 GeV primary protons are converted). By collecting antiprotons at 3 GeV and decelerating them rather than taking low-energy particles from the target, 4 to 7 orders of magnitude are gained in antiproton intensity.

Apart from the adiabatic emittance increase, other heating mechanisms which are governed by their proper time constants tend to spoil the quality of stored beams. Amongst them are:

- *Diffusion* due to betatron resonances driven by small errors of the electromagnetic guide fields of the ring (high-order resonances) together with various sorts of field ripple.

- *Scattering* on the residual gas. Pile-up effects of small-angle Coulomb scattering increase the beam emittance with a time constant τ_{ms}. Single scattering with angles larger than the acceptance causes losses with a typical time τ_{ss}. Both τ_{ms} and τ_{ss} are inversely proportional to the residual vacuum pressure P, and they depend strongly on the particle velocity β and the Lorentz factor γ:

$$\tau_{ms}, \tau_{ss} \propto \frac{\beta^3 \gamma^2}{P} . \tag{16}$$

- *Beam-beam interaction* via electromagnetic fields of a counter-rotating colliding beam.

- *Beam instabilities* due to the wake field of the particles in the beam environment.

Such mechanisms can be compensated if the cooling rate exceeds the blow-up rate. A cooling system with a strong damping rate $1/\tau_c = (1/E)(dE/dt)$ will lead to equilibrium values of emittance E_{eq} and momentum spread $(\Delta p/p)_{eq}$

$$E_{eq} = \tau_c \left.\frac{dE}{dt}\right|_{\text{blow-up}} , \quad \left(\frac{\Delta p}{p}\right)_{eq} = \tau_c \left.\frac{d(\Delta p/p)}{dt}\right|_{\text{blow-up}} . \tag{17}$$

They are small compared to the acceptance if τ_c is short enough. The remaining losses are then due to single scattering with angles or momentum deviations larger than the acceptance. Since large deviations occur with very small probability, cooling in all three directions is therefore expected to increase dramatically the lifetime of a circulating beam up to $\leq \tau_{ss}$. This effect was clearly observed in the cooling experiments at CERN.

4.3 Internal targets

The advantages of an internal target in conjunction with phase-space cooling have been recognized by Skrinsky et al.[3,7]. For LEAR, internal targetry has been promoted since the very beginning and the (machine-dependent) balance between cooling and single as well as multiple scattering on the target has been worked out[18]. LEAR has been designed to include three possible locations for the addition of gas targets. Recently, this idea has been taken up by Pollock et al.[19], who are working on a project based on the use of gaseous targets in an ion storage ring with electron cooling for nuclear physics.

The preferred operation of an internal target will be a cyclic mode with (fast) filling and (slow) depletion of the stack by interactions on the target. The interaction time constant is given by

$$\tau_i = (\nu t \sigma)^{-1} , \tag{18}$$

where ν is the revolution frequency of the stack, t the target thickness (particles per cm^2), and σ the total effective cross-section. The time τ_i should be larger than the cooling time constant τ_c if one wants to ensure that cooling makes a strong improvement in emittance and resolution. On the other hand a large reaction rate r requires a short interaction time constant τ_i, since r depends on τ_i and on the number n of stored particles:

$$r = n \frac{1}{\tau_i} = nt\nu\sigma . \tag{19}$$

So one has to balance reaction rate (statistical precision) against angular and energy resolution. Both can be improved simultaneously with increasing cooling speed.

Maximal count rate with optimal duty cycle (~ 1) is reached when r equals the available average particle flux. Equation (19) shows that this defines an upper limit for $\{nt\}$. At LEAR, where about 10^6 $\bar{p}$/s are available on the average, we get at ~ 800 MeV/c where $\sigma \approx 150$ mb with $n = 10^9$ $\bar{p}$ stored: $\{nt\} \approx \{10^9 \times 5 \times 10^{-9}$ g/$cm^2\}$. This tells us that it does not make sense to have internal proton targets thicker than 5×10^{-9} g/cm^2 in the case of LEAR.

Important properties and advantages of internal targets are:

a) Recirculation of all particles occurs with the high revolution frequency ν ($\geq 10^6$ s^{-1}) except for those which undergo the wanted reactions or scattering by an angle larger than Θ_{ap}, the acceptance limit of the machine (see Fig. 4).

The fast recirculation leads to high luminosity and allows economic use of rare particles. The efficiency for hadronic interaction is

$$\varepsilon \approx \frac{\sigma_h}{\sigma_h + \Delta\sigma_{Cb}} = \frac{\sigma_h}{\sigma} . \tag{20}$$

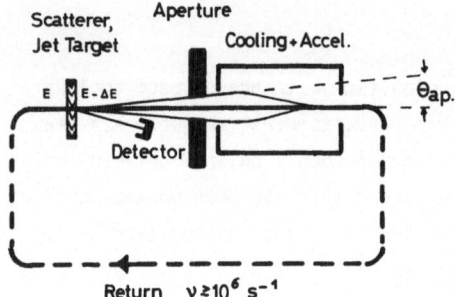

Fig. 4   Schematic view of components relevant for internal gas target operation. Particles which do not interact in the target can be recycled if they are scattered by less than the acceptance angle θ_{ap}. The energy loss ΔE in the target and single scattering below θ_{ap} (and therefore also multiple scattering) are compensated by a phase-space cooling system.

Here $\sigma = \sigma_h + \Delta\sigma_{Cb}$ is the above-mentioned effective total cross-section; σ_h is the total hadronic cross-section and $\Delta\sigma_{Cb}$ is the Coulomb cross-section for scattering with angles larger than the acceptance angle θ_{ap}

$$\Delta\sigma_{Cb} = \int_{\theta_{ap}}^{180°} \frac{d\sigma_{Cb}}{d\Omega} \, d\Omega \approx 261 \text{ mb} \left(\frac{Z_{beam} Z_{target}}{p^* \beta^* \theta^*}\right)^2 . \tag{21}$$

The star indicates c.m. quantities. In Fig. 5 we compare efficiencies for internal and external target operation for $p + p$. If high resolution is needed then external targets need to be thin and they become very inefficient as shown in Fig. 5.

b) Very good angular and energy definition is possible, limited only by the equilibrium between cooling strength and blow-up effects (mostly) on the target.

With electron cooling at LEAR one hopes to reach a momentum spread $\Delta p/p < 10^{-4}$. The corresponding energy resolution in the c.m. system $\delta\sqrt{s} = \delta p_{lab} m_{target} / \sqrt{s}$ is shown for $p + p$ in Fig. 6 and compared with external target operation.

In Fig. 7 we show as an example the equilibrium emittance expected for antiprotons in LEAR with a proton gas jet target which leads to 10^6 interactions per second with 10^9 stored antiprotons. Figure 7 gives an indication of the possible angular resolution $\delta\theta \approx E/\Delta x$, where Δx is the dimension of the beam-target overlap (typical < 1 cm).

Very good energy resolution is needed if one wants to find narrow structures in excitation functions, e.g. in intermediate nuclear compound systems. It may also be desirable in studies of spontaneous positron emission in heavy-ion interactions. Energy and angular resolution are crucial if one plans to do precise measurements close to thresholds[20].

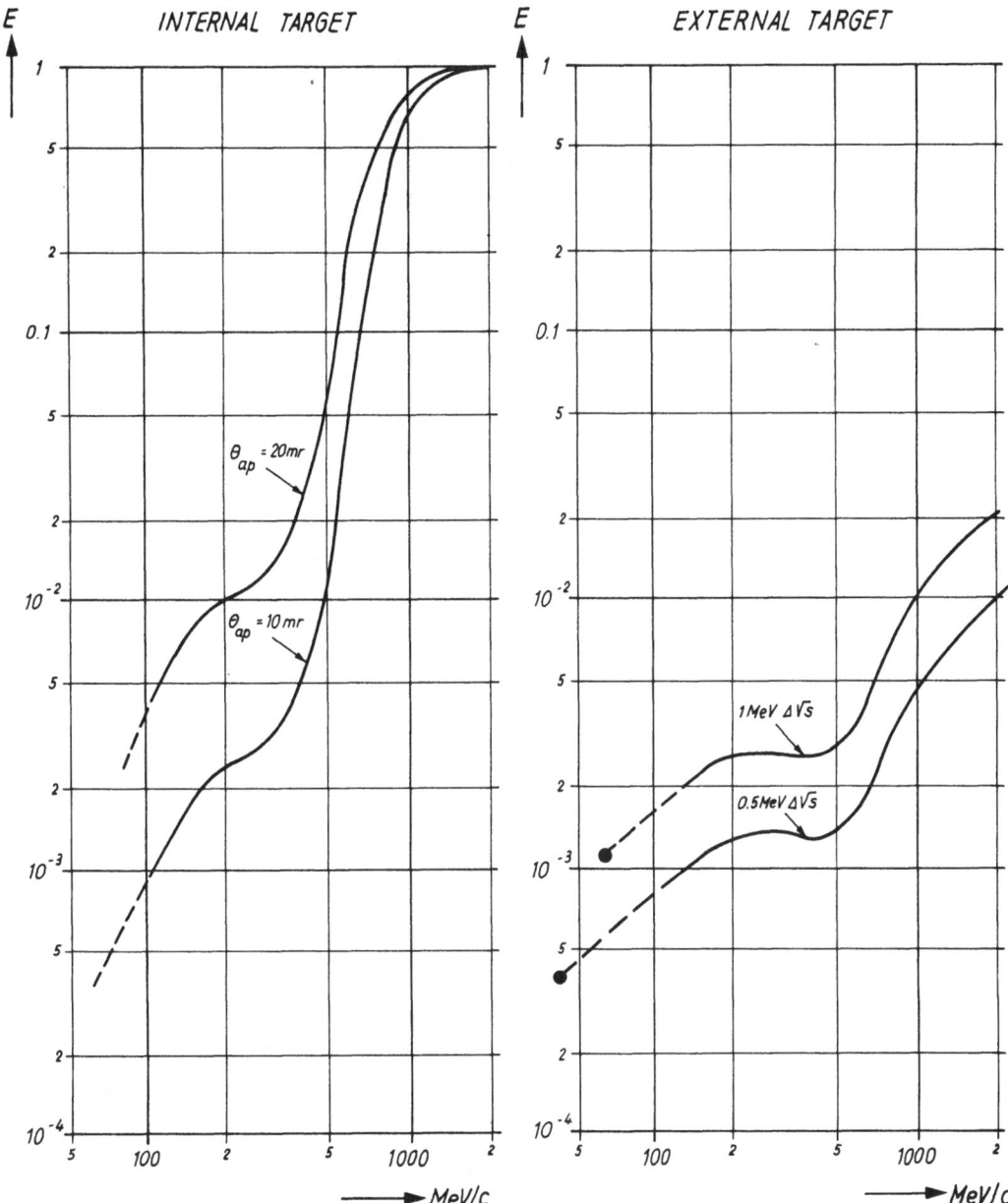

Fig. 5 Target efficiencies for proton proton interactions: a) on an internal hydrogen gas target with perfect phase-space cooling. The resolution here is defined by the cooling. b) On an external hydrogen transmission target with a thickness chosen to give a certain resolution $\Delta\sqrt{s}$.

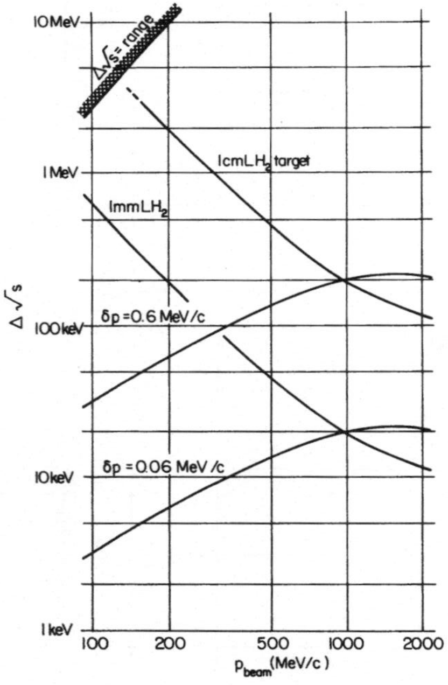

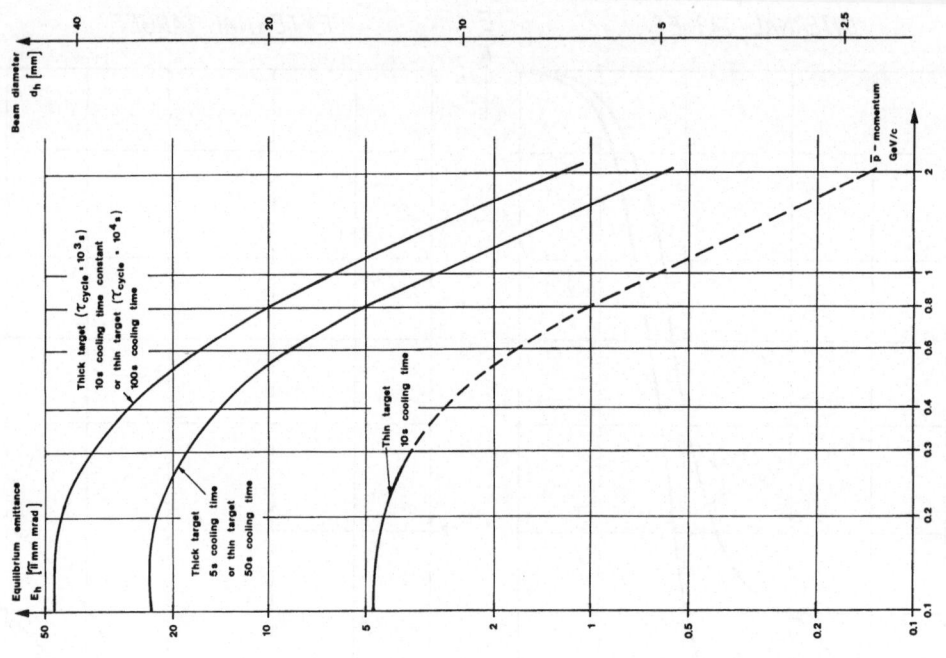

Fig. 7 Equilibrium beam size (horizontal).

Fig. 6 The c.m. energy resolution $\Delta\sqrt{s}$ as a function of the beam momentum in $\bar{p}p$ interactions for constant momentum spread $\delta p = 0.6$ MeV/c and $\delta p = 0.06$ MeV/c and for liquid hydrogen targets of 1 mm and 10 mm thickness.

It is worth noting that the quality of the circulating beam can easily be improved at the expense of "beam particle economy" [Eq. (20)] if one reduces artificially the machine acceptance angle θ_{ap}. This increases $\Delta\sigma_{Cb}$ and reduces the interaction time constant τ_i. This poses no problems if one has a fast enough cycling source (like a cyclotron). The advantage lies in the fact that large deviations in momentum and angle are cut out so that the beam temperature [θ in Eq. (1)] stays smaller and electron cooling becomes *faster*.

c) There are no multiple interactions in the extremely thin target. Hence all kinematical correlations, even for very heavy and/or slow recoil particles, are undisturbed. This is, of course, attractive for coincidence measurements, especially in heavy-ion interactions, where energy loss and angular straggling impose very stringent limits. It might also have applications for atomic physics because, for example, δ-electrons or Auger electrons can be seen undisturbed.

d) Full advantage can be taken of the high density of cooled circulating beams. This is because extraction and beam transport to an external target -- processes both liable to mis-steering and jitter -- are not needed.

e) A polarized hydrogen or deuterium atomic beam can be used[21] as a target with a high degree of polarization and virtually free from carbon and other contaminants. It still gives excellent luminosity in the recirculated beam if it reaches a (realistic) thickness of $\sim 10^{-12}$ g/cm^2.

f) Thin filaments ($\emptyset \lesssim 1$ μm) can be employed as internal targets[22]. Their apparent thickness is reduced to reasonable values owing to the small overlap with the beam. Their advantage is simplicity. Furthermore, they have extremely high brilliance for the reaction products, a very attractive feature for a subsequent magnetic spectrometer.

g) There is virtually no beam dump. Therefore the background conditions can be very clean. One has to worry only if $\Delta\sigma_{Cb}$ is large compared to σ_h. But also under these conditions the (uninteresting) Coulomb forward scattered particles can be absorbed on an aperture limiter, which is installed in the ring far away from the detectors.

4.4 Colliding beams

Hadronic interactions in colliding beams are under study so far only at CERN in the Intersecting Storage Rings and in the SPS $p\bar{p}$ collider. There the beam energies are so high (> 10 GeV) that Coulomb scattering is not important. In the low-energy range blow-up effects become crucial [Eq. (16)]. Here collider operation with protons and heavier ions will be feasible only if sufficient cooling can be applied.

The luminosity L in a collider with coasting co-rotating or counter-rotating (0° crossing angle) beams is

$$L = \frac{1}{4} \frac{N_1 N_2}{U_1 U_2} \frac{A_{min}}{A_1 A_2} \ell c |\beta_1 \mp \beta_2| .$$

(22)

Here ℓ is the length of the interaction region. $U_{1,2}$ are the circumferences, $N_{1,2}$ the number of particles stored, and $A_{1,2}$ the beam cross-sections for the storage rings 1 and 2, respectively. A_{min} is the overlap area of both beams. Ideally $A_1 \approx A_2 \approx A_{min}$. These beam areas and therefore the luminosity are constrained by disruptive effects of the space-charge field of one beam on the other[23]. Phase-space cooling will counteract this effect to some (so far not tested) extent. Furthermore $c\beta_{1,2}$ are the velocities of the stored particles. The + and − signs in Eq. (22) apply for counter-rotating and co-rotating beams, respectively.

With heavy ions losses will occur also owing to stripping or pick-up of electrons. This problem can be reduced when using fully stripped ions and/or machines with very high momentum acceptance Δp. Since $\Delta p/p \approx \Delta Z/Z$ one can store several ionization states which differ by $\sim \Delta Z$. Machines for this purpose have been frequently discussed[24].

If one foresees a storage ring with non-dispersive straight sections one can install there electron cooling, which equalizes the velocities of all particles, independently of their charge states. In the dispersive regions, where different charge states have separated orbits, stochastic cooling could be applied on these orbits individually. Clearly when working with partially stripped ions one has to foresee also dispersion-free interaction regions. One may consider the AA ring at CERN[9] as an example where these criteria (e.g. $\Delta p/p \approx 6\%$) are realized.

Finally, we wish to give some further examples of possible future applications of cooling, which are however quite exotic.

4.4.1 *Experiments with ultimate resolution*

A double-ring system with counter-rotating or co-rotating beams allows the study of a large variety of ion-ion interactions in a wide energy range. If one works with fully stripped ions, and only then, one can obtain optimal energy resolution. Only under these conditions can one avoid the degradation of resolution caused otherwise by ejection of δ-electrons (Landau tails) or by inelastic excitations of electronic states. The resolution of the invariant mass $\sqrt{s}$ in interactions of fully stripped beams depends on the momentum spread of the cooled beams

$$\partial \sqrt{s} = \partial p_1 \frac{(E_2/E_1)p_1 - \cos \theta_{12} p_2}{\sqrt{s}}$$

$$= \partial p_1 \frac{E_2(\beta_1 - \cos \theta_{12}\beta_2)}{\sqrt{s}} .$$

(23)

We see from Eqs. (23) that for ultimate resolution the interaction angle θ_{12} also has to be well defined by cooling.

4.4.2 *Studies of charge exchange in collisions of two ions*

A double-ring system as in Fig. 8 with two partially stripped ion beams (different in charge and/or mass) may be tuned to have nearly the same ion velocities $\beta_1 \approx \beta_2$. The excitation energy of the compound system can be made so small that it comes into the range of typical electronic excitation energies. Equation (23) shows that the energy resolution also gets better with $(\beta_1 - \beta_2)$. One can study excitation functions of atomic charge-exchange processes. The new charge states appear in well-defined directions, where they can be easily measured even in coincidence. In order to illustrate the energy resolution, consider a hypothetical case where, for example, Ca^{20+} and Ca^{10+} beams are stored with $\beta = 0.1$ (5 MeV/A) and $\beta = 0.1 (1 + 3 \times 10^{-3})$. This corresponds to an excitation energy of about 1 keV. Assuming a momentum resolution of $\Delta p/p \approx 10^{-5}$ one would obtain in this case a centre-of-mass resolution of 60 eV.

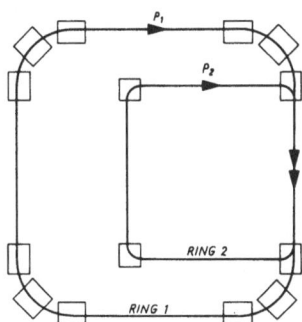

Fig. 8 Schematic view of the double ring system (for $\bar{d}$ production, see Ref. 25). The rings 1 and 2 overlap in a common straight section. Here $\bar{d}$ are produced and stored in ring 1.

4.4.3 *Accumulation of reaction products*

In a double-ring collider (Fig. 8) a given collision energy can be realized with different momentum combinations p_1 and p_2 of the stored beams. What changes is the centre-of-mass velocity of the compound system. For certain conditions one can arrange for two-body reactions of the type $m_1 + m_2 \to m_3 + m_4$ that the momentum vectors in the laboratory system are equal: $\vec{p}_1 \approx \vec{p}_3$ and $\vec{p}_2 \approx \vec{p}_4$ [25]. In the case of heavy ions one can arrange that $\vec{p}_1/Z_1 = \vec{p}_3/Z_3$ or $\vec{p}_2/Z_2 = \vec{p}_4/Z_4$, where Z denotes ionization

states. The consequence is that for such settings a fraction of the reaction products 3 (and) or 4 is permanently accepted in ring 1 (and) or 2. The new particles
have the correct magnetic rigidity but they can be distinguished by different revolution frequencies from their parent particles 1 and 2.

One might think of "breeding" and accumulating rare isotopes, etc. This trick
has been discussed in Ref. 25 for the production of antideuterons in the reaction
$\bar{p}\bar{p} \rightarrow \bar{d}\pi^-$. The limits of this method come with losses induced by elastic forward
Coulomb scattering.

4.4.4 *Production of fully or highly stripped heavy ions*

For phase-space cooling, a stored beam of (partially) stripped heavy ions over-
laps with an intense electron beam of nearly the same velocity. One can easily increase the relative electron-ion energy. Instead of pick-up we expect then knockout
of electrons from the ions, as long as enough energy is available in the collision.
What counts for the maximal obtainable ionization state is the velocity difference
between the ion and the stripper electron.

In order to get fully stripped Xe^{54+} (U^{92+}) on a fixed stripper an ion beam with
> 80 MeV/A (> 250 MeV/A) would be needed. An electron beam with an energy of only
> 40 keV (> 125 keV) overlapping with a stored slow ion beam can, in principle, do
the same.

The heavy ions could be kept at a convenient energy in a storage ring. In a non-
dispersive section the electron beam can overlap with all charge states. Large
$\Delta p/p \approx \Delta Z/Z$ would allow simultaneous storage of all charge states which are produced.
Injection and extraction is on separate orbits in dispersive regions. It could be
done in a pulsed or continuous way. Electron cooling could be performed in a second
non-dispersive straight section or stochastic cooling could be applied to the separated charge states in the dispersive regions. Cooling is needed to avoid blow-up and
to provide sufficient storage time to reach large ionization. As in electron beam
ionization sources (EBIS)[26] also in our example we have to worry about the time constants with which a given charge state is populated. For the production of an ion
with charge $i+1$ from ionization state i the time $\tau_{i \rightarrow i+1}$ is determined by the ionization cross-section $\sigma_{i \rightarrow i+1}$, the current density of the electron beam j, and the fraction η of the storage ring which is filled with the ionizing electron beam

$$\tau_{i \rightarrow i+1} = (\sigma_{i \rightarrow i+1} j \eta)^{-1}.$$

Table 4 gives order-of-magnitude information for time constants for the increase
of ionization of Xe and U ions. In Table 4 we assume $\eta = 3\%$, $j = 16$ A/cm$^2 \cong 10^{20}$ elec-
trons/cm^{2}s and for $\sigma_{i \rightarrow i+1}$ we use a conservative approximation, $\sigma_{i \rightarrow i+1} \gtrsim 2 \times 10^{-14}$ cm^2

Table 4

Order of magnitude of the time constant
for ionization of inner-shell electrons
in Xe and U ions by an electron current
density of 16 A/cm^2

Ion	$i \to i+1$	$\tau_{i \to i+1}$ (s)	E_i (keV)	$\sigma_{i \to i+1}$ (cm^2)
Xe	50 → 51	250	5.5	1.35×10^{-21}
	51 → 52	480	5.5	6.94×10^{-22}
	52 → 53	10^4	34.6	3.4×10^{-23}
	53 → 54	2×10^4	34.6	1.7×10^{-23}
U	88 → 89	3.8×10^3	21.8	8.7×10^{-23}
	89 → 90	7.7×10^3	21.8	4.5×10^{-23}
	90 → 91	1.1×10^5	115.6	3×10^{-24}
	91 → 92	2.2×10^5	115.6	1.5×10^{-24}

$1/(E_i/eV)^2$, which holds when the electron energy $E_e \approx 2E_i$ to $12E_i$ (see Ref. 27).
The separation energies E_i are taken from Ref. 28.

If we have n heavy ions of a charge state i stored, then we can get a continuous
flux of charge state i + 1 of the order $I_{i+1} \sim n/\tau_{i \to i+1}$. For example, with
10^6 Xe^{53+} (U^{91+}) stored we can get 50 Xe^{54+} (4.6 U^{92+}) per second. These higher
charge states can either be cooled and then extracted or they can be accumulated for
internal use in the machine or for further ionization. In order to reach fully
stripped ion states one may consider a cascade of stripper rings each for a certain
range of ionization ΔZ.

High-quality beams of highly or fully stripped heavy ions can be efficiently
handled in a given accelerator structure; such beams could have many applications,
for example:

- in atomic physics studies, where the high-quality beams could be decelerated without losses to very low energy.

- in studies of spontaneous positron emission from scattering of very heavy ions,
where well-defined kinematical parameters and K-shell vacancies already in the beam
particles would be an advantage.

5. CONCLUSIONS

We have discussed some properties and limitations of the new techniques of phase-
space cooling relevant in nuclear and heavy-ion physics. Many applications seem to

be possible in the future. We have tried to indicate some of them. But in order to take full advantage of cooled beams, detectors and experimental techniques will have to be adapted. Phase-space cooling is already applied on a large scale in the anti-proton projects at CERN.

REFERENCES

1) G.I. Budker, Atomnaya Energiya 22 (1967) 346, and Proc. Symposium on Electron-Positron Storage Rings, Saclay, 1966 (PUF, Paris, 1966), p. II-1-1.
2) S. van der Meer, Stochastic damping of betatron oscillations, internal report CERN ISR-PO/72-31 (1972).
3) G.I. Budker and A.N. Skrinsky, Sov. Phys. Usp. 124 (1978) 277;
   G.I. Budker et al., Part. Acc. 7 (1976) 197.
4) S. van der Meer, D. Möhl, G. Petrucci and L. Thorndahl, Phys. Rep. 58 (1980) 73.
5) F.T. Cole and F. Mills, Ann. Rev. Nucl. Part. Sci. 31 (1981) 295.
6) G. Carron et al., Phys. Lett. 77B (1978) 353.
7) G. Budker et al., CERN report 77-08 (1977).
8) M. Bregman et al., Phys. Lett. 78B (1978) 174.
9) E. Jones, The antiproton accumulator AA, a performance report, to be published in Proc. Workshop on Physics with Cooled Low Energy Antiprotons, Erice, 1982.
10) M. Bell et al., Phys. Lett. 87B (1980) 275; see also ref. 7 above.
11) C. Rubbia et al., Proc. Neutrino Conf., Aachen, 1976 (Vieweg, Braunschweig, 1977), p. 683.
    AA-study group, Design study of a proton-antiproton colliding beam facility, internal report CERN PS/AA/78-3 (1978).
    LEAR study group, Design study of a facility for experiments with low energy antiprotons, internal report CERN PS/DL/80-7 (1980).
    For a more recent summary, see for instance J. Gareyte, The CERN p-p̄ complex, Proc. 11th Int. Conf. on High Energy Accelerators, Geneva, 1980, Experientia Supplementum 40 (Birkhäuser, Basle, 1980), p. 79.
12) H.G. Hereward, Artificial damping in the CERN proton storage ring, Proc. Symposium on Electron-Positron Storage Rings, Saclay, 1966 (PUF, Paris, 1966), p. VIII-3-1.
13) See, for instance D.R. Bates and A. Dalgarno in: Atomic and molecular pro-cesses (R.D. Bates, ed.) (Academic Press, New York-London, 1962), p. 245;
    or H.A. Bethe and E.E. Salpeter, Quantum mechanics of one- and two-electron atoms (Springer, Berlin, 1957).
14) M. Bell and J.S. Bell, Capture of cooling electrons by cool protons, preprint CERN TH-3054 (1981), to be published in Particle Accelerators.
15) M.J. Seaton, Mon. Not. Royal Astr. Soc. 118 (1958) 477.
    See also Bates, ref. 13.
16) P. Lefèvre, Construction of the LEAR-facility, a status report, to be published in Proc. Workshop on Physics at LEAR with Cooled Low Energy Antiprotons, Erice, 1982.
    D. Möhl, Phase-space cooling techniques and their combination in LEAR, same Proc.
17) R. Cappi, W. Hardt and C. Steinbach, Ultraslow extraction with good duty factor, Proc. 11th Int. Conf. on High Energy Accelerators, Geneva, 1980, Experientia Supplementum 40 (Birkhäuser, Basle, 1980), p. 335.
    R. Cappi, R. Giannini and W. Hardt, Ultraslow extraction from LEAR, Status report, to be published in Proc. Workshop on Physics at LEAR with Cooled Low Energy Antiprotons, Erice, 1982.
18) R. Klapisch, Chairman's report on the CERN Workshop on Intermediate Energy Physics, CERN-PSC-DI-77-50 (1977).
    K. Kilian, U. Gastaldi and D. Möhl, Deceleration of antiprotons for physics experiments at low energy (a low-energy antiproton factory), Proc. 10th Int. Conf. on High-Energy Accelerators, Protvino, 1977 (IHEP, Serpukhov, 1977), Vol. 2, p. 179.

K. Kilian and D. Möhl, Gas jet target in LEAR, CERN p̄ LEAR Note 44 (1979).

K. Kilian, Antiproton interactions above threshold, Proc. 5th European Symposium on Nucleon-Antinucleon Interactions, Bressanone, 1980 (CLEUP, Padua, 1980), p. 681.

K. Kilian, D. Möhl, J. Gspann and H. Poth, Internal targets for LEAR, to be published in Proc. Workshop on Physics at LEAR with Cooled Low Energy Antiprotons, Erice, 1982.

19) R.E. Pollock, D.W. Miller and P.P. Singh, Proposal to the National Science Foundation: The IUCF Cooler-Tripler, Proposal for an advanced light-ion physics facility, Indiana University, Phys. Dept., Bloomington, Indiana (1980).

20) P.D. Barnes et al., Threshold studies at LEAR, to be published in Proc. Workshop on Physics at LEAR with Cooled Low Energy Antiprotons, Erice, 1982.

21) L. Dick, J. Jeanneret, W. Kubischta and J. Antille, The CERN polarized atomic hydrogen beam target, Proc. Conf. on High-Energy Physics with Polarized Beams and Polarized Targets, Lausanne, 1980 (eds. C. Joseph and J. Soffer), Experientia Supplementum 38 (Birkhäuser, Basle, 1980), p. 212.

22) K. Kilian and D. Möhl, Internal hydrogen or solid targets and polarization experiments at LEAR, to be published in Proc. Workshop on Physics at LEAR with Cooled Low Energy Antiprotons, Erice, 1982.

23) M. Sands, internal report SLAC 121 (1970);
E. Keil, Beam-beam interactions in p-p storage rings, in Theoretical aspects of the behaviour of beams in accelerators and storage rings, CERN 77-13 (1977), p. 314.

24) G. Hortig, Nucl. Instrum. Methods 45 (1966) 347.
J.G. Cramer, Nucl. Instrum. Methods 130 (1975) 121.
B. Franzke, K. Blasche and B. Franczak, Stripsy oder Recycling von Stripper abfällen, in Gesellschaft für Schwerionenforschung Jahresbericht 1977, GSI-J-1-78, p. 157.
M.G. Nagaenko, E.M. Reshetvikova, Yu.P. Severgin and I.A. Shukeilo, Beam storage for increasing heavy ion synchrotron intensity, Proc. 11th Int. Conf. on High Energy Accelerators, Geneva, 1980, Experientia Supplementum 40 (Birkhäuser, Basle, 1980), p. 283.

25) K. Kilian, D. Möhl, H. Pilkuhn and H. Poth, Nucl. Instrum. Methods 202 (1982) 427.

26) J. Arianer and R. Geller, The advanced positive heavy ion sources, Ann. Rev. Nucl. Part. Sci. 31 (1981) 19.

27) C.J. Powell, Cross-sections for ionization of inner-shell electrons by electrons, Rev. Mod. Phys. 48, No. 1 (1976) 33.

28) C.M. Lederer and V. Shirley (editors), Tables of Isotopes, 7th ed. (J. Wiley and Sons, Inc., New York, 1978).

FAST ON-LINE SPECTROSCOPY OF EXOTIC NUCLEI

O. Klepper and E. Roeckl

GSI Darmstadt

D-6100 Darmstadt

1. INTRODUCTION

It is a typical feature in nuclear science to pursue phenomena of the nuclear structure
and decay over a range as wide as possible in proton and neutron number. Therefore
all paths are intensively explored to produce new or rare species of nuclei in sufficient
quantity to facilitate detailed spectroscopy. There are many areas of the nuclear chart
where reactions between heavy ions, which are considered mainly in this review, are
superior in producing nuclei far away from stability compared to methods using the fission process or the spallation and/or the fragmentation of heavy nuclei by high-energy
protons. Spectroscopy of such reaction products is generally faced with the problem,
that the nuclei of interest represent only a small branch of the total reaction
cross-section. In addition, the decay of nuclei far away from the beta stability line
often becomes increasingly complex. These facts, which make such nuclei "exotic species", often prevent their identification or their detailed investigation right at the
production site with its high-level of background radiation. In these "in-beam" techniques only sophisticated multiple coincidence conditions sometimes allow to pick out a
reaction channel with a small relative cross-section.

In order to achieve a separation of the irradiation area and the measuring position, and
maybe simultaneously some selectivity in respect to the wanted nuclei, a variety of
techniques has evolved adapting themselves to the different production methods and
the individual spectroscopic needs. Some of them are sketched in fig.1. In all these
cases the target is kept thin so that the recoiling reaction products escape. They can
be stopped in a catcher foil or a gas and then be transported mechanically together
with the foil or in a gas stream to a shielded measuring position. If one wants to
implant the recoiling nuclei directly into a detector system, the beam has to be suppressed in order to keep the number of particles reaching the detectors small. Such
recoil separators allow to investigate nuclei with very short half-lives and very small
production cross-sections. If further selectivity concerning the nuclear charge or
mass number is indispensable, chemical and/or mass separation can be combined with
the methods above sacrificing in general separation speed and efficiency. As an example fig. 1 shows how in principle the reaction products stopped in a hot catcher are
volatilized, reionized and then collected from a mass-separated beam in front of a detector. The other example demonstrates how volatile atoms, that are sprayed by a gas jet
onto a hot foil, will be selectively desorbed from this and then condense on a cold
catcher.

Samples provided with these methods should have saturation activities of at least a few hundred decays/s of the desired nuclei if some detailed spectroscopy, like for instance γ-γ coincidences, is wanted. On the other hand, one decay within a few days may be enough in an enforced hunt of a rare species of nuclei. In addition, for facilitating spectroscopy of charged particles, the separation techniques should produce small thin sources.

Simple transport systems

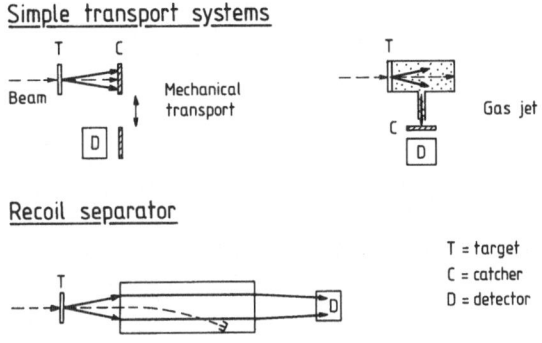

Recoil separator

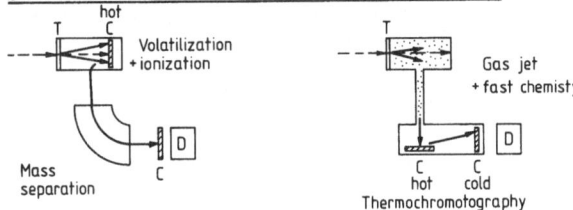

T = target
C = catcher
D = detector

On line selection of mass and / or element number

Figure 1. Principles of some separation methods facilitating the spectroscopy of short-lived products from heavy-ion reactions.

In the following we will present a few examples for those methods sketched in fig. 1 and allowing spectroscopy of nuclei with half-lives in the minute region or shorter. We will not strive for any completeness in this wide field of separation techniques and also omit "in-beam" methods. In order to show the performance and specific strength of the presented method results from a recent experiment will be described shortly. Lately reviews on studies of nuclei far from beta stability by means of heavy-ion beams have been given by Hamilton[1] and Roeckl[2].

2. SIMPLE TRANSPORT SYSTEMS

Already simple devices are often sufficient to transport the reaction products from the target area with its intensive background radiation to a shielded area and to achieve a close geometry for the sample and a complicated detector arrangement. As an example a wheel transport-system[3] is shown in fig.2. Here residual nuclei recoiling from the target are slowed down by a gold foil and captured on a tantalum foil mounted on a wheel. By appropriately choosing the thicknesses of all the foils some selectivity concerning the collection of the wanted reaction products can be achieved. In addition, the slowing down in the gold foil provides a shallower implantation and therefore a thinner source facilitating particle spectroscopy.

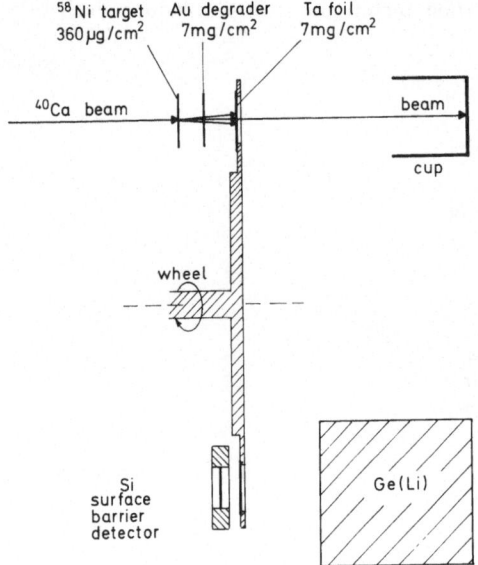

Figure 2. Schematic drawing of a wheel transport-system. After irradiation of the target, the wheel transports the collected activity of recoiling atoms within 0.5 s by a 180° rotation into a position between a particle and a gamma detector. The distance between collection and measuring point is 14 cm. During the measurement the irradiation of the target is interrupted. (From ref. 3)

With this system very neutron-deficient evaporation residues from the reaction 135-MeV ^{40}Ca on ^{58}Ni have been investigated and the long-lived isomer ^{95}Pd^m has been identified[3] (fig. 3). Out of the total cross-section of about 250 mb for evaporation residues, roughly 2 % feed the n2p channel populating this spin-gap isomer. A source strength of about 450 ^{95}Pd^m atoms per second was produced at 5 particle•nA of ^{40}Ca ions, which allowed proton and gamma singles as well as coincidence measurements.

In order to achieve a larger separation of the detection area from the highly radioactive production site than in the experiment described above, one can apply pneumatic transfer ("rabbit") systems, or gas-jet transport. Both techniques are reviewed in ref. 4.

An example for the gas-jet technique is presented in fig. 4. The reaction products recoil out of three targets and are stopped in helium gas containing aerosol particles ("clusters"), e.g. ethylene glycol or NaCl. The thermalized atoms stick to the clusters and are then transported along with the carrier gas through capillaries to the counting chamber. The multiple-capillary system allows to extract short-lived reaction products efficiently from a large gas volume. The 3-target 12-capillary system of fig. 4 was especially designed for reactions with ^{3}He beams. The number of targets and capillaries maybe reduced for heavier projectiles because of their higher stopping power. In the counting chamber the reaction products are collected on an aluminium catcher in front of a telescope consisting of three surface barrier detectors. Collimators and a magnetic field of 0.4 T serve to reduce the number of β-particles reaching the detector. Using the β-delayed α-particle emitter ^{20}Na(446 ms) mean transport times between 180 and 540 ms have been found[6] depending on the number and inner diameter of the capillaries. Overall transport efficiencies were typically above 10 %.

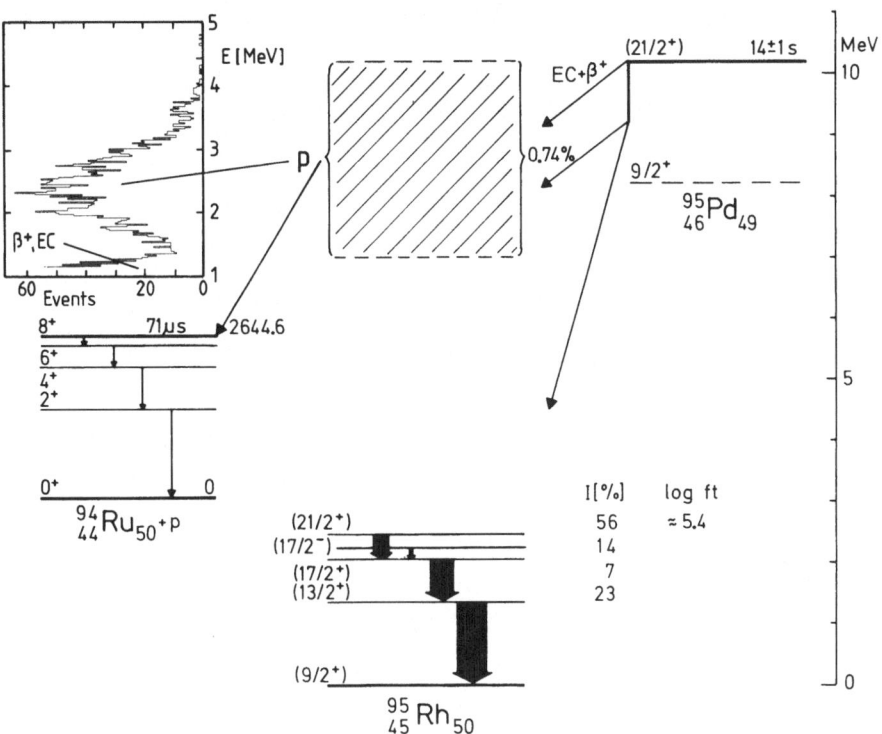

Figure 3. Spectrum of β-delayed protons and β-decay scheme of the 14-s isomer ^{95}Pd^m. The decay scheme was deduced from gamma and proton multispectrum analyses and pγ coincidences applying the wheel transport-system of fig. 2. Additional γγ coincidences were taken by an in-beam technique (After ref. 3).

This helium-jet system and the reaction ^{24}Mg(^{3}He,p4n)^{22}Al were employed to discover[5] the first odd-odd T_z=-2 nuclide ^{22}Al. Fig. 5 shows the observed β-delayed proton spectrum and the deduced decay scheme for ^{22}Al with a half-life of about 70 ms. This nucleus lies very close to the proton drip line. The proton separation-energy is estimated to be 155 keV only[5]. As indicated in fig. 5 ^{22}Al can decay by β-delayed two-proton emission to ^{20}Ne, and there is indeed preliminary experimental evidence for two-proton coincidences populating the first excited state of ^{20}Ne (ref. 7).

The production of T_z=-2 nuclides is always accompanied by the production of other strong β-delayed proton emitters, like ^{21}Mg and ^{25}Si in this case. Here the high proton energies, occurring in the decay of the odd-odd nucleus ^{22}Al, make the application of

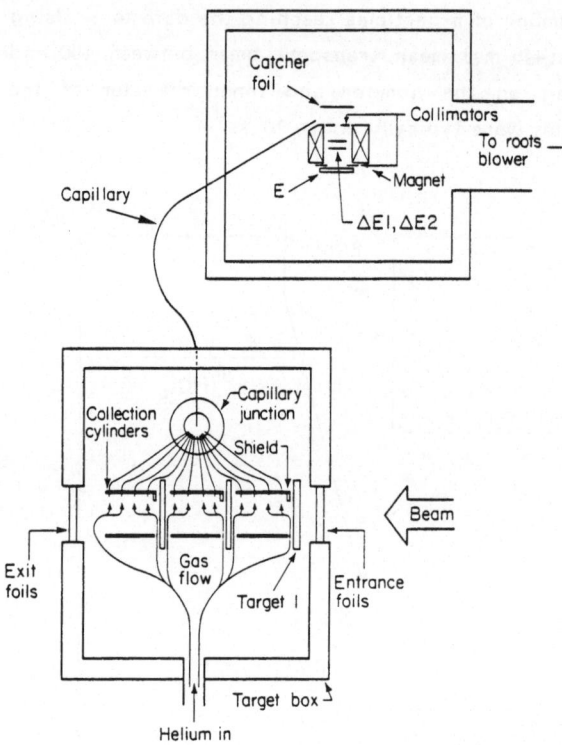

Figure 4. Schematic diagram of the target and the counting chamber of a gas jet system. The connecting main capillary of stainless steel is 1.1 m long and has an inner diameter of 1.6 mm (Ref. 5).

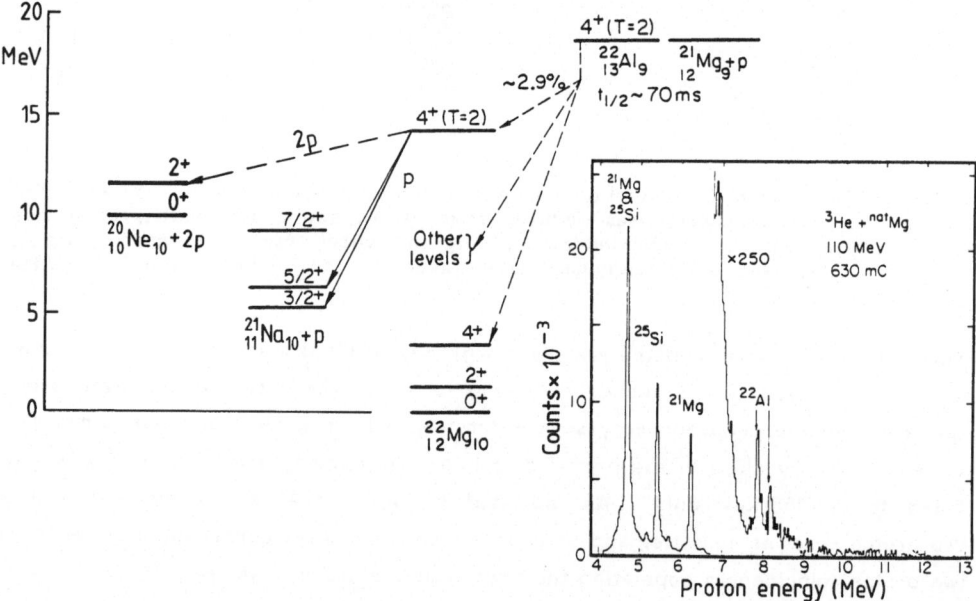

Figure 5. Spectrum of β-delayed protons and the proposed β-decay scheme of ^{22}Al. The two proton lines at 7.839 and 8.149 MeV can be attributed to the isospin forbidden proton decay of the lowest T=2 state in ^{22}Mg fed by superallowed β-decay of ^{22}Al. The effective production cross-section for these β-delayed protons is about 1nb (After refs. 5,7).

the simple helium-jet method possible. For similar investigations[8] of the even-even $T_z=-2$ nuclei ^{20}Mg, ^{24}Si and ^{36}Ca, however, with proton energies smaller than 4.5 MeV the helium-jet had to be coupled to a mass separator, as will be dicussed later.

As most of the reaction products stopped in a gas remain ionized in the presence of an electrostatic field, one can employ such fields instead of differential pumping for the collection process[9]. In a recent application[10] of this method the α-particle emitting evaporation residues from the reaction ^{40}Ca on ^{147,148}Sm are directly deposited on the active surface of a solid state detector resulting in a detection efficiency of 45 %. The collection yield for a 0.5-s isotope is about 4 %. The identification of the new α-particle emitter $^{184}_{82}$Pb(0.55 s) by excitation functions was confirmed by the observation of a time correlation between the new α-particle line and α-particles of $^{180}_{80}$Hg.

2. RECOIL SPECTROMETERS

In the experiments discussed so far the reaction products were slowed down and thermalized before being transported to the detector. However, the strong recoil momentum, generally inborn to products of heavy-ion reactions, can be used for the separation and transport process. As the distribution of recoil velocities depends strongly on the type of reaction employed, different so-called recoil or kinematic spectrometers have been developed. (For a review see ref.11.) In the following we will discuss two examples that allow to separate projectile fragments or evaporation residues of fusion reactions from the primary beam and to implant them within micro seconds directly into surface barrier detectors for spectroscopic investigation.

In the fragmentation of heavy projectiles of several 100 MeV/u on light targets, the fragments move almost with beam velocity and are strongly peaked at 0°. This kinematic advantage is used in the spectrometer shown in fig. 6. A magnetic analysis suffices to separate isotopes according to their A/Z value. Therefore exotic nuclei are separated from the more abundantly produced ones closer to stability and from the projectiles, and can be directly implanted into a multielement solid-state telescope to provide charge and mass identification via their energy losses and ranges. Within 2 ms of seeing a specific isotope, the beam is interrupted and the detector electronics are switched to a higher gain necessary to detect low-energy electrons. A period of 2 s is allowed to observe the β-decay from the embedded fragment. Half-lives between about 1 ms and 1 s are accessible with this technique. In fig. 7a we compare the cross-sections for the production of sodium isotopes by heavy-ion fragmentation and by spallation[14] from protons on uranium. Concerning the heavier sodium isotopes only the neutron-rich ^{48}Ca fragmentation reaches the cross-sections obtained by spallation. Fig. 7b demonstrates the clear mass resolution between ^{31}Al and ^{32}Al achieved with the apparatus shown in fig. 6.

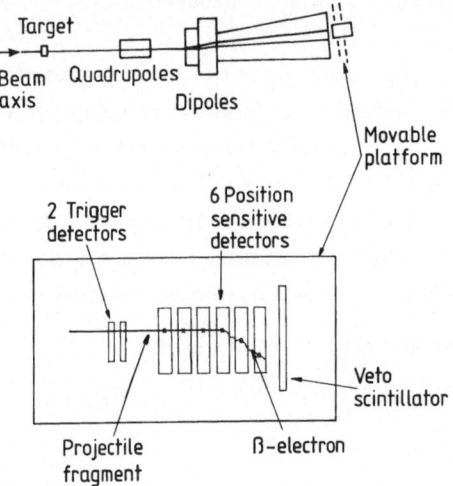

Figure 6. Layout of the LBL recoil spectrometer used to identify neutron-rich nuclei from projectile fragmentation and to measure their β-decay lifetimes. The solid-state telescope on the movable platform stops and identifies the fragments. It consists of two Si(Li) detectors (ϕ=5 cm; 800 μm thick) used as a particle trigger and of six position sensitive Si(Li) detectors (ϕ=7 cm; 5 mm thick). The latter allow a position correlation in one dimension between the origin of the β-particles and the site of the implanted fragment (After ref. 12).

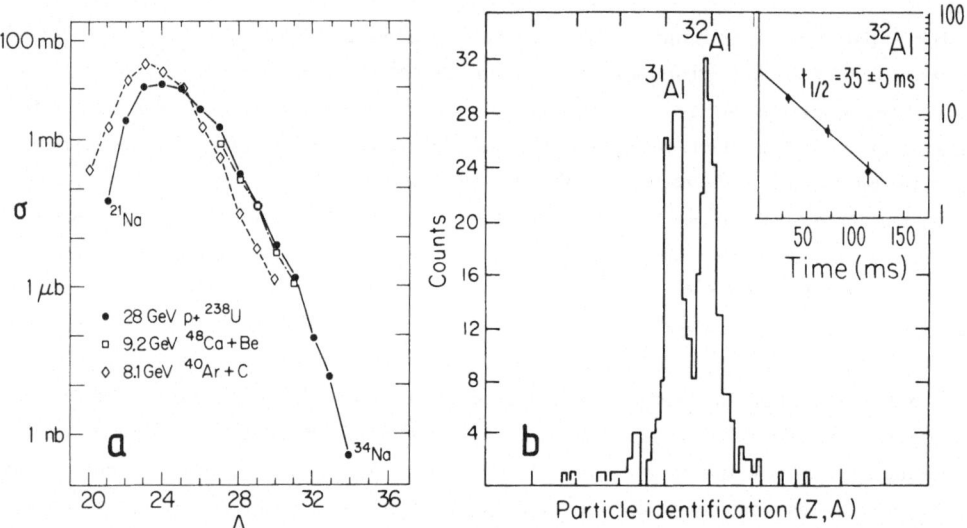

Figure 7. a) Comparison of production cross-sections for sodium isotopes produced in ^{48}Ca on ^{9}Be, ^{40}Ar on C and protons on ^{238}U reactions (After ref. 12).
b) Particle-identification spectrum for aluminium isotopes produced by ^{40}Ar fragmentation on ^{9}Be at 11.4 GeV. The inset shows the half-life measurement of ^{32}Al (After ref. 13).

As a second example for a recoil spectrometer the velocity filter SHIP[15] (Separator for Heavy Ion reaction Products) at GSI is presented in fig. 8. It is especially designed to separate the fusion reaction products under 0° from the faster heavy-ion projectiles. The SHIP consists of two consecutive velocity selectors each with spatially separated

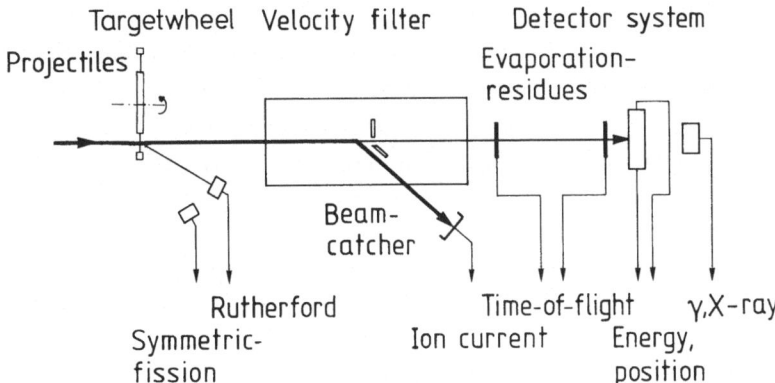

Figure 8. Experimental set-up to investigate rare events in the production of heavy elements with the velocity filter SHIP.

electric and magnetic fields and with opposite deflection senses. The velocity dispersion between these two sections allows to separate the reaction products from the primary particles with suppression factors 10^{11} - 10^7. The better values are obtained for the production of heavy compound systems with relatively light ions. The accepted angle for recoiling atoms is $\pm 1.5°$ and the accepted relative charge and velocity widths are $\pm 10\%$ and $\pm 5\%$, respectively. For residues after neutron evaporation the overall-efficiency of the filter is at least 20%. The separation time is of the order of 10^{-6}s.

The high background suppression generally allows to implant the reaction products directly into an array of seven position- sensitive surface-barrier detectors with a total area of 2000 mm². From the energy deposited in the detector and the time of flight a rough estimate ($\pm$ 10%) of the nuclear mass is possible. The decay of the evaporation residue is investigated by correlating the event of the implantation in position and time with the following decay products like α-particles or fission fragments. As the evaporation residues are implanted close to the detector surface, only about 50% of the emitted α-particles are registered with their full energy.

As the SHIP has no selectivity in mass or nuclear charge, it is well suited for survey experiments. For example, the α-decay of neutron-deficient isotopes in the region from terbium to platinum was explored[16]. The time and position correlation between α-particles of new and already known α-emitters were used to establish parent-daughter relationships and by this to identify new isotopes. Also a search[17] for proton radioactivity was performed in the region of rare earth and neighbouring elements. Fig. 9a shows a proton line with a half-life of 85 ms observed in the reaction ^{58}Ni on ^{96}Ru. From cross bombardments, from measurements of excitation functions and angular distributions of evaporation residues the mass 151 of this proton emitter was obtained.

From the systematics of proton separation-energies the decay was then assigned to the ground state or a low-lying isomeric state of ^{151}Lu. After establishing this "fixpoint" beyond the proton drip line, another very similar proton emitter ^{147}Tm (see fig. 9b), that differs from ^{151}Lu just by one α-particle, was identified[18] with the GSI on-line separator, as discussed below. (This proton line had actually also been seen in an earlier survey experiment by the SHIP.) By comparing calculated partial proton-decay half-lives with the measured total half-lives as lower limits, one can speculate that both proton decays involve an angular momentum change of 5 ℏ and represent transitions from a $\pi h_{11/2}$ single-proton state to the 0^+ ground state of the even daughter nuclei ^{150}Yb and ^{146}Er, respectively.

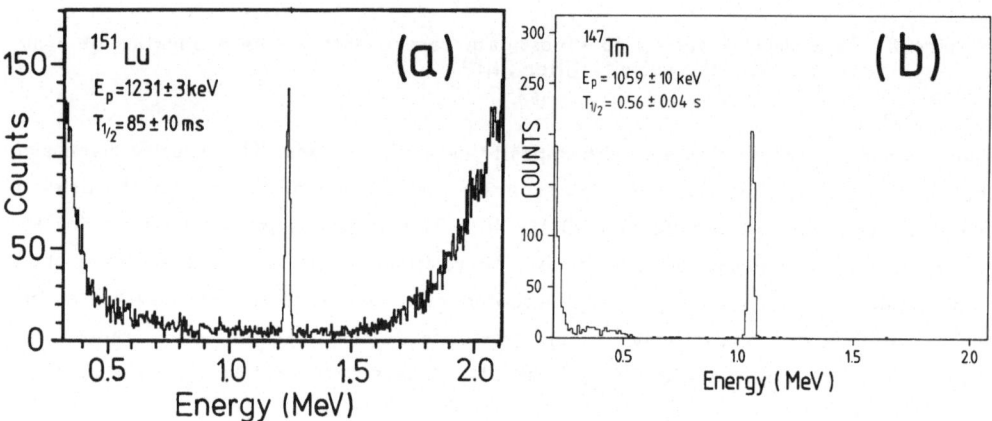

Figure 9. Energy spectra with proton lines assigned to the decay of ^{151}Lu and ^{147}Tm. Spectra (a) and (b) were obtained from exploring ^{58}Ni + ^{96}Ru reactions (ref. 17) at the velocity filter SHIP and from ^{58}Ni + ^{92}Mo reactions (refs. 18,19) at the GSI on-line mass separator.

Extreme sensitivity and background suppression is necessary for the investigation of isotopes of elements with Z ≥ 100. First isotopes of the known elements fermium (Z=100), mendelevium (Z=101), 104, and 105 were produced[20] in order to explore the cold fusion reactions of neutron-rich projectiles with targets near the doubly magic ^{208}Pb and to establish a base of spectroscopic data on α-decay and spontaneous fission in elements with Z ≤ 105. The next aim was then to produce odd-odd isotopes of the elements 107 and 109 with a low probability for spontaneous fission and to identify the isotopes unambiguously by correlated α-decay chains to known transitions. Cold fusion of ^{54}Cr and ^{58}Fe projectiles with ^{209}Bi targets were employed to produce isotopes of the elements Z=107 and Z=109, respectively, via one-neutron evaporation. In the ^{54}Cr experiment[21] six events were detected and assigned to the α-decay of ^{267}107. One of the most significantly registered events is presented in fig. 10; the others are partly incomplete, as 50% of the α particles escape the detector in backward direction. This

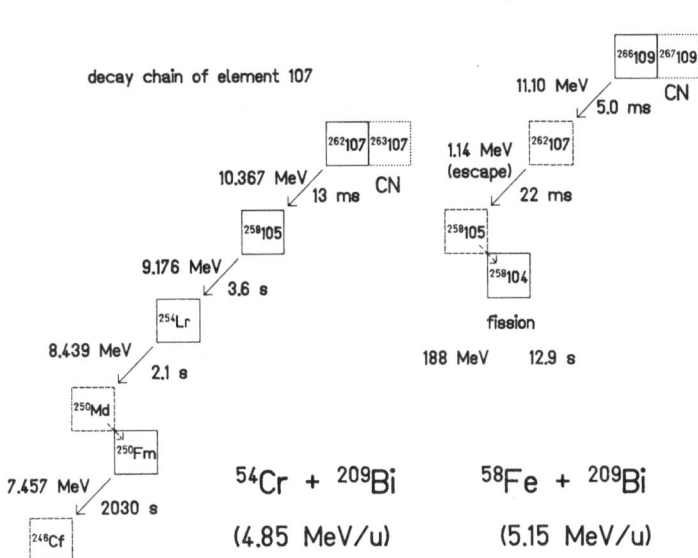

Figure 10. To the left side the most complete α-decay chain with the observed α-ener-
gies and time differences is presented that was detected in the decay of
^{262}107. This correlation of the new α-particle activity of 10.367 MeV with
known α-emitters of lower Z elements identifies the isotope ^{262}107 unambiguously. On the right side the interpretation of the only observed decay chain
in the reaction ^{58}Fe on ^{209}Bi is given (After refs. 21,22).

holds also for the one and only one event, shown in fig. 10, that was observed in the 9
days experiment[22] with the ^{58}Fe beam. The two correlated α-decays and a subsequent
spontaneous fission are interpreted as the decay of ^{266}109, the heaviest nucleus ever
made. The observed event rates correspond to production cross-sections of the order
of 200 and 10 pb for the isotopes ^{262}107 and ^{266}109, respectively.

4. ON-LINE CHEMISTRY AND ISOTOPIC MASS SEPARATION.

Often the above mentioned techniques are not sufficient for an unambiguous assignment
of mass and atomic number to a newly found activity, therefore fast chemistry and/or
mass separation have to be applied alone or in combination with the described methods.
In order to be applicable to nuclei with short half-lives, the rather time-consuming
conventional radiochemical procedures have to be speeded up by automation (For a
review see ref. 23).

From the few cases, where these techniques have been employed so far in heavy-ion
reactions, we mention only a recent search[24] for volatile superheavy elements which

combined gas-jet and thermochromatography (compare fig. 1). Reaction products from damped collisions of uranium on uranium were transported with a gas-jet to an oven and caught on a quartz-wool plug at a temperature of $\simeq 1000°C$. From here volatile elements were continuously evaporated and condensed on a cooled copper wheel that transported the collected activity into positions in front of surface- barrier detectors. In a first experiment with an integrated uranium beam intensity of $8 \cdot 10^{14}$ particles and collection times of 200 s no fission event was observed. From this an upper limit for the production cross-section of $\simeq 2$ nb for fissioning volatile superheavy elements in the half-life region from 10^2 to 10^4 s was deduced.

The gas-jet can also be coupled to a mass separator. For instance in the RAMA-system[6,25] (Recoil Atom Mass Analyser) the main capillary of the helium-jet, shown in fig. 4, transports the clusters with the reaction products direct to an ion source after skimming off the helium gas by a strong vacuum pump. Ion- source efficiencies for RAMA range from 0.1-0.5% for elements such as sodium, magnesium, silicon, calcium, indium, tellurium, cesium, holmium, dysprosium and astatine. As mentioned earlier, for the discovery[8] of the three short-lived ($\simeq$100 ms) T_Z= -2 beta-delayed proton emitters ^{20}Mg, ^{24}Si and ^{36}Ca the mass separation of RAMA was clearly necessary. The effective production cross-section for these observed proton activities is of the order of $1\mu b$. The current status of on-line mass separation (ISOL) has been discussed at a recent conference[26] and in a review article by Ravn[27]. In many laboratories with heavy-ion accelerators on-line mass separators are in operation or under construction, e.g. at Berkeley (RAMA, OASIS), Chalk River, Daresbury, Darmstadt, Grenoble, Louvain (LISOL), Oak Ridge (UNISOR), Orsay (ISOCELE). As an example we describe in the following the GSI separator[28] on-line to the heavy-ion accelerator UNILAC. Even though the ISOLDE facility at CERN is currently only using proton and helium beams, and therefore does not quite fit into the scope of this review article, it should be mentioned here, as pioneering work concerning the ISOL technique and spectroscopic methods has been done in this laboratory.

Facing the task to prepare short-lived reaction products as mass separated samples in a fast, efficient and possibly chemically selective way, the crucial element of an ISOL facility is the catcher ion-source system. At the GSI on-line separator (fig. 11) generally two types of ion sources are used[29]: a discharge ion source of the FEBIAD-type and a thermoionizer of the cavity-type. The former one is less chemically selective and isotopes of the elements from chromium (Z=24) to rubidium (Z=37), from rhodium (Z=45) to cesium (Z=55) and from mercury (Z=80) to francium (Z=87) have been identified in the separator with varying overall separation efficiencies. With the thermal ion source the elements indium, cesium, barium, all lanthanides from neodymium to lutetium except gadolinium, hafnium and thallium have been separated. The separation efficiencies often depend quite sensitively on the choice of catcher material[29]. In general tantalum and graphite prove to be better than niobium and molybdenum.

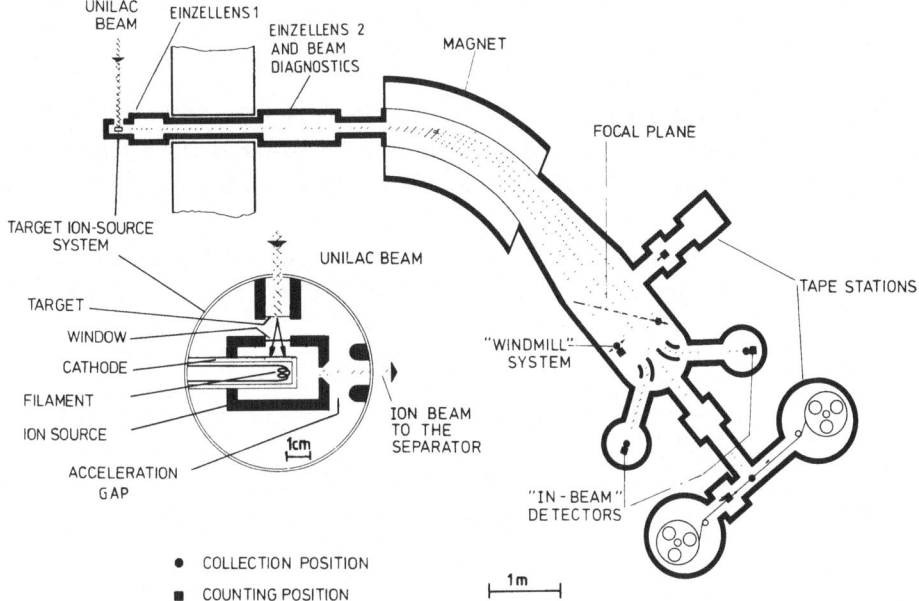

Figure 11. Layout of the GSI mass-separator facility. The inset shows schematically, how recoiling reaction products enter the ion source through a thin tantalum window and are caught on the hot tantalum cathode. A thin foil of graphite or of a refractory metal can be brought in front of the cathode in order to vary the catcher material. In case of targets with high melting points the window and/or the catcher can be made of this material and serve as targets as well. The implanted atoms have to be released from the catcher, re-ionized, extracted from the ion source and accelerated into the separator. Positions where the mass separated beams are collected and assayed are indicated (Ref. 28).

As an example the overall efficiencies and source strengths for two nuclei at the proton drip line, ^{110}I and ^{114}Cs, and the direct proton emitter ^{147}Tm are given in the table.

TABLE:

OVERALL EFFICIENCIES η AND SOURCE STRENGTHS

NUCLEUS	$T_{1/2}$ (s)	η (%)	ATOMS/S
^{110}I	0.6	2.5	120
^{114}Cs	0.6	10	700
^{147}Tm	0.6	≥ 0.4	≥ 1

As mentioned in the previous chapter, the direct proton decay of 0.56-s ^{147}Tm has been identified[18] with this separator. The other proton emitter[17] 85-ms ^{151}Lu has not been seen by mass separation, because the separation efficiency for a lutetium nucleus with this short half-life is rather low.

The shortest-lived nucleus identified so far in the separator is the α-emitter 35-ms ^{192}Po. The smallest production cross-section detected so far is 90 nb for the α-decay of ^{114}Cs. The two very short-lived nuclei[30] 60-μs ^{106}Te and 3.6-ms ^{107}Te have been identified as α-daughters of the separated isotopes ^{110}Xe and ^{111}Xe, respectively, using the same correlation technique as mentioned earlier for the identifying of new elements. The discovery of ^{106}Te, the lightest known α-emitter, is illustrated in fig. 12.

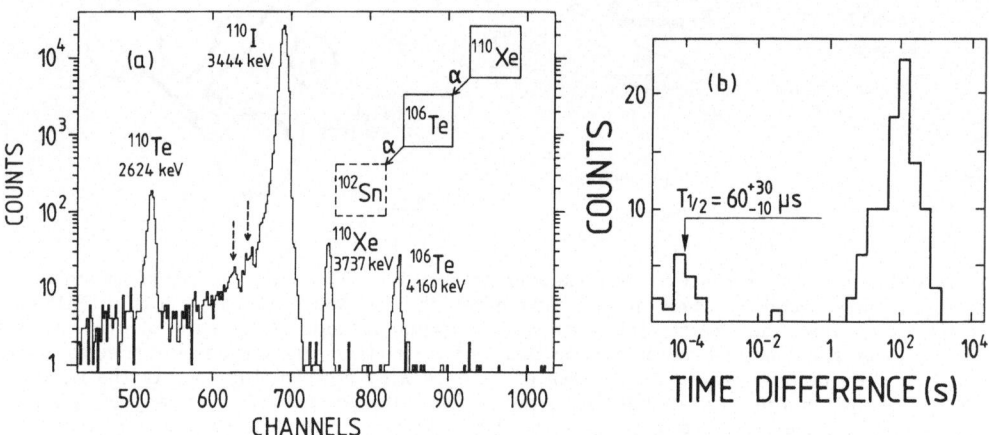

Figure 12. a) α-spectrum obtained at mass 110 from the reaction ^{58}Ni + ^{58}Ni
b) Distribution of time differences between α-decay events of ^{110}Xe (3737 keV) and of the newly observed line at 4160 keV. The peak at long time-differences is due to random correlations, whereas the one at short time-differences represents a parent-daughter correlation, here ^{110}Xe-^{106}Te, with a daughter half-life of 60 μs. Note the logarithmic time scale (Ref. 30).

So far only fusion-evaporation reactions have been exploited with the GSI on-line separator leading naturally to the investigation of very neutron-deficient isotopes. Recently also multinucleon transfer-reactions have been applied to produce new neutron-rich isotopes outside the well explored region of fission fragments. Employing ^{nat}W/Ta targets and two beams of 9 MeV/u ^{76}Ge and ^{136}Xe the β- and γ-decays of the following new isotopes have been investigated: in the first experiment[31] projectile-like nuclei as ^{62}Mn, ^{63}Fe, $^{71-73}$Cu and in the second[32,33] target-like nuclei as ^{179}Yb and $^{181-183}$Lu. The weakest separation yield of ≃ 5 atoms/s was observed for the isotope ^{183}Lu. This corresponds to a production cross-section of ≃ 50 μb assuming that ^{183}Lu is produced only from ^{186}W via three-proton stripping reactions and that the separation efficiency is of the order of 10%.

5. <u>OUTLOOK</u>

We have selected a few examples illustrating how with present techniques the broad spectrum of reaction products from different types of heavy-ion reactions is utilized for spectroscopy of exotic nuclei far from β-stability. Some methods have been cultivated to such a degree that in searching for new elements at the very end of the periodic table multi-coincidence experiments are feasible even at production cross-sections of the order of 10 pb. On the other hand, much effort has been devoted in developing rapid on-line chemistry for nuclei with half-lives in the subminute region, these techniques, however, have only scarcely been applied so far for heavy-ion reaction products, especially for neutron-deficient nuclei.

Most investigations of products of heavy-ion reactions concern only relatively simple spectroscopy and often are only of survey type. It is impressive, on the other hand, to consider the recent developments made at ISOLDE, like e.g. collinear laser spectroscopy; they are facilitated by the much higher source strengths (up to 10^{11} decays/s) than the ones achieved with ISOL facilities at heavy-ion accelerators. One can speculate, however, that with the advent of new heavy-ion accelerators and further development of experimental techniques more such refined spectroscopy studies will be possible also with heavy ions. Indeed, collinear laser experiments are in preparation at some heavy-ion based ISOL facilities (e.g. Louvain, Oak Ridge, Darmstadt). With so rich assets at hand and prospects ahead, the field of spectroscopy of nuclei far from β-stability will probably continue to be one of the major forefronts in nuclear research.

REFERENCES

1. J.H. Hamilton, in: Heavy Ion Collisions, Vol. 3, ed R. Bock, North Holland Pub.Co., Amsterdam, in press
2. E. Roeckl, Proceedings of the Intern. Conf. on Nucleus-Nucleus Collisions, Michigan State University, 1982; to be published in Nucl. Phys. A
3. E. Nolte, H. Hick, Z. Physik A 305, 289 (1982) and private communication
4. R. D. Macfarlane, Wm. C. McHarris, in: Nuclear Spectroscopy and Reactions, Vol. A, ed. J. Cerny (Academic, New York and London, 1974) p. 243
5. M. D. Cable, J. Honkanen, R. F. Parry, H. M. Thierens, J. M. Wouters, Z. Y. Zhou, Joseph Cerny, Phys. Rev. C 26, 1778 (1982)
6. D. M. Moltz, J. M. Wouters. J. Äystö, M. D. Cable, R. F. Parry, R. D. von Dincklage, J. Cerny, Nucl. Instr. Meth. 172, 519 (1980)
7. J. Cerny, contribution to the International Workshop XI, Hirschegg, Austria, January 1983
8. J. Äystö, M. D. Cable, R. F. Parry, J. M. Wouters, D. M. Moltz, Joseph Cerny, Phys. Rev. C 23, 897 (1981)
9. J.P. Dufour, R. Del Moral, A. Fleury, F. Hubert, Y. Llabador, M. B. Mauhourat, R. Bimbot, D. Gardes, M. F. Rivet, Proc. 4th Intern. Conf. on Nuclei Far From Stability, Helsingør 1981, CERN Report 81-09, p. 711

10. J. P. Dufour, A. Fleury, F. Hubert, Y. Llabador, M. B. Mauhourat, R. Bimbot, D. Gardes, Z.f. Physik A294, 107, 1980

11. H.A. Enge, Nucl. Instr. Meth. 186, 413 (1981)

12. T.J.M. Symons, CERN Report 81-09, p. 668, (cf. ref. 9)

13. M.J. Murphy, T.J.M. Symons, G.D. Westfall, H. J. Crawford, Phys. Rev. Lett. 49, 455 (1982)

14. C. Thibault, R. Klapisch, C. Rigaud, A. M. Poskanzer, R. Prieels, L. Lessard, W. Reisdorf, Phys. Rev. C12, 644 (1975)

15. G. Münzenberg, W. Faust, F. P. Hessberger, S. Hofmann, W. Reisdorf, K.H. Schmidt, W.F.W. Schneider, H. Schött, P. Armbruster, K. Güttner, B. Thuma, H. Ewald, Nucl. Instr. Meth. 186, 423 (1981)

16. S. Hofmann, G. Münzenberg, W. Faust, F.P. Hessberger, W. Reisdorf, J.R.H. Schneider, P. Armbruster, K. Güttner, B. Thuma, CERN Report 81-09, p. 190 (cf.ref.9)

17. S. Hofmann, W. Reisdorf, G. Münzenberg, F.P. Hessberger, J.R.H. Schneider, P. Armbruster, Z. f. Physik A305, 111 (1982)

18. O. Klepper, T. Batsch, S. Hofmann, R. Kirchner, W. Kurcewicz, W. Reisdorf, E. Roeckl, D. Schardt, G. Nyman, Z. f. Physik A305, 125 (1982)

19. D. Schardt, Lecture Notes in Physics 168, 'Heavy Ion Collisions', Springer Verlag, Berlin 1982, p. 256

20. G. Münzenberg, S. Hofmann, F. P. Hessberger, W. Reisdorf, K.H.Schmidt, W. Faust, P. Armbruster, K. Güttner, B. Thuma, D. Vermeulen, C.-C. Sahm, CERN Report 81-09, p. 755 (cf. ref. 9)

21. G. Münzenberg, S. Hofmann, F. P. Hessberger, W. Reisdorf, K.-H. Schmidt, J.H.R. Schneider, P. Armbruster, C.-C. Sahm, B. Thuma, Z. f. Phys. A300, 107 (1981)

22. G. Münzenberg, P. Armbruster, F. P. Hessberger, S. Hofmann, K. Poppensieker, W. Reisdorf, J.R.H. Schneider, K.-H. Schmidt, W.F.W. Schneider, C.-C. Sahm, D. Vermeulen, "Observation of One Correlated α-Decay in the Reaction ^{58}Fe on ^{209}Bi → ^{267}109", to be published

23. G. Herrmann, N. Trautmann, to be published in Ann. Rev. of Nucl. and Particle Science, Vol. 32 (1982)

24. H. Dornhöfer, W. D. Schmidt-Ott, W. Fan, H. Gäggeler, K. Sümmerer, G. Dersch, D. Hirdes, P. Lemmertz, N. Greulich, N. Trautmann, GSI Scientific Report 1981, p. 216

25. D.M. Moltz, J. Äystö, M. D. Cable, R. F. Parry, P. E. Haustein, J. M. Wouters, Joseph Cerny, Nucl. Instr. Meth. 186, 141 (1981)

26. H.L. Ravn, E. Kugler, S. Sundell, eds., Proc. of the 10th Intern. Conf. on Electromagnetic Isotope Separators and Techniques Related to Their Applications, Nucl. Instr. Meth. 186 (1981)

27. H. L. Ravn, Phys. Rep. 54, 201 (1979)

28. C. Bruske, K. H. Burkard, W. Hüller, R. Kirchner, O. Klepper, E. Roeckl, Nucl. Instr. Meth. 186, 61 (1981)

29. R. Kirchner, K. H. Burkard, W. Hüller, O. Klepper, Nucl. Instr. Meth. 186, 295 (1981)

30. D. Schardt, T. Batsch, R. Kirchner, O. Klepper, W. Kurcewicz, E. Roeckl, P. Tidemand-Petersson, Nucl. Phys. A368, 153 (1981)

31. E. Runte, W.-D. Schmidt-Ott, P. Tidemand-Petersson, R. Kirchner, O. Klepper, W. Kurcewicz, E. Roeckl, N. Kaffrell, P. Peuser, K. Rykaczewski, M. Bernas, P. Dessagne, M. Langevin, "Decay Studies of Neutron-Rich Products from ^{76}Ge-Induced Multi-Nucleon Transfer Reactions Including the New Isotopes ^{62}Mn ^{63}Fe and $^{71-73}$Cu", to be published

32. R. Kirchner, O. Klepper, W. Kurcewicz, E. Roeckl, E. F. Zganjar, E. Runte, W.-D. Schmidt-Ott, P. Tidemand- Petersson, N. Kaffrell, P. Peuser, K. Rykaczewski, Nucl. Phys. A378, 549 (1981)

33. K. Rykaczewski, R. Kirchner, W. Kurcewicz, D. Schardt, N. Kaffrell, P. Peuser, E. Runte, W.-D. Schmidt-Ott, P. Tidemand- Petersson, K. L. Gippert, "The New Neutron-Rich Isotope ^{183}Lu", submitted for publication in Z. f. Physik A

Springer Series in Chemical Physics

Editors: V. I. Goldanskii,
R. Gomer, F. P. Schäfer,
J. P. Toennies

Volume 1
I. I. Sobelman

Atomic Spectra and Radiative Transitions

1979. 21 figures, 46 tables. XII, 306 pages
ISBN 3-540-09082-7

Contents: Elementary Information on Atomic Spectra: The Hydrogen Spectrum. Systematics of the Spectra of Multi-electron Atoms. Spectra of Multielectron Atoms. – Theory of Atomic Spectra: Angular Momenta. Systematics of the Levels of Multielectron Atoms. Hyperfine Structure of Spectral Lines. The Atom in an External Electric Field. The Atom in an External Magnetic Field. Radiative Transitions. – References. – List of Symbols. – Subject Index.

Volume 7
I. I. Sobelman, L. A. Vainshtein, E. A. Yukov

Excitation of Atmos and Broadening of Spectral Lines

1981. 34 figures, 40 tables. X, 315 pages
ISBN 3-540-09890-9

Contents: Elementary Processes Giving Rise to Spectra. – Theory of Atomic Collisions. – Approximate Methods for Calculating Cross Sections. – Collisions Between Heavy Particles. – Some Problems of Excitation Kinetics. – Tables and Formulas for the Estimation of Effective Cross Sections. – Broadening of Spectral Lines. – References. – List of Symbols. – Subject Index. – Errata for volume 1 of this series.

Volume 13
I. Lindgfren, J. Morrison

Atomic Many-Body Theory

1982. 96 figures. XIII, 469 pages
ISBN 3-540-10504-2

Contents: Angular-Momentum Theory and the Independent-Particle Model: Introduction. Angular-Momentum and Spherical Tensor Operators. Angular-Momentum Graphs. Further Developments of Angular-Momentum Graphs. Applications to Physical Problems. The Independent-Particle Model. The Central-Field Model. The Hartree-Fock Model. Many-Electron Wave Functions. – Perturbation Theory and the Treatment of Atomic Many-Body Effects: Perturbation Theory. First-Order Perturbation for Closed-Shell Atoms. Second Quantization and the Particle-Hole Formalism. Application of Perturbation Theory to Closed-Shell Systems. Application of Perturbation Theory to Open-Shell Systems. The Hyperfine Interaction. The Pair-Correlation Problem and the Coupled-Cluster Approach. – Appendices A-D. – References. – Author Index. – Subject Index.

Springer-Verlag
Berlin
Heidelberg
New York
Tokyo

Lecture Notes in Physics